Rudiments of Signal Processing and Systems

Tom J. Moir

Rudiments of Signal Processing and Systems

 Springer

Tom J. Moir
Department of Electrical and Electronic Engineering
School of Engineering, Computer and Mathematical
Sciences
Auckland University of Technology
Auckland, New Zealand

ISBN 978-3-030-76949-9   ISBN 978-3-030-76947-5   (eBook)
https://doi.org/10.1007/978-3-030-76947-5

Cover illustration: Auckland Harbour bridge at night, *courtesy* James Moir

This Springer imprint is published by the registered company Springer Nature Switzerland AG
The registered company address is: Gewerbestrasse 11, 6330 Cham, Switzerland

To James and Callum.

Preface

The author has been a researcher and teacher of electrical engineering subjects for nearly 38 years. I began with control systems and taught that for the first 16 years of my life along with various electrical engineering topics. My research drifted slowly over from control systems to signal processing, and when I later emigrated from Scotland to New Zealand, I found myself teaching the subject. Signal processing interested me more at the time since I could do basic experiments on a computer almost anywhere and did not need to be in a laboratory with hardware. By that time, personal computers were getting powerful enough to do routine research. Prior to that, we had to use Apollo or Sun workstations on a UNIX platform or specialised minicomputers that were very expensive. Twenty years later, I wrote a book on control systems but felt afterwards that I also needed to write one for signal processing. This is the cumulation of 20 years of teaching signal processing. I decided to concentrate most of the book on the digital side of signal processing since it is becoming widely used by now. The theory has been around since the 1960s or even earlier, but the hardware has taken some time to catch up.

My interest in electronic and electrical engineering began when I was around 5 years old. I was always peering into the back of televisions or radios and eventually pulled the old ones apart. Reading books from the Mackie Academy school library and UK electronics magazines, I quickly found out how to make my own simple radios and amplifiers. Imagine my surprise nowadays that I can buy a universal radio receiver on a tiny integrated circuit that works mainly using software algorithms. This is all down to digital signal processing and the many great people who have contributed to this area over the years. Signal processing does not come on its own, you must know about the systems that signals drive into or come from. Therefore, the book is not entirely about signals. As a student, I would often have to spend many hours trying to understand a variety of textbooks on subjects I was studying, and many were written in a way a beginner could not understand. I have written this book in a similar manner as I teach, and nothing is taken for granted. Therefore, I have used the word rudiments in the title. Like learning a musical instrument, we all need the absolute basic theory first. If we neglect it, we end up having to go back again at some point to fundamentals anyway and it is a false economy. I do assume a basic level of understanding of mathematics, without which no graduate-level engineer can exist. It is not too advanced, and any non-standard theory is covered in the book. I am increasingly of the opinion that as we advance in technology, we should use that technology in learning. When I was a student, most calculations were done by hand or using a slide rule. Then, we had the calculator and eventually computers. We have symbolic mathematical algorithms that can solve standard integrations that are available online for free. In a mathematics class, we must do the work by hand, since that is the whole object of the exercise, but in this subject, we can use tools where possible to speed up mathematical problem-solving. This is not cheating, but a natural progression from earlier methods. It also removes unnecessary clutter from a text since the emphasis is on learning the subject and not churning through pages of integrals.

This book is aimed at undergraduate second-year engineers and onwards. It takes the subject from the fundamentals which is common to most universities, to more research-oriented topics nearer the end of the book. One can buy several books to cover the topics that are covered in this book. The emphasis is always on examples and working software demos using MATLAB where possible. I try to avoid MATLAB at the earlier stages where it is not really needed and bring it in as required. The book starts with a historical context on electronic engineering to show us where we are in history and how we got here. Then, all the fundamentals are introduced from time domain to frequency domain, sampling theory and discrete theory. Chapters 1–8 should be enough for years 2 and 3 of an electrical engineering major. Beyond Chap. 8, we move into master's level topics on optimal and adaptive filters. One method that has not appeared in any DSP book to date is the Toeplitz method of designing optimal filters. Here, we exploit the similarity of polynomials and lower triangular Toeplitz matrices. I hope you enjoy the book and learn from my experience. It was written during and after COVID-19 lockdown in New Zealand.

Auckland, New Zealand   A/Prof. Tom J. Moir

Contents

Chapter 1
Introduction and Basic Signal Properties

1.1 Brief History of the Subject from a Personal Perspective

Rather than diving straight into the theory, I always feel it better to reflect on how we got here in the first place. To fix our place in history is always a good thing, as it leads to an appreciation of how hard it has been to even get to this level of human achievement in the first place.

The industrial revolution is usually placed around the mid-eighteenth century to a little later and was the success it was mainly due to the birth of the machine. The improved steam-engine of James Watt [1] started a series of inventions and discoveries that has never stopped to the day you are reading this text. There are whole books devoted to this subject, so perhaps it is best to confine ourselves to the specialities of this book, namely electronic signals, systems and the mathematical descriptions and manipulation of them. After the famous kite experiment by Benjamin Franklin in 1752, a use needed to be made of this new discovery. Many see this as the beginning point in our journey. However, it should be remembered that William Gilbert, an English physicist had already discovered static electricity by 1600, though the connection to lightening had not been realised. Electricity that occurred naturally in fish (the electric eel) was known about up to two millennium earlier [2]. By the eighteenth century, some debate emerged when the knowledge of electric fish became known. The debate was a rather obvious one which asked the question whether the shock obtained from an electric eel was the same as obtained from a Leyden jar, a kind of early capacitor for storing static electricity. From the point of view of signals, the most exciting discovery was made by Hans Christian Orsted, a Danish Physicist and Chemist. In 1820, Orsted published his discovery that electricity and magnetism were related. He discovered this when a wire powered by an early battery was accidently placed closed to a compass and the needle spun around. This was a major discovery. It had only been in 1799 that Alessandro Volta (the unit of voltage is named in his honour) had published his work on the first battery, the voltaic pile. Hence Orsted had a voltaic pile available for his experiments at the time of the discovery. When the battery was connected to the wire the needle spun

© The Author(s), under exclusive license to Springer Nature Switzerland AG 2022
T. J. Moir, *Rudiments of Signal Processing and Systems*,
https://doi.org/10.1007/978-3-030-76947-5_1

around and history was made. Soon after Michael Faraday in England formalised the law of electromagnet induction. Mankind could then generate electricity or used stored electricity in batteries and move things (the basic motor invented by Faraday). When things move, we have the basis of signalling and the invention of a basic on–off signal.

In 1825 an English inventor William Sturgeon invented the electromagnet. A coil of wire around an iron former was all that was needed. When the coil was energised it became magnetic and attracted metallic objects. It did not have much power however since the wire used was not insulated and the wire could not be overlapped. Around the same time Georg Simon Ohm of Germany discovered Ohm's law. Ten years later Joseph Henry invented a more powerful electromagnet by using insulated wire. An electromagnet is just a small step away from a solenoid or relay and this in turn was used as the first electric signalling device, the Telegraph. Invented by William Cook and Charles William Wheatstone in 1837, the early Telegraph used needles to point at letters at a receiving station. These needles were easily moved by using the earlier discoveries already mentioned. A picture of such a machine is shown in Fig. 1.1.

Although this was the first practical telegraph, the theory had been previously suggested by two giants of science 10 years previously. The two were Pierre-Simon Laplace (the inventor of Laplace Transforms and other mathematical methods) and Andre-Marie Ampere (of whom the unit of current is now named after). Even earlier in 1753 a Scotsman wrote a letter to the *Scots magazine* (the oldest magazine in the world still being published) describing a Telegraph system using static electricity. However, as with many theoretical ideas, the old engineers saying that "Saying is

Fig. 1.1 Cooke and Wheatstone 5 needle Telegraph (Wikimedia commons)

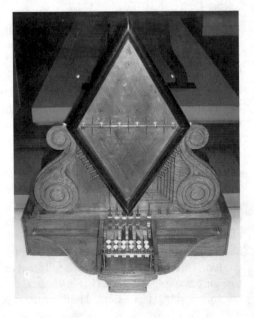

much easier than doing" comes to mind. This is born out time and time again in technological inventions.

As regards the telegraph, it is history to say that the original method above was later replaced by the later method due to Samuel Morse in the USA in the early 1840s [3]. The Morse Code (a series of short and long on and off currents) that was sent as the coding method for letters is often thought to be the first binary signal. However, spaces also need to be sent and therefore the code as described later by the founder of digital signals (Claude Shannon) is thought to be more a ternary signal.

With the invention of electric communication, the race was on to find if it was possible to send speech signals instead of just a code. However, this had to wait until after the invention of the microphone and the telephone. The telephone is generally accepted to have been invented by Scottish American Alexander Graham Bell in 1876 (Fig. 1.2).

The theory that underpinned the later invention of wireless comes from Scottish Physicist James Clark Maxwell. He discovered that electricity, light and magnetism are all related under a linked set of equation nowadays known as Maxwells equations [4]. Perhaps not so well known is the fact that the present form of these equation in vector notation is due to the later work of English engineer Oliver Heaviside [5]. A German physicist Hertz (the SI unit of frequency is named in his honour) later verified Maxwells theory. Gugliemo Marconi in 1895 managed to transmit wireless electromagnetic signals over 2 miles. The following year he obtained a patent in London UK for his wireless apparatus. However, Nikola Tesla also filed a wireless patent in 1897 beating Marconi to the chase in the USA at least. There ensued legal wrangling's between Tesla and Marconi in the USA. However, Tesla appeared to be

Fig. 1.2 Alexander Graham Bell who patented the first telephone. (Wikimedia commons)

more interested in the transmission of electrical power via wireless rather than as a communication device. Marconi's had a rather primitive on–off spark transmitter that was only capable of handling Morse-code. What was needed was the merging of the earlier telephone invention with wireless technology. This was achieved by using a carbon microphone as earlier used in telephony. A Canadian engineer Reginald Fessenden made the breakthrough and, in the process, also invented Double-sideband Amplitude Modulation.

Amplitude modulation using a carrier frequency had to wait for the invention of the sinusoidal oscillator. This in turn needed the invention of the thermionic valve in 1908 (or vacuum tube) for amplification by Lee De Forest in the USA. All these inventions were quite close to one another but there is a chain or sequence that cannot be broken, a technology needs a previous step to progress. With the invention of amplification with the thermionic valve, the first signal generators could be invented. A stable oscillator needs an amplifier (unless it is of the switching type) to work and this in itself was a major inventive step. A theory of feedback was produced by Bode [6] of Bell labs, not so much to produce oscillators, but to stop oscillation when feedback is put around an amplifier. This was a common problem when Black in 1934, also of Bell Labs suggested improving amplifier design by using negative feedback [7]. A great innovation was the ability to generate stable and low distortion sinusoids which were needed for wireless and for testing (frequency response methods). The breakthrough came by 1942 by William Hewlett by using an incandescent lamp to stabilise the feedback loop. He later teamed up with Dave Packard and the combined HP company has now become a household name.

Signals are not just confined to one dimension, however. Images too can be thought of as 2D signals and the invention of television is a separate story. The invention of television involves many intentions, but there are key inventions which define the subject. Surprisingly, the theory for electronic television pre-dates mechanical television. Alan Archibald Campbell Swinton was a Scottish engineer who published in Nature the theory of how television would work. This was in 1908, some 20 years prior to 1925 when fellow Scotsman John Logie Baird Fig. 1.3 first publicly demonstrated the transmission of moving images from a dummy model. Bairds invention was of course mechanical scanning and impractical by today's standards and soon afterwards Philo Farnsworth in 1927 patented electronic television (Fig. 1.4).

Usually Farnsworth is credited as being the first to produce the first electronic transmission of moving images, but Kenjiro Takayanagi of Japan also produced electronic television by 1925. Baird formed his own company, hired engineers, and built his own electronic system. He also started to work with 3D television, but it was never publicly demonstrated.

The invention of semiconductor devices enabled technology to get much smaller and consume less power (also producing less wasteful heat). Transistor radios appeared by the late 1950s and early 1960s and eventually televisions also became transistorised. In 1969 a man first walked on the moon and TV pictures (black and white and of low resolution) were transmitted live from space. The timescale of events is not very long, and technology advanced very quickly in the twentieth

Fig. 1.3 John Logie Baird. First to demonstrate mechanical television (picryl.com, public domain)

Fig. 1.4 Philo Farnsworth and his early television apparatus (with permission, University of Utah Marriott Library)

century. Soon after the moon walk, integration of analogue and digital devices became commonplace and much larger circuitry could be built in a smaller space.

For signals, small signal generators could be produced, and oscilloscopes dropped in size to become more portable. Amplifiers began to be manufactured on a single integrated circuit. Back in 1948 Claude Shannon had developed a theory of digital communications including the theory of sampling of analogue signals [8]. This was a long time before digital signals became commonplace. The invention of the computer on a chip (microprocessor) and later the microcontroller enabled complex systems to be vastly simplified and the hardware to be fitted in a small physical space. Everyday machines like washing machines became controlled by microcontrollers. With faster microprocessors it became possible to implement outside of the laboratory, sampled signals and digital systems using Shannon's theory.

Analogue filters had been invented first with passive components (resistors, capacitors and inductors). Credited with the first invention was an American inventor Campbell [9] in 1910. With the advent of the thermionic valve and later the transistor, active filters could be produced which produce voltage gain as well as filter at the same time. Some of the designs found their way into later electronic topologies using operational amplifier integrated circuits. Notable among these was the so-called Sallen-key variety first invented at MIT Lincoln Labs in 1955.

With the advent of fast processors and analogue to digital convertors it became possible to implement digital filters. Filters closer to the ideal filter could be produced using this method with steep roll-offs and stable operation. Eventually this led to the evolution of radio from a hardware system to a purely software-based system. Software-defined radio (SDR) was born as early as 1984 by a company called E-Systems. Practical single-chip SDRs radios were not being used until around the early 2000s. Part of the technology that made some of the SDR systems practical was the advent of the field-programmable gate-array or FPGA for short. Arrays of switching gates that can be re-configured and programmed using software and can run at high clock rates. Particularly of use when parallel operations need to be done, they can be used to created highly stable oscillators using directly digital synthesis (DDS). Oscillators which once required incandescent lamps to be stable, or that used manipulation of square waves (function generators) could be replaced by devices that can generate almost any mathematically defined signal that is needed. They are rock stable both in amplitude and in frequency. With the reduction size and cost, consumer electronic items become ever cheaper.

It is perhaps with some degree of irony that the author recalls his boyhood when hand-held communications devices that could see the picture of the caller were lusted over. There was much talk of the feasibility of such technology in the 1960s and 1970s. By the 1980s or 1990 some degree of video-telephony had been produced but it was a niche market and expensive. By the early 2000s, smartphones had been produced which could enable the user to see and talk simultaneously to another person. Around the same time 3D televisions were being manufactured for the consumer market. In the case of the smartphone, most people prefer to use texting to communicate rather than seeing the video, and 3D televisions are becoming obsolete due to lack of demand. Engineers often produce technology that they feel others should use, but in the end, they opt for another solution due to market forces.

A whole heap of technology needed to be developed at the same time as our electronic technologies. Software had to change from the old-fashioned approach of basic mathematical programming using say FORTRAN, PASCAL to new object-oriented approaches. From a user's point of view only, an old-style program would execute and ask for data input step by step and then produce an output. With the object-orientated approach the program will just sit and wait until you pull down a menu option. Things can be done in different orders and new user interface devices were introduced like the invention of the Mouse. In a Word processor for example, you can make letters bold and then do a spell-check or do things the other way around. From the point of view of the structure, it is a subject with a great many innovations based on the paradigm of objects.

The subject of speech reignition is a method which is only in recent times is becoming a practical reality, though in terms of research it is an old idea dating back to the 1950s and 60s. From a signal processing point of view, noise is a major problem in speech recognition and algorithm development has been prolific in noise cancellation or reduction. Smartphones have basic noise cancelling algorithms built into the software. Speech recognition is the convergence of many areas of mathematical signal processing (for example Digital Fourier Transforms), machine learning, feature extraction, fast computer hardware and linguistics.

As well as signals, the systems or machines that have been invented have also gone through generations of change. Outside of mechanical engineering which has the internal combustion engine and other engine technologies, in electrical engineering the invention of the motor has led to one of the most used devices both in the home and in industry. The science and laboratory demonstrations were of course started by Michael Faraday in 1830, but the first practical dc motor was made by Thomas Davenport in 1834. Even earlier in 1832 Frenchman Hippolyte Pixii developed an ac generator (or alternator) based on Faradays experiments. He also used a commutator to produce a dc voltage. A great many inventors contributed to the invention of the electric motor (or electric machine as the general term is commonly used). Perhaps notably for ac motors was Galileo Ferraris in 1885 and Nikola Tesla in 1882 (though there remains some debate about who was first). Dolivo-Dobrovolosky [10] invented the first three-phase induction motor in 1890 to go with his three-phase generator and transformer. In the USA, Tesla is known for doing similar independent work. In the past, ac motors were used to control fans and pumps whilst dc motors generated greater torque at low speed and were favoured for speed-control and position-control servomechanisms. It was much harder to control the speed of an ac motor whilst keeping torque high at low speeds. In more recent times however due to the increase in embedded computing power and fast MOSFET power electronics, brushless dc motors and flux-vector drives are all but making dc motors obsolete. The idea is not new, dating back to the late 1960s from the company Seimens. Modern robotic systems will use brushless dc or ac motor drives. Almost 200 years of discovery is converging on ac motor technology (Fig. 1.5).

Fig. 1.5 NASA robot nicknamed Surge (Jet propulsion laboratory)

Applications to many disciplines are too numerous to mention in this brief introduction. Aerospace has been a great innovator in signal processing for navigation purposes and control-systems. Processing biological signals coming from the human heart and brain as well as muscles, is a discipline in its own right. Drug delivery systems, heart monitors, heart–lung machines and endless equipment found in hospitals. The automobile industry, railway systems, home automation, sound recording (analogue and digital) and power systems all have major contributions.

1.2 Basic Signals and Properties

In the field of electrical-electronic engineering there are several waveforms of special interest and their properties are worth knowing. The first of these is the sinewave (or sinusoidal waveform).

This signal has several uses but mainly for finding the frequency–response of systems. By varying the frequency and keeping the amplitude constant we measure the output amplitude and phase and then draw two graphs: Ratio of output to input amplitude (or Gain) versus frequency and phase versus frequency. From this we can deduce the type of system that we are dealing with. Alternatively, we can check the purity of the output of a system with a sinewave input and this tells us how much distortion there is in say an amplifier design. The purity is checked using a spectral analyser, the theory of which is covered later in this book.

Define the waveform as

$$f(t) = V_p \sin(\theta) \tag{1.1}$$

Fig. 1.6 Sinewave

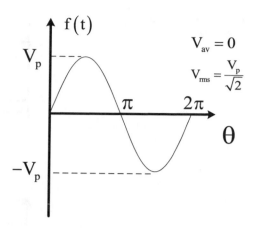

Then its average or dc value is given by the area of the signal in Fig. 1.6. It has an equal area which is positive and negative and so the average value is zero. we can verify by calculating

$$V_{av} = \frac{V_p}{2\pi} \int_0^{2\pi} \sin(\theta)d\theta = 0 \tag{1.2}$$

Its root-mean square (rms) value is therefore needed as some kind of measure. This is covered in most introductory books on circuit analysis. We find the rms value from the square root of the mean (or average) squared value. It is easiest to find the rms value squared and then take the square root.

$$V_{rms}^2 = \frac{V_p^2}{2\pi} \int_0^{2\pi} \sin^2(\theta)d\theta = \frac{V_p^2}{2} \tag{1.3}$$

From which

$$V_{rms} = \frac{V_p}{\sqrt{2}} \tag{1.4}$$

Note that we don't dwell too much on the intricacies of the basic calculus problem. This is a particularly simple integral to calculate by substituting $\sin^2(\theta) = \frac{1}{2}(1 - \cos(2\theta))$ and going down that path. A simpler approach is to use the tools available at the time, and we have many computer algebra software packages available including web pages that solve such problems.

This is the celebrated rms value of a sinewave that is used by all electrical engineers. A little less commonly known is what happens to a square wave. We consider two types of square wave. See Fig. 1.7.

Fig. 1.7 a, b Two different
square waves

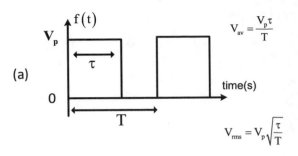

(a)

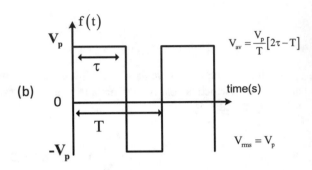

(b)

Figure 1.7a is a waveform that is commonly met in many areas. One possibility is that if the width τ varies with time but the period T stays constant, that we have pulse width modulation or PWM. This is used in power electronics for controlling the average (or dc) voltage from say a microprocessor. Microprocessors or any digital device usually only have +5 V (or lower) outputs and zero. A digital output port can only switch between the two and cannot take on an intermediate value of voltage. So, to get around this we vary the width and this in turn varies the average value. Since the average value is indeed the dc value, we can control dc. The average value of Fig. 1.7a is easily calculated without calculus as it is just the area of a rectangle divided by the period of the waveform (since the remainder of the waveform is zero area) We have that

$$V_{av} = \frac{V_p \tau}{T} \qquad (1.5)$$

By squaring the amplitude and taking the area we can find that

$$V_{rms}^2 = \frac{V_p^2 \tau}{T} \qquad (1.6)$$

So that

$$V_{rms} = V_p \sqrt{\frac{\tau}{T}} \qquad (1.7)$$

From Fig. 1.7b the average value of a bipolar square wave can be calculated. Find the areas in the positive and negative cycle and add giving

$$\begin{aligned} V_{av} &= \frac{V_p}{T}[\tau - (T - \tau)] \\ &= \frac{V_p}{T}[2\tau - T] \end{aligned} \qquad (1.8)$$

The rms value squared is just the amplitude of the two parts of the waveform squared (positive and negative) divided by the period. Hence

$$\begin{aligned} V^2_{rms} &= \frac{V_p^2}{T}[\tau + (T - \tau)] \\ &= V_p^2 \end{aligned} \qquad (1.9)$$

The rms value becomes

$$V_{rms} = V_p \qquad (1.10)$$

So, the rms value of a square-wave which swings an equal amount positive and negative is just its peak value.

Now consider the example below of a square wave with a dc level.

Example 1.1 Find the average and rms values of the following waveform.

Figure 1.8 Shows a square-wave type waveform. Find the average and rms values.

The average or dc value is just the total area divided by the period of the waveform which is 10 s.

$$V_{av} = \frac{2 \times 6 - 4 \times 1}{10} = 0.8 \qquad (1.11)$$

The rms value squared is the sum of the squared amplitudes divided by the period.

Fig. 1.8 Example 1.1 waveform

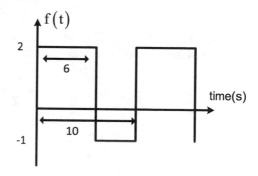

$$V_{rms}^2 = \frac{4 \times 6 + 1 \times 4}{10} = 1.4 \qquad (1.12)$$

Take the square root

$$V_{rms} = 1.183 \qquad (1.13)$$

Finally, a triangular waveform is considered. Consider two different types. A wide range of applications exist for these types of waveform. They can be used to test linearity for instance (the ramp part of the waveform) or to generate a hardware PWM type signal. For hardware PWM generation the signal to be modulated is compared to a triangular waveform and the output (since it is hard-limited) is PWM. In any case this is a common enough waveform to analyse.

For Fig. 1.9(a), the area is the area under a triangle. (0.5 × base × height). The average value is then

$$V_{av} = \frac{V_p T}{2}\frac{1}{T} = \frac{V_p}{2} \qquad (1.14)$$

The rms value is found by squaring the equation for the signal, dividing by the period T and taking the square root. Define the line as having equation

$$f(t) = \left(\frac{V_p}{T}\right)t \qquad (1.15)$$

where t is the independent variable time in seconds. Then

$$V_{rms}^2 = \frac{1}{T}\int_0^T \left(\frac{V_p}{T}t\right)^2 dt$$

$$= \frac{V_p^2}{3} \qquad (1.16)$$

$$V_{rms} = \frac{V_p}{\sqrt{3}} \qquad (1.17)$$

Waveform (b) in Fig. 1.9 is a little harder. The average is easy however and is found from the area divided by the period.

$$V_{av} = \frac{V_p t_1}{2}\frac{2}{T}$$

$$= V_p \frac{t_1}{T} \qquad (1.18)$$

Fig. 1.9 Two different
triangular waveforms

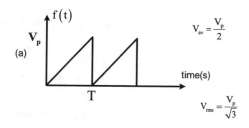

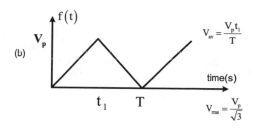

The rms can be found by squaring and integrating the equation of the line of the first triangle up to t_1 and the second from t_1 to T. For the first part we have a straight line at the origin with slope $\frac{V_p}{t_1}$ and therefore

$$V_{rms}^2 = \frac{1}{T} \int_0^{t_1} \left(\frac{V_p}{t_1} t\right)^2 dt \quad 0 \le t \le t_1$$

$$= \frac{V_p^2 t_1}{3T}$$

(1.19)

For $t_1 \le t \le T$ the slope of the line (call it m) is negative going

$$m = -\frac{V_p}{(T - t_1)}$$

(1.20)

Two points on the line are say (x_0, y_0) and are at $(T,0)$ and (t_1, V_p). The equation of a straight line going through a point (x_0, y_0) is

$$f(t) - y_0 = m(t - x_0)$$

(1.21)

Using the coordinate $(T,0)$, the equation of the line is therefore

$$f(t) = -\frac{V_p}{T - t_1}(t - T)$$

(1.22)

The rms value squared is

$$V_{rms}^2 = \frac{1}{T} \int_{t_1}^{T} \frac{V_p^2(T-t)^2}{(T-t_1)^2} dt$$

$$= \frac{V_p^2}{3T}(T-t_1)$$

(1.23)

Adding the two parts we have the sum of squares of two rms values over the two different time ranges. The total rms value squared is

$$V_{rms}^2 = \frac{V_p^2 t_1}{3T} + \frac{V_p^2}{3T}(T-t_1)$$

$$= \frac{V_p^2}{3}$$

(1.24)

The rms value is now

$$V_{rms} = \frac{V_p}{\sqrt{3}}$$

(1.25)

The previous signals all repeat with period T and are called periodic signals. Signals that do not repeat are known as *aperiodic* signals. Such a signal is shown below (Fig. 1.10).

This signal has different spectral (frequency) properties to a periodic signal.

Signals can also be *random*. Another word commonly used is *stochastic*. This means that unlike say a periodic wave where we know exactly what the next values in time will be based on previous ones, this is no longer true. Periodic waves are also known as *deterministic* signals. A good example of a random signal is the signal obtained from a person speaking. This is usually known as a speech signal. An example is shown below (Fig. 1.11).

We cannot tell what the next sample of the signal will be based on previous records. Instead, we can only determine that there is a *probability* that the signal will have a particular value. What governs or defines a random signal is usually its probability distribution and statistical properties. More of this will be studied in a later chapter.

Fig. 1.10 Example of an
aperiodic signal

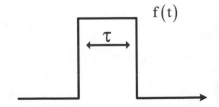

Fig. 1.11 Example of a
random or stochastic signal.
Section of speech

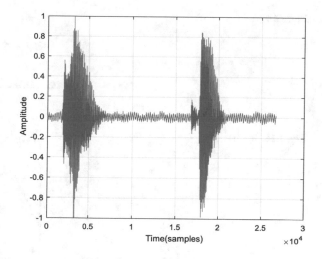

1.3 Signal Operations

A signal can be shifted to the right or left with a simple operation. For a signal f(t),
then $f(t - \tau)$ shifts the signal to the right in time by τ seconds. Also $f(t + \tau)$ shifts
the signal to the left by τ seconds. Performing the operation f(2t) reduces the length
of the signal by a half (if it is aperiodic) and f(0.5t) doubles the length of the signal.
If the signal is periodic then it either doubles or halves the frequency.

Example 1.2 Signal operations.

Consider a signal f(t) (Fig. 1.12).
Now we see what happened when t = –t. That is f(–t) (Fig. 1.13).
The signal just runs backwards in time. Put simply it is flipped in the vertical axis
like a mirror.
Figure 1.14 shows f(0.5t). The time duration of the finite-length signal is doubled.
In a similar manner f(2t) has the length halved from f(t) (Fig. 1.15).
Getting a little more complicated consider f(2t + 1) (Fig. 1.16).
For f(2t + 1), equate the argument within brackets to zero giving 2t + 1 = 0 and
solving gives t = -0.5. This corresponds to the starting point of the waveform. The

Fig. 1.12 f(t)

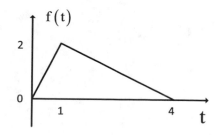

Fig. 1.13 f(–t)

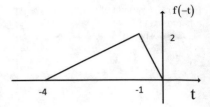

Fig. 1.14 f(0.5t)

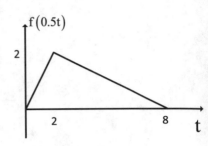

Fig. 1.15 f(2t)

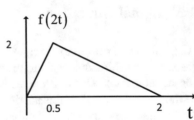

Fig. 1.16 Shifting and
scaling of a waveform from
Fig. 1.12

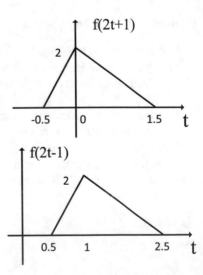

Fig. 1.17 Shift and scale a
waveform from Fig. 1.12

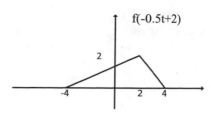

waveform has half the duration due to the 2t term and therefore the rest follows. Likewise, f(2t − 1) has a shift to the right by 0.5 (not 1 which would be a common mistake), and the duration is also halved in size.

Finally consider f(−0.5t + 2). Equate the term in brackets to zero giving −0.5t + 2 = 0 or t = 4. Time 4 s is the starting point and due to the negative term i.e. −0.5t. the waveform must go backwards in time and have twice the duration of the original.

A good way to check your results is to substitute values of time into −0.5t + 2. This should give the original point in time from f(t). So, from Fig. 1.17 substitute t = 4 into −0.5t + 2 and get 0. Substitute t = 2 and get 1 in the original. Finally substitute t = −4 and get 4. Hence the salient points in the transformed signal all map onto the original at the right point in the waveform.

1.4 Signal Symmetry

Of some use to us later in Fourier series, we need to know the difference between waveforms which are classified as even or odd. In mathematics they are called even or odd function, but here we stick to the engineering term odd or even signals.

For a signal f(t), it is classified as being *even* if f(− t) = f(t). A good example of this is a cosine waveform. For example, $\cos(-\omega t) = \cos(\omega t)$ is even. A signal is classified as being *odd* if f(− t) = − f(t). A good example of an odd signal is a sine waveform. Clearly $\sin(-\omega t) = -\sin(\omega t)$ is an odd waveform. A waveform that does not satisfy either the odd or even conditions is classified as being neither odd nor even. This is best illustrated with an example.

Example 1.3 Symmetry in waveforms.

Consider the following waveform in Fig. 1.18.

It is symmetric about the vertical axis and therefore f(− t) = f(t) making the signal even. The best way to recognise even symmetry is to think of the vertical axis as a mirror and the reflection of the waveform in positive time becoming the waveform in negative time.

Such a signal is easily made to be odd by shifting the vertical axis to the left or right. Figure 1.19 is an example of odd symmetry.

The waveform is odd because f(−t) = −f(t).

Fig. 1.18 An even
symmetry signal

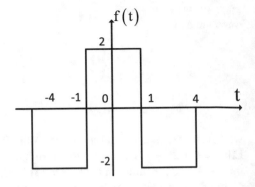

Fig. 1.19 Example of an
odd symmetry signal

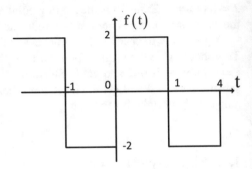

Now consider the signal in Fig. 1.20.

At first sight the waveform looks odd, but on closer inspection we see that f(- t) $\neq$ - f(t) since the amplitude goes up to +2 but down to only −1.

In real life there is no such thing as an even or odd waveform since there is no reference to a y (vertical) access like there is on paper. When a waveform generator is switched on, how we view the waveform or capture it all depends on a relative measurement in time. Time zero can be any part of a periodic waveform you choose as a starting point. Here the concept is a mathematical convenience.

Fig. 1.20 A signal that is
neither odd nor even

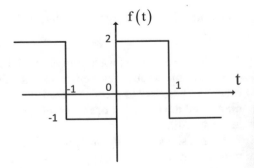

1.4.1 *Writing Any Signal as the Sum of Even and Odd Signals*

Any signal can be written as a sum of an odd and even waveform. For any signal $f(t)$ we can write

$$f(t)=[f(t) + f(-t)]/2+[f(t) - f(-t)]/2 \qquad (1.26)$$

The first part of (1.26) $[f(t) + f(-t)]/2$ is always an even signal and the second part $[f(t) - f(-t)]/2$ will always be odd. To illustrate this, consider the waveform shown in Fig. 1.21a. It is neither even nor odd.

However, now look at Fig. 1.22.

Fig. 1.21 a Shows a signal that is neither odd or even and **b** its time-reversed version

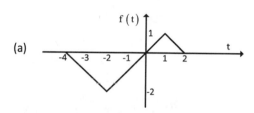

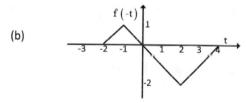

Fig. 1.22 Shows the decomposition into an even (**a**) and odd (**b**) waveform

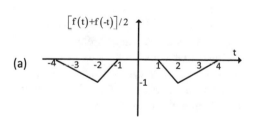

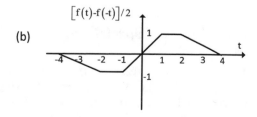

Fig. 1.23 Examples of
a causal, **b** aniticausal and
c noncausal signals

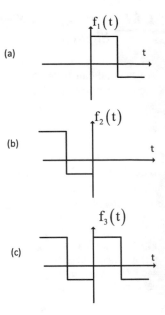

The waveform has been decomposed into two waveforms, one even and one odd. Adding (a) and (b) from Fig. 1.22 will result in the original waveform of Fig. 1.21a.

1.5   Causal and Non-causal Signals

In our physical world, there are basic laws of physics that forbid certain things from happening. One such thing are signals that go backwards in time or that start in negative time. Although in this book and in general we study such signals, this is more of a mathematical convenience, the signals do not appear in the physical world. A signal that goes backwards in time is known as an *anticausal* signal. A signal that is found in everyday life has positive time only as the independent variable and is known as a *causal* signal. Signals with both positive and negative time in their waveform are known as *noncausal* signals. All three are illustrated in Fig. 1.23a–c.

1.6   Signals of Special Importance

There are several signals that are used or seen commonly in electrical engineering. Perhaps the best known is the sinewave (Fig. 1.6). This is used as a test signal for determining the frequency response of amplifiers or electro-mechanical control-systems. It is also used in communication systems as a carrier frequency.

Fig. 1.24 **a** Step, **b** ramp and **c** pulse waveforms

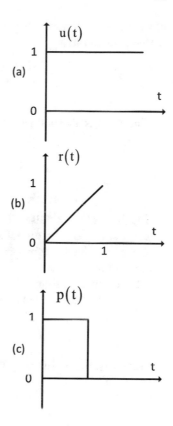

Figure 1.24a shows a step waveform (usually referred to as a step function). It is just the positive edge of a square wave. It looks like a doorstep and hence its name. The drop to zero is omitted since it is used to find the transient response of systems as a test signal and the negative going drop to zero gives us similar information (it is just a negative going step). So the step response of a system is a system with input of a step u(t), with usually unit magnitude. Practically we use a function generator set to a square wave but set its frequency to be very low (one Hz or less). We then apply this to an unknown system and record the output from the positive going edge (or step). We define a ramp mathematically as

$$u(t) = \begin{cases} 1, t \geq 0 \\ 0, t < 0 \end{cases} \tag{1.27}$$

Figure 1.24b represents a ramp waveform (of unit slope). This is generated in the laboratory from a very low frequency triangular or sawtooth wave. It is used for a number of applications including similar time-domain method of finding whether a system is linear. A ramp r(t) is a good way of finding the linearity of say an amplifier. For a ramp at the input, the output should also be a straight-line ramp. If it curves upwards or downwards then we have linearities in the system. A ramp is defined mathematically as

$$r(t) = \begin{cases} t, & t \geq 0 \\ 0, & t < 0 \end{cases} \tag{1.28}$$

The ramp is also defined as the integral of a step.

$$r(t) = \int_0^t u(t)dt \tag{1.29}$$

Figure 1.24c is a pulse signal. It is found in a great number of systems since it can be thought of as a digital signal. It can be used as an input to a digital latch to enable a line which performs some other action. It's duration in time depends on the application.

The rectangular signal shown in Fig. 1.25a is just a pulse of width T seconds but has a formal mathematical definition.

$$\prod(t) = A.\text{rect}(t/T) \tag{1.30}$$

Similarly, there is a triangular pulse shown in Fig. 1.25b.

$$\text{triang}(t) = f(t) = A[1 - |t|/T] \tag{1.31}$$

These are just shortcut methods of describing signals of this nature.
Finally, consider the impulse signal (or function) as shown in Fig. 1.26.

Fig. 1.25 Rectangular and triangular signals

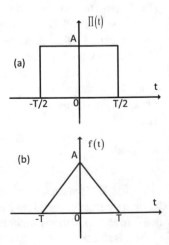

Fig. 1.26 Impulse or delta signal

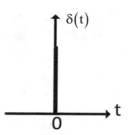

The impulse or Dirac delta signal $\delta(t)$ is defined as a rectangular pulse of width epsilon and height $1/\epsilon$ for some small value of epsilon. It therefore has an area which is unity. In the limit we let $\varepsilon \to 0$ and the theoretical Dirac impulse has height infinity and zero width. Hence

$$\int_{-\infty}^{\infty} \delta(t)dt = 1 \tag{1.32}$$

Practically, an impulse does not exist, though we can approximate one by using a function generator to have a low square wave frequency (to create one pulse every say 10 s) and reduce the width of the pulse to value as small as possible. This gives us a train of impulses of course and not a single one, but this doesn't matter. A more practical theoretical impulse is given by the *unit impulse*. It is defined as

$$\delta(t) = \begin{cases} 1, t = 0 \\ 0, t \neq 0 \end{cases} \tag{1.33}$$

It has a value of unity when $t = 0$ and otherwise is zero. This gives rise to an interesting mathematical proposition. It means that for any signal $f(t)$

$$\int_{-\infty}^{\infty} f(t)\delta(t - \tau)dt = f(\tau) \tag{1.34}$$

This property has often been referred to as the *sifting* property. In (1.34) above $\delta(t - \tau)$ is an impulse shifted in time so that it has a value of unity when $t = \tau$ and is otherwise zero. Therefore, when multiplied by an arbitrary signal and integrated, the value of f(t) at $t = \tau$ is given out as the solution. The other terms in the integration must be zero. We also note that the step signal is found from the impulse accordingly.

$$u(t) = \int_{-\infty}^{t} \delta(t)dt \tag{1.35}$$

Note there is another definition of impulse, namely the Kronecker impulse covered in sampled-data signals.

References

1. J.W. Klooster, *Icons of Invention: The Makers of the Modern World from Gutenberg*, vol. 1 (Greenwood press, London, UK, 2009)
2. C. Wu, Electric fish and the discovery of animal electricity. Am. Sci. **72**(6) (1984)
3. L. Coe, *A history of Morses's invention and its predecessors in the United States* (McFarland, Jefferson, North Carolina, 1993)
4. J.C. Maxwell, A dynamical theory of the electromagnetic field. Philos. Trans. R. Soc. Lond. **155**, 459–512 (1865)
5. T.K. Sarkar, R. Mailloux, A.A. Oliner, M. Salazar-Palma, D.L. Sengupta, *History of Wireless* (Wiley, New Jersey, 2006).
6. H.W. Bode, Relationship between attenuation and phase in feedback amplifiers. Bell Syst. Tech. J. **19**, 421–454 (1940)
7. H.S. Black, Stabilized feedback amplifiers. Bell Syst. Tech. J. **13**(1), 1–18 (1934)
8. C.E. Shannon, A mathematical theory of communications. Bell Syst. Tech. J. **27**, 379–423 (1948)
9. G.A. Campbell, in *The Froehlich/Kent Encyclopedia of Telecommunications*, vol. 2, ed. by F. E. Froehlich and A. Kent (Marcel Dekker Inc, NY, USA, 1991), p. 241
10. P.E. Valivach, Basic stages of the history of electrical engineering and possible prospects for its development. Russ. Electr. Eng. **80**(6), 350–358 (2009)

Chapter 2
Dynamic Systems Introduction

2.1 Definition of a Linear System

A system is just a generic term used to define some mechanism or circuitry or combination of both that has an input and an output. (often many inputs and many outputs). For example, consider Fig. 2.1.

The input x(t) is some signal that is applied to the system and the output y(t)is known as the response.

A system with one input and output is called a single-input, single-output system or SISO system. One with multiple inputs and outputs is known as a multi-input multi-output system or MIMO system. For MIMO systems the term Multivariable is often used in the discipline of control engineering and Multivariate in statistics and mathematics. For the basic definitions, a SISO system will be used.

2.1.1 Linear System Definition

A linear system originated in mathematics under the topic of linear differential equations (D.Es). This is a little different from the engineering description but means the same thing. Since dynamical systems are represented by differential equations, then for a differential equation in y it can be written in the form

$$a_n(x)\frac{d^n y}{dx^n} + \cdots + a_1(x)\frac{dy}{dx} + a_0(x)y = b(x) \qquad (2.1)$$

Here y and all its derivatives exist and are to the first power only and the coefficients $a_i(x), i = 0,1,2...n$ are functions of x only (or a constant). For example, $\left(\frac{dy}{dx}\right)^2$ cannot exist in a linear differential equation as the derivative is a power of 2. As an example, suppose we have the D.E

Fig. 2.1 Generic system
with one input and one
output

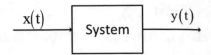

$$\frac{d^2y}{dx^2} + \cos(x)\frac{dy}{dx} + y = \sin(x) \tag{2.2}$$

Since y and its two derivatives are all of power one, and the coefficients of these derivatives are all functions of x, the system is linear.

Likewise

$$\frac{d^2y}{dx^2} + y\frac{dy}{dx} + y = \sin(x) \tag{2.3}$$

y and its derivatives are all of power one but one of the coefficients is not a function of x. Therefore, it is a nonlinear D.E. However

$$x^3\frac{d^2y}{dx^2} + \frac{dy}{dx} + y = 2 \tag{2.4}$$

Although the cubic makes it look nonlinear, this system is in fact linear because y and all its derivatives are one and the coefficients are all powers of x. Finally,

$$\frac{d^2y}{dx^2} + x^2\sqrt{\frac{dy}{dx}} + y = 3 \tag{2.5}$$

Must be nonlinear because the derivative of y is a square root and not order one.

Usually for our engineering applications the independent variable x is time.

In electrical engineering the definition is usually given that a linear system must satisfy *superposition*. Superposition is usually studied in a first-level electronics or circuit-theory course and goes like this:

Superposition

Suppose a system has one input and one output. A test for the superposition principle is as follows. We apply two inputs separately to the system, say $x_1(t)$ and $x_2(t)$ with each giving rise to a unique output $y_1(t)$ and $y_2(t)$ respectively. If we then add the two inputs $x_1(t) + x_2(t)$ and apply that signal instead, the output will be the same as the summed outputs $y_1(t) + y_2(t)$ obtained from the individual inputs.

This is more practical than the mathematical approach since system models are not always known in advance and tests need to be performed to obtain it. Figure 2.2 illustrates this.

Fig. 2.2 Superposition
principle

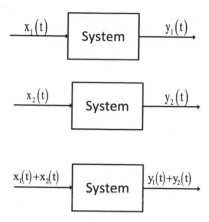

Linear system definition and testing

A linear system is a system that satisfied the superposition principle. To practically
test a real electronic system, we apply a signal (say x_1, a sinewave of frequency f1)
to the system and measure the output amplitude and frequency. We then remove x_1
and apply a signal x_2 at a frequency f2. We note the amplitude and frequency of this
output. In both cases there should only be one frequency present at the output. Now
if we add both x_1 and x_2 and apply to the same system we should get the sum of
the two previous signals. Usually if a high frequency sinewave and low frequency
sinewave are used then the sum is easy to spot as the low frequency one with the high
frequency sinewave superimposed on top. For a unity-gain amplifier (a buffer) we
would see something like the inputs [(a) and (b)] and output (c) shown in Fig. 2.3.
Of course, the scaling depends on the gain and in this case, it is assumed to be unity.
Also, no phase-shift is shown and in a real system the phases would not be aligned.
Often this is termed *Additivity* and the more generally definition of linearity is taken
to satisfy *Homogeneity* as well as *Additivity*. Homogeneity is the simple property
that if you scale the input by some factor that the output must also be scaled by the
same quantity.

Suppose the system was a square-law device, then the output would look as shown
in Fig. 2.4.

The output will then not only consist of the original two sinewaves, but
cross-product terms as well. We can verify this mathematically as follows. For
two cosine inputs we have $x_1(t) = a\cos(\omega_1 t), x_2(t) = b\cos(\omega_2 t)$. If the system
is linear with say a gain K then the output becomes (ignoring phase-shift),
$y(t) = K a\cos(\omega_1 t) + K b\cos(\omega_2 t)$ or the sum of the scaled outputs found when
inputs are taken one at a time. Hence the system must satisfy both *Additivity and
Homogeneity*. With the square-law device we have as output

$$y(t) = [a\cos(\omega_1 t) + b\cos(\omega_2 t)]^2 \qquad (2.6)$$

Expanding we get

Fig. 2.3 Sum of two
sinewaves at output of
system with unity gain

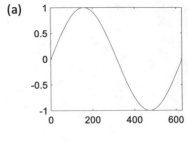

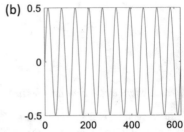

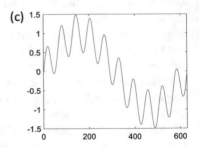

Fig. 2.4 Output of system
when the system is a
square-law device

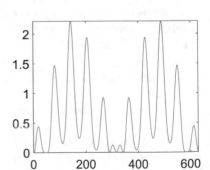

$$y(t) = a^2\cos^2(\omega_1 t) + b^2\cos^2(\omega_2 t) + 2ab\cos(\omega_1 t)\cos(\omega_2 t) \qquad (2.7)$$

And straight away it can be seen that a cross-product term appears plus the squared
individual terms. Since by two trig identities

Fig. 2.5 Time-invariance.
a original input and output.
b delayed input and output

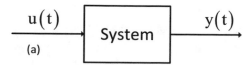

$$\cos^2(x) = \frac{1}{2}[1 + \cos(2x)] \tag{2.8}$$

and

$$\cos(\omega_1 t)\cos(\omega_2 t) = \frac{1}{2}\cos(\omega_1 + \omega_2)t + \frac{1}{2}\cos(\omega_1 - \omega_2)t \tag{2.9}$$

then (2.6) becomes

$$y(t) = \frac{a^2}{2}[1 + \cos(2\omega_1 t)] + \frac{b^2}{2}[1 + \cos(2\omega_2 t)] + ab[\cos(\omega_1 + \omega_2)t + \cos(\omega_1 - \omega_2)t]$$
$$= \frac{a^2 + b^2}{2} + \left(\frac{a^2}{2}\right)\cos(2\omega_1 t) + \left(\frac{b^2}{2}\right)\cos(2\omega_2 t) + ab\cos(\omega_1 + \omega_2)t + ab\cos(\omega_1 - \omega_2)t \tag{2.10}$$

Equation (2.10) gives a dc term $\frac{a^2 + b^2}{2}$, two frequencies at double the input frequencies $2\omega_1$ and $2\omega_2$, and two sum and difference signals. In fact, we do not get the original frequencies at all in the output.

2.1.2 Examples of Linear and Nonlinear Systems

Example 1 Consider a static system which has a straight-line representation from input to output given by.

$$y(t) = k_1 x(t) + k_2 \tag{2.11}$$

Determine if this system is linear or nonlinear.

Solution

For an input $x_1(t)$, using (2.11) the output is $y_1(t) = k_1 x_1(t) + k_2$. Likewise, for an input $x_2(t)$ the output is $y_2(t) = k_1 x_2(t) + k_2$. The sum of these (superposition) that we would expect to get out were the system linear is therefore $y(t) = k_1 (x_1(t) + x_2(t)) + 2k_2$.

Now calculate the output for the summation of the two inputs $x(t) = (x_1(t) + x_2(t))$ substituted directly into (2.11) and get $y(t) = k_1 (x_1(t) + x_2(t)) + k_2$. Now clearly this is not the equation to our previously found output, it differs. The two are only the same when $k_2 = 0$. The system is therefore nonlinear.

Example 2 System is given by.

$$y(t) = tx(t) \tag{2.12}$$

Two outputs are $y_1(t) = tx_1(t), y_2(t) = tx_2(t)$. Adding for superposition gives

$$y(t) = tx_1(t) + tx_2(t) \tag{2.13}$$

Consider both inputs added together applied to (2.12). The output is.
$y(t) = t[x_1(t) + x_2(t)]$ which is the same as (2.13). Therefore, the system is linear.

2.2 Definition of Time-Invariant System

For a time-invariant system, we just simply delay the input $u(t)$ by some amount in time, say τ to give $u(t - t)$. If the original output with no delay was $y(t)$, then when delayed the output should become $y(t - \tau)$. The system responds the same no matter at what time the input is applied.

2.2.1 Examples of Time-Invariance

Suppose a system is defined by

$$y(t) = tu(t) \tag{2.14}$$

Delaying the input only gives an output $tu(t - \tau)$. But if there is a delay the output must be delayed according to. $(t - \tau)u(t - \tau)$.

The system is therefore not time invariant since a time-delay at the input does not give the same result as a time-delayed output. In terms of differential equations, this means that any coefficients must be constant and not vary with respect to the independent variable.

Interestingly the squared function as a system which was previously shown to be nonlinear.

$$y(t) = (u(t))^2 \tag{2.15}$$

is clearly time-invariant since delaying the input and output give the same results of $(u(t - \tau))^2$.

A system given by

$$y(t) = Ku(t) \tag{2.16}$$

Is also time-invariant since a delay in the input and output yield the same result. This system is also linear as explained with a previous example (Fig. 2.5).

2.3 Linear Time Invariant (LTI) Systems

Put the two earlier concepts of a linear and time-invariant system together and we arrive at the definition for a linear time-invariant system or LTI for short. Such a system must satisfy both the linearity and time-invariance properties. In terms of differential equations, an LTI system is defined in terms of differential equations with constant coefficients and no powers of y (where y is the output) or its derivatives greater than 1. (or less than 1 as the case may be). For example

$$\frac{d^2y(t)}{dt^2} + a_1\frac{dy(t)}{dt} + a_2y(t) = u(t) \tag{2.17}$$

Is a second-order linear differential equation with constant coefficients. It is a LTI system with output y and input u, both functions of time. Variations of this are also possible when the input also has derivatives and is represented by a linear differential equation with constant coefficients. For example

$$\frac{d^2y(t)}{dt^2} + a_1\frac{dy(t)}{dt} + a_2y(t) = u(t) + b_1\frac{du(t)}{dt} \tag{2.18}$$

Is also a LTI system.

Consider a simple RC circuit as shown in Fig. 2.6.

Summing voltages around the loop (Kirchhoff's voltage law) gives the linear first-order differential equation.

$$RC\frac{dy(t)}{dt} + y(t) = u(t) \tag{2.19}$$

Fig. 2.6 RC Circuit. Gives
first order ODE

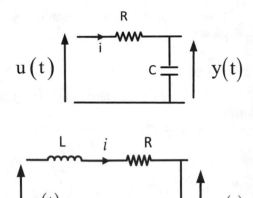

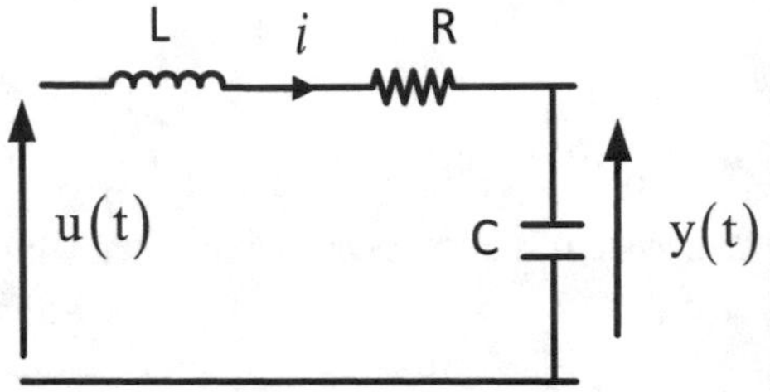

Fig. 2.7 RLC Circuit gives
second-order ODE

The coefficients of y(t) and its derivative are all constants indicating a linear differential equation. To find the output of this system to any arbitrary input the equation could be solved using classical methods found in basic calculus and goes back to the times of Isaac Newton and Leibniz, the inventors of calculus. This sort of equation is usually commonly known as an ordinary differential equation (ODE). Closed-form solutions for y(t) are readily available for most input types. This is a first order ODE because there is one derivative of y(t) in (2.19). In engineering terms, it is also first order due to the physics of the circuit. Namely, it has one energy storage device, the capacitor. Adding an inductor will increase the order to 2 (Fig. 2.7).

The differential equation is found by summing voltages around the loop as before. This gives a second-order ODE with constant coefficients.

$$LC\frac{dy^2(t)}{dt^2} + RC\frac{dy(t)}{dt} + y(t) = u(t) \tag{2.20}$$

The theory for solving such equations has been known for centuries but is a little cumbersome. In this textbook, since it is an engineering one, the method of Laplace transforms is preferred. This is studied later in the chapter.

As the number of energy storage devices increases, so does the order of the system. We can have high order systems formed by the interconnection of several lower order ones. The systems must be in some way isolated, however. For example, the RC circuit output can be connected to a second RC and a second-order system will result. The system will not be just two of the same, this is because an interaction from one to the other occurs when current flows from the first circuit into the second. This is shown in Fig. 2.8(a).

If a unity-gain buffer amplifier is inserted between the systems, the systems are said to be in cascade and we get two identical first-order systems, the output of the first going into the input of the second. Otherwise, the system dynamics are different

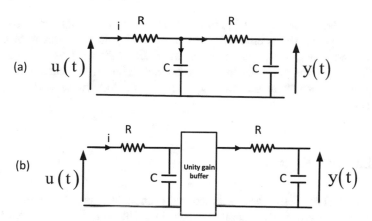

Fig. 2.8 Two first-order systems **a** has interaction from one to the other whilst **b** has two duplicate systems

Fig. 2.9 Mass-spring damper system

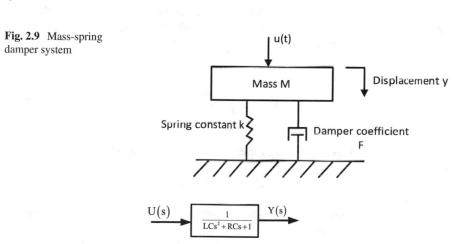

Fig. 2.10 Transfer function representation as a block diagram

due to a flow of current from the first into the second. The buffer acts as an isolator and enables the two circuits to behave independently (Figs. 2.9 and 2.10).

Finally consider a mechanical system consisting of a mass M, spring of stiffness k and a damper with coefficient of factor F. The displacement is y.

Using a free-body diagram it is simple to show that the differential equation of this system is second order and given by

$$M\frac{dy^2(t)}{dt^2} + F\frac{dy(t)}{dt} + ky(t) = u(t) \qquad (2.21)$$

where u(t) is the applied force and y(t) is the displacement. The system has two energy storage elements, the mass and spring and the system is second order. Clearly from earlier definitions Eq. (2.21) is linear because y(t) and its derivatives are of order one only, and the coefficients are constant. The same applies in mechanical systems as with electrical, if we connect two together the net system is not always two isolated systems one after the other (cascade), but in this case only so if the forces from one are isolated from that of the second. Otherwise, multiple systems will interact.

2.4 Laplace Transforms

The Laplace Transform was invented by French Mathematician Pierre-Simon Laplace when it arose in probability theory [1]. The technique is a great tool for engineers in solving LTI differential equations for various input types. The method probably lay unused until a similar method was proposed by engineer Oliver Heaviside. It is an operator method which transforms differential equations into algebraic equations. In popularising the method and extending it, Heaviside also introduced the idea of *transfer-functions*. The Laplace transform (LT) for a signal f(t) is defined as

$$F(s) = \int_0^\infty f(t)e^{-st}dt \tag{2.22}$$

This is also written using the shortcut $\mathcal{L}$ (script L for Laplace) notation thus

$$\mathcal{L}\{f(t)\} = F(s) \tag{2.23}$$

The usual convention is that upper case letters are used for LT and lower-case for time-domain, but not always.

Sometimes for mathematical convenience the LT is defined as the *bilateral* LT, where the lower limit on the integration is minus infinity instead of zero. This means that non-causal signals or anticausal signals can be transformed. For simple signals, the solution to (2.22) is readily available in tables.

2.4.1 *Laplace Transform of Common Signals.*

To calculate the LT of a *unit step* let f(t) = 1. Then

$$F(s) = \int_0^\infty e^{-st}dt = \frac{1}{s} \tag{2.24}$$

The LT of a *ramp* with slope unity is found when $f(t) = t$

$$F(s) = \int_0^\infty te^{-st}dt = \frac{1}{s^2} \tag{2.25}$$

The LT of a *sinusoid* is found when $f(t) = \sin(\omega t)$

$$F(s) = \int_0^\infty \sin(\omega t)e^{-st}dt = \frac{\omega}{s^2 + \omega^2} \tag{2.26}$$

The LT of a *unit impulse* is found when $f(t) = \delta(t)$

$$F(s) = \int_0^\infty d(t)e^{-st}dt = 1 \tag{2.27}$$

To find the LT of a *time-delayed* signal say $f(t - \tau)$.

$$\mathcal{L}\{f(t - t)\} = \int_0^\infty f(t - t)e^{-st}dt \tag{2.28}$$

Substitute $t - \tau = x$, $dt = dx$. When $t = 0$, $x = -\tau$ and when $t = \infty$, $x = \infty$. We then obtain

$$\mathcal{L}\{f(t - t)\} = \int_{-\tau}^\infty f(x)e^{-(x+t)s}dx = e^{-st}\int_{-t}^\infty f(x)e^{-xs}dx \tag{2.29}$$

If $f(x)$ is zero for negative time we get

$$\mathcal{L}\{f(t - \tau)\} = e^{-st}\int_0^\infty f(x)e^{-xs}dx$$

$$= e^{-st}F(s) \tag{2.30}$$

This is also called the *shifting* property of the LT.
The LT of a *decaying exponential* $f(t) = e^{-at}$ is found from

$$\mathcal{L}\{e^{-at}\} = \int_0^\infty e^{-at}e^{-st}dt$$

$$= \int_0^\infty e^{-(s+a)t} dt = \frac{1}{s+a} \tag{2.31}$$

2.4.2   Laplace Transform of a First Derivative

$$\mathcal{L}\left\{\frac{dy(t)}{dt}\right\} = \int_0^\infty \frac{dy(t)}{dt} e^{-st} dt \tag{2.32}$$

This can be integrated by parts to yield

$$\mathcal{L}\left\{\frac{dy(t)}{dt}\right\} = sY(s) - y(0) \tag{2.33}$$

where $y(0)$ is the initial condition, that is $y(t)$ at time zero.

2.4.3   Laplace Transform of a Second Derivative

$$\mathcal{L}\left\{\frac{dy^2(t)}{dt^2}\right\} = \int_0^\infty \frac{dy^2(t)}{dt^2} e^{-st} dt \tag{2.34}$$

This can be found to be

$$\mathcal{L}\left\{\frac{dy^2(t)}{dt^2}\right\} = s^2 Y(s) - sf(0) - f(0) \tag{2.35}$$

2.4.4   Laplace Transform of a System and Its Transfer-Function

Consider the previously used RLC circuit and its Eq. (2.21).

$$LC\frac{dy^2(t)}{dt^2} + RC\frac{dy(t)}{dt} + y(t) = u(t)$$

Taking LTs with zero initial conditions yields

$$LCs^2Y(s) + RCsY(s) + Y(s) = U(s) \tag{2.36}$$

This is an algebraic equation which can be written as

$$\left[LCs^2 + RCs + 1\right]Y(s) = U(s) \tag{2.37}$$

Then we write the ratio of the output to input

$$\frac{Y(s)}{U(s)} = \frac{1}{LCs^2 + RCs + 1} \tag{2.38}$$

Known as the *Transfer-function* of the system. We can draw a block-diagram illustrating this method, which is far more intuitive than differential equations.

The originator of this work (Oliver Heaviside) used the letter $p = d/dt$ rather than s as an operator to represent differentiation. When taking transfer functions from differential equations the initial conditions are always zero. When finding the response of a system to a given input, the initial conditions may or may not be non-zero.

2.4.5 Table of Laplace Transforms

Tables of commonly met LTs are readily found in the literature. Here is a table of commonly met LTs. We use tables to convert from the time-domain into Laplace, then multiply by the LT of the input. To find the output we then inverse LT back to the time-domain. To find the inverse LT the tables are used directly, and if the function does not appear on the tables it must be split into smaller parts using a partial-fraction expansion. Inverse LT has $\mathcal{L}^{-1}$ as the usual notation used. For example, $\mathcal{L}^{-1}\{\frac{1}{s^2}\} = t$, is read directly from Table 2.1 third row down by finding the Laplace value in the third row and reading off the time-domain equivalent in the same row.

2.4.6 Multiple Connected Systems

Systems can be connected to yield higher-order systems. Figure 2.11a shows two systems in series (or cascade) and (b) two systems connected in parallel. Here we relax the necessity to use upper-case letter for LT variables.

For Fig. 2.11a, the output of $G_1(s)$ is given by $x(s) = G_1(s)u(s)$ and the output of $G_2(s)$ is given by $y(s) = G_2(s)x(s)$. Substituting means that for two systems in series $y(s) = G_1(s)G_2(s)u(s)$. For (b), the two parallel systems are simply added giving $y(s) = [G_1(s) + G_2(s)]u(s)$.

For systems in series, we therefore multiply their Laplace transfer functions. Theoretically it makes no difference which systems comes first since $G_1(s)G_2(s) =$

Table 2.1 Common laplace transforms

Time-Domain	Laplace Domain
$\delta(t)$ Unit Impulse at $t=0$	1
$u(t)=1$, unit step function	$\frac{1}{s}$
t unit ramp	$\frac{1}{s^2}$
t^n	$\frac{n!}{s^{n+1}}$
e^{-at} exponential decay	$\frac{1}{s+a}$
$\sin(\omega t)$ Sine wave	$\frac{\omega}{s^2+\omega^2}$
$\cos(\omega t)$	$\frac{s}{s^2+\omega^2}$
$e^{-at}\sin(\omega t)$	$\frac{\omega}{(s+a)^2+\omega^2}$
$e^{-at}\cos(\omega t)$	$\frac{s+a}{(s+a)^2+\omega^2}$
$\frac{dy(t)}{dt}$ 1st derivative	$sy(s)-y(0)$
$\frac{d^2y(t)}{dt^2}$ 2nd derivative	$s^2y(s)-sy(0)-\frac{dy(0)}{dt}$
$u(t-\tau)=1$, unit delayed step	$\frac{e^{-s\tau}}{s}$
$y(t-\tau)$, delayed signal	$Y(s)e^{-s\tau}$

Fig. 2.11 Two systems in **a** series or cascade, **b** in parallel

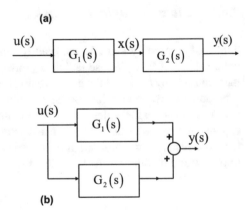

$G_2(s)G_1(s)$. However, they are usually physical restraints on which system comes first. For example, the first system could be a low power amplifier and the second high-power. The high-power one would then have to be second in the chain.

2.4.7 Example of Step Response, First-Order System

For the first-order RC circuit of Fig. 2.6, find the response to a unit step input. The DE of the system is Eq. (2.19)

Fig. 2.12 Step response of first order system. Shown with three different time-constant values

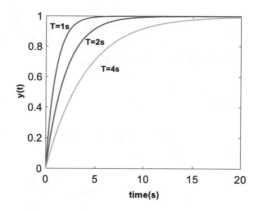

$$RC\frac{dy(t)}{dt} + y(t) = u(t)$$

Define $T = RC$ as the *time-constant* of the system. Taking LTs, the transfer function becomes

$$\frac{y}{u}(s) = \frac{1}{1 + sT} \tag{2.39}$$

For a unit step input $u(s) = 1/s$ and

$$y(s) = \frac{1}{s(1 + sT)} \tag{2.40}$$

However, $\frac{1}{s(1+sT)}$ does not appear on the LT table. It must be split using partial fractions in smaller pieces. First write $\frac{1}{s(1+sT)} = \frac{1/T}{s(s+1/T)}$. Then we have a partial fraction expansion

$$\frac{1/T}{s(s + 1/T)} = \frac{A}{s} + \frac{B}{s + 1/T} \tag{2.41}$$

where A and B must be determined. A few ways exist to do this. The simplest is by multiplying out (2.41) from the left by $s(s + 1/T)$ to give

$$1/T = A(s + 1/T) + Bs \tag{2.42}$$

Now compare coefficients of powers of s on both sides of (2.42).
Powers of s^0 gives $1/T = A/T$ from which $A = 1$.
Powers of s^1 gives $0 = A + B$ from which $B = -1$.
Then

$$\frac{1/T}{s(s+1/T)} = \frac{1}{s} - \frac{1}{s+1/T} \tag{2.43}$$

From (2.40), the output y(s) is

$$y(s) = \frac{1}{s} - \frac{1}{s+1/T} \tag{2.44}$$

Reading the LT table in reverse (that is looking at the Laplace value and reading the time-domain value) gives

$$y(s) = \mathcal{L}^{-1}\left\{\frac{1}{s} - \frac{1}{s+1/T}\right\} \tag{2.45}$$

where $\mathcal{L}^{-1}$ represents *inverse* Laplace transform.
From the table of LTs (the unit step and the exponential decay rows)

$$y(t) = 1 - e^{-t/T} \tag{2.46}$$

The time-constant T (units in seconds s) determines how long the system takes to reach the final value, in this case voltage as it is an RC circuit. It is usually taken as a rough measure that by $5 \times T$ the system has reached steady state. Steady state is defined as when the output has reached a constant value and does not change. Mathematically speaking, steady state is when time goes to infinity:

$$\begin{aligned} y(t)|_{t\to\infty} &= 1 - e^{-t/T}\big|_{t\to\infty} \\ &= 1 \end{aligned} \tag{2.47}$$

Any dynamic system always has a transient time before reaching steady state. The transient time is where the capacitor is charging as in Fig. 2.11. To get a measure of this we usually consider a time when t = T from which (Fig. 2.12)

$$\begin{aligned} y(T) &= 1 - e^{-1} \\ &= 0.632 \end{aligned} \tag{2.48}$$

The time-constant can therefore be directly measured by finding the time the system takes to reach 0.632 of its final value (unity in this case). See Fig. 2.13.
It can also be seen that by time 20 s ($5 \times T$) the system has reached steady state.

Fig. 2.13 Illustration of how to measure time constant from first order response. T = 4 s

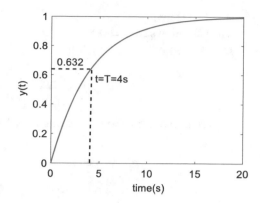

2.4.8 Higher Order Systems

Systems can be either naturally higher order than 1 or 2 or created as such by cascading other systems together. In such cases the method of Laplace is still possible (though simulation is preferred). The same method using LT tables is used but care must be taken when using partial fraction expansions. A few examples are shown for the step input of some higher order systems. Suppose the system TF is G(s) and a unit step is applied with Laplace value u(s) = 1/s. Then the output y(s) = G(s)u(s).

Example 1 If the TF is second order.

$$G(s) = \frac{1}{(s+1)(s+2)}$$

Then the output is found from

$$y(s) = \frac{1}{s(s+1)(s+2)}$$
$$= \frac{A}{s} + \frac{B}{s+1} + \frac{C}{s+2}$$

Multiplying out gives

$$1 = A(s+1)(s+2) + Bs(s+2) + Cs(s+1)$$

Compare coefficients of s on both sides of the equation.
s^0 1 = 2A giving A = 0.5.
s^1 0 = 3A + 2B + C = 1.5 + 2B + C.
s^2 0 = A + B + C = 0.5 + B + C.
Now solve simultaneously for B and C. We try subtraction first for the last two equations.

$0 = 1 + B$ giving $B = -1$.
Then C follows as $C = 0.5$
Using these values

$$y(s) = \frac{0.5}{s} - \frac{1}{s+1} + \frac{0.5}{s+2}$$

Inverse Laplace transforming using the table gives

$$y(t) = \mathcal{L}^{-1}\left\{\frac{0.5}{s} - \frac{1}{s+1} + \frac{0.5}{s+2}\right\}$$
$$= 0.5 - e^{-t} + 0.5e^{-2t}$$

The steady-state value when t goes to infinity is $y(t) = 0.5$.

Example 2 Suppose $G(s) = \frac{1}{s^2+\sqrt{2}s+1}$.

The polynomial $s^2+\sqrt{2}s + 1$ has complex roots and cannot be factorised. The output is

$$y(s) = \frac{1}{s\left(s^2+\sqrt{2}s+1\right)}$$
$$= \frac{A}{s} + \frac{Bs+C}{s^2+\sqrt{2}s+1}$$

From which we must find the constants A, B, C. Note the numerator of the second part has $Bs + C$, which is degree one less than the denominator. If the system was third order, we would need a quadratic in the numerator instead. Multiply out

$$1 = A\left(s^2+\sqrt{2}s + 1\right)+(Bs + C)s$$

Compare coefficients on both sides of the equation.
s^0 $1 = A$.
s^1 $0 = \sqrt{2}\,A + C$ giving $C = -\sqrt{2}$.
s^2 $0 = A + B$ giving $B = -1$.
Putting this together yields

$$y(s) = \frac{1}{s\left(s^2+\sqrt{2}s+1\right)}$$
$$= \frac{A}{s} + \frac{Bs+C}{s^2+\sqrt{2}s+1}$$
$$= \frac{1}{s} - \frac{s+\sqrt{2}}{s^2+\sqrt{2}s+1}$$

The second fraction could be a problem, but we recognise from the Table of LTs that

$$e^{-at}\cos(\omega t) = \mathcal{L}^{-1}\left\{\frac{s+a}{(s+a)^2+\omega^2}\right\}, \quad e^{-at}\sin(\omega t) = \mathcal{L}^{-1}\left\{\frac{\omega}{(s+a)^2+\omega^2}\right\}$$

Therefore, we re-arrange $\frac{s+\sqrt{2}}{s^2+\sqrt{2}s+1} = \frac{(s+\sqrt{2}/2)+\sqrt{2}/2}{\left(s+\frac{\sqrt{2}}{2}\right)^2+\frac{1}{2}}$.

and

$$y(s) = \frac{1}{s} - \frac{\left(s+\sqrt{2}/2\right)+1/\sqrt{2}}{\left[\left(s+\frac{\sqrt{2}}{2}\right)^2+\frac{1}{2}\right]}$$

$$= \frac{1}{s} - \frac{\left(s+\sqrt{2}/2\right)}{\left[\left(s+\frac{\sqrt{2}}{2}\right)^2+\frac{1}{2}\right]} - \frac{1/\sqrt{2}}{\left[\left(s+\frac{\sqrt{2}}{2}\right)^2+\frac{1}{2}\right]}$$

Using the above formula to inverse LT

$$y(t) = 1 - e^{-t/\sqrt{2}}\cos\left(\frac{1}{\sqrt{2}}t\right) - e^{-t/\sqrt{2}}\sin\left(\frac{1}{\sqrt{2}}t\right)$$

$$= 1 - e^{-t/\sqrt{2}}\left[\cos\left(\frac{1}{\sqrt{2}}t\right) + \sin\left(\frac{1}{\sqrt{2}}t\right)\right]$$

But $\cos\left(\frac{1}{\sqrt{2}}t\right) + \sin\left(\frac{1}{\sqrt{2}}t\right) = \sqrt{2}\cos\left(\frac{1}{\sqrt{2}}t - \pi/4\right)$.
Therefore

$$y(t) = 1 - e^{-t/\sqrt{2}}\sqrt{2}\cos\left(\frac{1}{\sqrt{2}}t - \frac{\pi}{4}\right)$$

The second term on the RHS of the equation is a decaying sinusoid of frequency $\frac{1}{\sqrt{2}}$ rad/s which goes to zero as time increases. This makes the steady-state value of $y(t) = 1$ as can be seen in Fig. 2.14.

2.5 Impulse Response

We have seen examples of dynamic systems and two methods of representing them. They are the differential equation and the Laplace Transform or Transfer Function. Another method that can act as a mathematical representation of a LTI system is the impulse response. As the name suggests, the impulse response of a system is its

Fig. 2.14 Step response of example 2.
$G(s) = \frac{1}{s^2 + \sqrt{2}s + 1}$

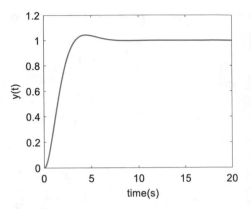

response to an impulse. We assume the impulse is a unit impulse for mathematical convenience. We will encounter the impulse response again when dealing later with convolution. The impulse response can be measured by applying a very narrow pulse (or series of pulses) to a system and recording the output. Usually for electrical systems this is not something encountered very often since the frequency response and step response is of more interest and we can glean most information from that. However, there are certain systems that the step response or frequency response cannot be used. One such example is the acoustics of a room. We can record the dynamics of the room by firing a starting pistol or bursting a balloon to act as the impulse and recording the sound from the response. This will give us a record of the room dynamics and we later use it for other purposes. More generally any LTI system has an impulse response we can calculate if we have the differential equation or transfer function. Here we prefer the use of the transfer function. The impulse response is just the *inverse* LT of the transfer function. It is a time-domain measure and always dies out to zero if the system is stable. For systems that are unstable they may will oscillate or the output will grow until saturation is reached. For example, consider the following examples.

Example 1 $G(s) = \frac{1}{s+2}$.

We denote the impulse response as $g(t) = \mathcal{L}^{-1}\left\{\frac{1}{s+2}\right\}$.
From tables this gives

$$g(t) = e^{-2t}$$

Which decays to zero as time increases.

Example 2 $G(s) = \frac{1}{(s+1)(s+2)}$

We denote the impulse response as $g(t) = \mathcal{L}^{-1}\left\{\frac{1}{(s+1)(s+2)}\right\}$.
Using partial fractions

$$g(t) = \mathcal{L}^{-1}\left\{\frac{A}{(s+1)} + \frac{B}{(s+2)}\right\}$$

Solving for the constants

$$g(t) = \mathcal{L}^{-1}\left\{\frac{1}{(s+1)} - \frac{1}{(s+2)}\right\}$$
$$= e^{-t} - e^{-2t}$$

This decays to zero as time increases.

Example 3 $G(s) = \frac{1}{s^2 + \sqrt{2}s + 1}$

We write this as.

$$G(s) = \frac{1}{s^2 + \sqrt{2}s + 1}$$
$$= \frac{1}{\left(s + \sqrt{2}/2\right)^2 + 1/2}$$

From the Table of LTs

$$e^{-at}\sin(\omega t) = \mathcal{L}^{-1}\left\{\frac{\omega}{(s+a)^2 + \omega^2}\right\}$$

Manipulate $\frac{1}{\left(s + \sqrt{2}/2\right)^2 + 1/2} = \sqrt{2}\frac{1/\sqrt{2}}{\left(s + \sqrt{2}/2\right)^2 + 1/2}$.
Then

$$g(t) = \mathcal{L}^{-1}\left\{\sqrt{2}\frac{1/\sqrt{2}}{\left(s + \sqrt{2}/2\right)^2 + 1/2}\right\}$$
$$= \sqrt{2}e^{-t/\sqrt{2}}\sin\left(\frac{1}{\sqrt{2}}t\right)$$

Which also decays to zero as shown in Fig. 2.15.

Fig. 2.15 Impulse response
of a second order system
with TF $G(s) = \frac{1}{s^2+\sqrt{2}s+1}$

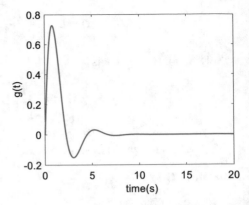

Reference

1. P.S. Laplace, *Des Fonctions Generatrices (on Generating Functions), in La Théorie analytique Des probabilités* (Libraire pour les mathematiques et la marine, Paris, 1812)

Chapter 3
Further Introductory Topics in Signals and Systems

3.1 Convolution Introduction

The mathematics of convolution goes back to at least the eighteenth century. In more recent times it was known in German as *Faltung* which means folding. It has a wide range of applications in the sciences, but in engineering it is usually most used in the signal processing field. The best way to see convolution in our applications is to see it as a time-domain version of Laplace transforms (LTs). Whereas LTs can be used to find the response of a linear system to a given input, convolution can also perform this task. However, convolution works entirely in the time-domain. As with LTs, convolution only applies to LTI systems. There are extensions to nonlinear systems, but it is not usually called convolution in these fields. Usually we convolve a signal with a system. The signal is already expressed in the time-domain and the system must be expressed by its impulse response. For a signal u(t) driving a system with transfer function G(s) or impulse response g(t), we can write its output as:

$$y(t) = \int_0^t u(t - \tau)g(\tau)d\tau \tag{3.1}$$

The limits on the integral are usually as shown for a causal system and signal, g(t) = 0, t < 0, u(t) = 0, t < 0 but the limits can be written as plus or minus infinity if needed without losing generality. This is because for the upper limit, the integral is taken with respect to τ and not t and hence u(t − τ)=0 for τ > t. If τ is negative then g(τ) = 0, so the lower limit can be less than zero without losing generality too. In some examples the signals can be noncausal, and we arrange the limits as necessary.

By substitution of t − τ = x, we can show that (3.1) also can be written as

$$y(t) = \int_0^t g(t - \tau)u(\tau)d\tau \tag{3.2}$$

© The Author(s), under exclusive license to Springer Nature Switzerland AG 2022
T. J. Moir, *Rudiments of Signal Processing and Systems*,
https://doi.org/10.1007/978-3-030-76947-5_3

We say that convolving u with g is the same as convolving g with u. The shortcut way of writing convolution is with a star symbol *. We write

$$y(t) = u(t)*g(t) = \int_0^t u(t - \tau)g(\tau)d\tau \tag{3.3}$$

And commutativity is satisfied.

$$u(t) * g(t) = g(t) * u(t) \tag{3.4}$$

More than one convolution can be used for multiple systems or signals, so we can have for example a distributive law satisfied too.

$$u(t) * \big[g(t) + h(t)\big]$$
$$= u(t) * g(t) + u(t) * h(t) \tag{3.5}$$

3.1.1 Examples of Convolution

Example 1 Consider the unit step response of the first order system $G(s) = \frac{1}{s+1}$.

The impulse response $g(t)$ is found from $g(t) = \mathcal{L}^{-1}\{\frac{1}{s+1}\} = e^{-t}$ Note that $g(t - \tau) = e^{-(t-\tau)}$.
For a unit step $u(t) = 1$. Therefore, the convolution integral is

$$y(t) = \int_0^t g(t - \tau)u(\tau)d\tau$$

$$= \int_0^t e^{-(t-\tau)}d\tau$$

$$= e^{-t} \int_0^t e^{\tau}d\tau$$

$$= e^{-t}\big[e^{\tau}\big]_{\tau=0}^{\tau=t}$$

$$= e^{-t}\big[e^t - 1\big]$$

$$= 1 - e^{-t}$$

We can also reverse the order in the convolution integral since $u(t-\tau) = 1$, $g(\tau) = e^{-\tau}$

$$y(t) = \int\limits_0^t u(t - \tau)g(\tau)d\tau$$

$$= \int\limits_0^t e^{-\tau}d\tau$$

$$= -\left[e^{-\tau}\right]_{\tau=0}^{\tau=t}$$
$$= -\left[e^{-t} - 1\right]$$
$$= 1 - e^{-t}$$

Example 2 Convolve $g(t) = \sqrt{2}e^{-t/\sqrt{2}} \sin\left(\frac{1}{\sqrt{2}}t\right)$ with a unit step $u(t) = 1$. This is the same example as considered in Chap. 2 with Laplace transforms. The transfer function is $G(s) = \frac{1}{s^2+\sqrt{2}s+1}$.

Take

$$y(t) = \int\limits_0^t u(t - \tau)g(\tau)d\tau$$

$$= \int\limits_0^t \sqrt{2}e^{-\tau/\sqrt{2}} \sin\left(\frac{1}{\sqrt{2}}\tau\right)d\tau$$

To integrate requires the product rule. This is a standard integral though the intricacies are a little messy and time consuming. We can easily integrate with computer algebra or any available tool. The result is

$$y(t) = 1 - e^{-t/\sqrt{2}}\left[\cos\left(\frac{1}{\sqrt{2}}t\right) + \sin\left(\frac{1}{\sqrt{2}}t\right)\right]$$

The integrations soon become quite formidable even for relatively low order systems. However, with the digital equivalent of convolution this problem is easily overcome. The computations are still quite numerically demanding though, even for digital equivalents and the fast-Fourier Transform (FFT) can be used to speed up the calculation. This enables acoustic impulse response data captured from real environments to be successfully used as applied to for instance processing of music audio. In this way, sound recorded in a studio can be made to sound as if it was recorded in a large theatre.

3.2 Convolution by Parts (or Piecewise Convolution)

If convolution was just solving the convolution integration, then it would be a relatively easy task. However, there are signals that the convolution integral cannot be applied to directly without careful consideration of various regions of integration. The integration must be done in parts, or piecewise.

Returning to the basic convolution equation

$$y(t) = \int_0^t u(t - \tau)g(\tau)d\tau$$

If it is examined closely, we see that the integration is done with respect to τ. This means that $u(t - \tau)$ is flipped in time (or folded to the left) if looked at graphically. As τ is then increased, $u(t - \tau)$ moves or slides to the right until it overlaps that of $g(\tau)$. The two are then multiplied together and integrated (summed). The process can be summarised as time-reversal, shifting, multiplying and summation. Consider two waveforms $u(t)$ and $g(t)$ as seen in Fig. 3.1a.

3.2.1 Example 1. Convolution by Parts

We change the independent variable from time to τ and draw $u(t - \tau)$ as seen in Fig. 3.1b. The negative τ means a flip around the vertical axis. This makes times 1

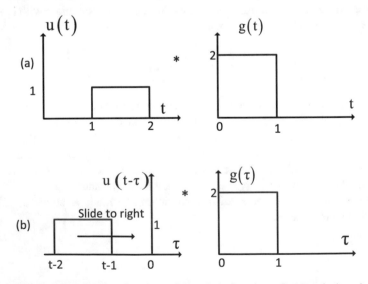

Fig. 3.1 **a** Original waveforms as functions of time. **b** As functions of $g(\tau)$ and $u(t - \tau)$

Fig. 3.2 a and b Two
regions of overlap

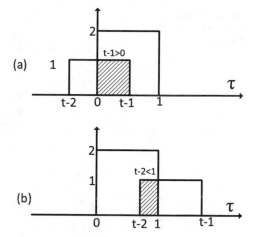

and 2 change to −1 and −2 on the τ scale. Then we add t to the scale and the limits
on the flipped waveform are t − 1 and t − 2. We also change the independent variable
from time to τ giving g(τ). We then put on the same graph and slide u(t − τ) to the
right until it overlaps g(τ). There are four regions of interest, the first and last are
zero and represent regions where u is outside and does not overlap g in any way. This
must happen first when t − 1 < 0 on Fig. 3.1b or t < 1. That is, the leading-edge t
1 of u is less than the trailing edge of g at time 0. Then when they overlap there are
two regions as shown in Fig. 3.2a, b.

For Fig. 3.2a t − 1 > 0 so t > 1. We write this as a region and find the area of
the *product* of the two waveforms. Since one has amplitude 1 and the other two, the
product is 2 and the area is.

Region 1

$$1 \le t \le 2$$

From the convolution integral, its limits are shown in the overlapping graph
Fig. 3.2a:

$$y(t) = 2 \int_0^{t-1} d\tau$$

$$= 2[\tau]_{\tau=0}^{\tau=t-1} = 2(t-1) \tag{3.6}$$

This is the equation of a ramp (straight line) of slope + 2 and it must start at time
t = 1 where it has value y (1) = 0. At the end of the region 1, t = 2 s and y(t) = 2.

The second region of overlap is shown in Fig. 3.2b when t − 2 < 1 giving t < 3.
Then the region is.

Region 2

Fig. 3.3 Result of
convolution by parts
example 1

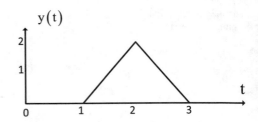

$$2 \leq t \leq 3$$

$$y(t) = 2 \int_{t-2}^{1} d\tau$$

$$= 2[\tau]_{\tau=t-2}^{\tau=1} = 2(3 - t) \tag{3.7}$$

At the beginning of region 2, it must have the same value as at the end of region 1.
We find $y(2) = 2$ which is the same. At the end of region 2, $y(3) = 0$. It has a slope of
-2 and ramps down to zero from a maximum of 2. Although the equations of these
lines clearly would be longer if defined on their own, the regions restrict their length
(Fig. 3.3).

Simple way to find regions of overlap

There is a simpler way defined by Ambarder [1] to find the regions of overlap. It is
more automatic and requires less thought. To do this we first write the limits of $g(t)$
and $u(t)$ side by side like this.

$$u(t) = \{1, 2\}, g(t) = \{0, 1\}$$

Then   add   in   turn   each   value   of $u(t)$ to   each   of $g(t)$ giving
$\{1 + 0, 1 + 1, 2 + 0, 2 + 1\} = \{1, 2, 2, 3\}$.

Keep only one of any duplicates and the result is $\{1, 2, 3\}$ in ascending order. This
tells us that the convolved waveform has regions 1–2 and 2–3 in time.

3.2.2   Example 2. Convolution by Parts

Consider the convolution of the two waveforms as shown in Fig. 3.4a.

$u(t)$ has width 1 s and is defined as having limits $\{5,6\}$. Whereas $g(t)$ has width 4 s
and has limits $\{0,4\}$. Using the shortcut method to find the regions we get $\{5 + 0,5$
$+ 4,6 + 0,6 + 4\} = \{5,9,6,10\}$. Re-arrange in ascending order and get $\{5,6,9,10\}$.
This gives us 3 regions of overlap. Time 5–6 s, 6–9 s and finally 9–10 s. Any other
region is zero.

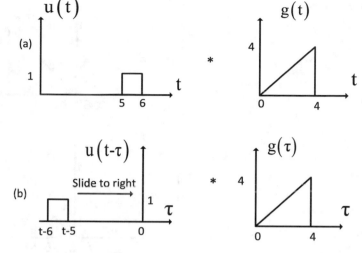

Fig. 3.4 a Original waveforms. **b** Folded and expressed with τ as independent variable

The equation of $g(\tau) = \tau$ is a straight line through the origin with unity slope. $u(t - \tau) = 1$ at all times.

For the three regions shown respectively in Fig. 3.5a–c we have:

Region 1

Just to double check the limits of this region, note that from Fig. 3.5a, t-5 > 0 so that t > 5. Also, $t - 6 < 0$ so $t < 6$. Therefore, as per the shortcut method

$$5 \leq t \leq 6$$

The convolution integration is

$$y(t) = \int_{\tau=0}^{\iota=\iota-5} \tau d\tau$$

$$= \frac{1}{2}[\tau^2]_{\tau=0}^{\tau=\tau-5} = \frac{1}{2}(t-5)^2$$

$$= \frac{1}{2}t^2 - 5t + 12.5 \tag{3.8}$$

We note that $y(5) = 0$ and $y(6) = \frac{1}{2}$.

Region 2.

A double check of the shaded region in Fig. 3.5b yields that $t - 6 > 0$ and $t - 5 < 4$. Then as per the shortcut method

$$6 \leq t \leq 9$$

Fig. 3.5 a–c Regions of
overlap

(a)

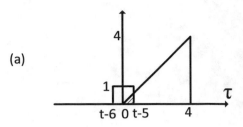

(b)

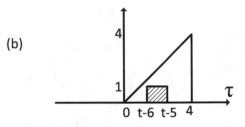

(c)

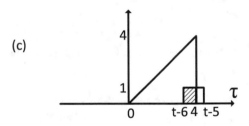

The convolution integration is

$$y(t) = \int_{\tau=t-6}^{\tau=t-5} \tau d\tau$$

$$= \frac{1}{2}[\tau^2]_{\tau=t-6}^{\tau=t-5} = t - 5.5 \qquad (3.9)$$

We note that $y(6) = \frac{1}{2}$ which has to be correct because it is the same as the last value of Region 1. We find $y(9) = 3.5$

Region 3

Here $t - 6 < 4$ so $t < 10$. From the shortcut method the region is

$$9 \le t \le 10$$

The convolution integration is

Fig. 3.6 Convolved waveform example 2

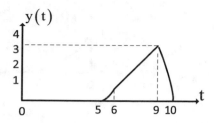

$$y(t) = \int_{\tau = t-6}^{\tau = 4} \tau \, d\tau$$

$$= \frac{1}{2} \left[\tau^2 \right]_{\tau = t-6}^{\tau = 4} = -1/2t^2 + 6t - 10 \qquad (3.10)$$

We note that $y(9) = \frac{1}{2}$ which has to be correct because it is the same as the last value of Region 1. We find $y(9) = 3.5$ which corresponds to the last value of Region 2, and $y(10) = 0$.

Drawing the convolved signal gives the graph of Fig. 3.6.

3.2.3 Example 3. Pulse of Duration 1 s Applied to an RC Network

A more practical example is shown in Fig. 3.7a pulse of duration 1 s is applied to an RC network.

The transfer function of the RC network can be found either from its differential equation (DE) as found in the previous chapter, or instead directly by recognising it is an impedance divider network. Another mathematical convenience which allows us to completely skip DEs is to use the impedance method with Laplace Transforms as used in circuit analysis. It was Oliver Heaviside who first pioneered this shortcut method of obtaining transfer functions. Recognise that for a capacitor its impedance is 1/sC and for a resistor it is just R. The transfer function is then found directly from

$$\frac{y}{u}(s) = \frac{1/sC}{R + 1/SC}$$

Fig. 3.7 Pulse applied to an RC network

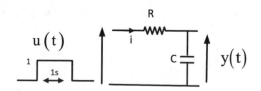

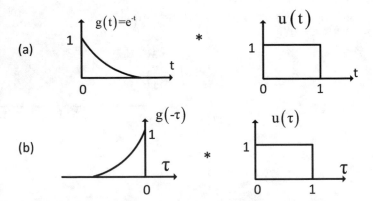

Fig. 3.8 a Impulse response in terms of time t. **b** in terms of τ with g flipped

$$= \frac{1}{1 + sT} \tag{3.11}$$

where the time-constant $T = CR$ The impulse response is then the inverse LT and found from tables to be

$$g(t) = \mathcal{L}^{-1}\left\{\frac{1/T}{s + 1/T}\right\}$$

$$= \frac{1}{T}e^{-t/T} \tag{3.12}$$

To get some numbers, consider the case when $T = 1$ s/ Then $g(t) = e^{-t}$. The input as just the pulse. Both are shown side by side in Fig. 3.8.

First method, flip g(t)

Either waveform can be flipped, so in this case $g(t)$ is chosen. The integral $y(t) = \int_0^t g(t - \tau)u(\tau)d\tau$ now applies. Figure 3.8b shows $g(t)$ flipped and both waveforms expressed in terms of τ. Figure 3.9a shows the two on the same graph and Fig. 3.9b, c shows the overlap regions.

There are two overlapping regions.

Region 1

In Fig. 3.9b the shaded area is found when $t > 0$ and $t < 1$. Therefore

$$0 \leq t \leq 1$$

$$y(t) = \int_0^t g(t - \tau)u(\tau)d\tau$$

Fig. 3.9 a Before sliding and **b, c** sliding through the two regions

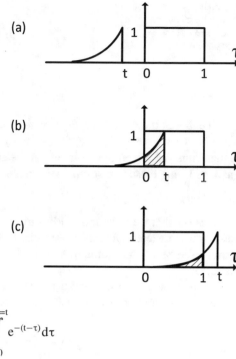

$$= \int_{\tau=0}^{\tau=t} e^{-(t-\tau)}d\tau$$

$$= e^{-t} \int_{\tau=0}^{\tau-t} e^{\tau}d\tau$$

$$= 1 - e^{-t}$$

Region 2

In Fig. 3.9c the shaded area is when $t \geq 1$

$$y(t) = \int_{0}^{t} g(t-\tau)u(\tau)d\tau$$

$$= \int_{\tau=0}^{\tau=1} e^{-(t-\tau)}d\tau$$

$$= e^{-t} \int_{\tau=0}^{\tau=1} e^{\tau}d\tau$$

$$= e^{-t}(e-1)$$

Fig. 3.10 RC circuit output
for a pulse of duration 1 s
input as found by
convolution

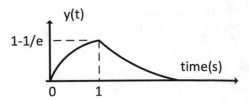

Note that at the end of region 1, when $t = 1$ s, $y(t) = 1 - e^{-1}$ Also in region 2 when $t = 1$ s we have $y(t) = e^{-1}(e - 1) = 1 - e^{-1}$. This makes sense since the two regions must be joined and coincide at the end of one region and the beginning of the next. Using these two results we can plot the result in Fig. 3.10.

Second method, flip u(t)

Just to show that we only chose arbitrarily to flip g(t), let us now repeat the process and flip u(t) instead. Quickly summarised, flipping the pulse u(t) gives us a pulse starting at time t-1 and going to time t. This will then slide to the right as shown in Fig. 3.11a.

For this case $g(\tau) = e^{-\tau}$ and $u(t - \tau) = 1$ during the overlap periods.

Region 1 From Fig. 3.11b $t \geq 0$ and the convolution integral becomes

$$y(t) = \int_0^t u(t - \tau)g(\tau)d\tau$$
$$= \int_{\tau=0}^{\tau=t} e^{-\tau}d\tau$$
$$1 - e^{-t}$$

Fig. 3.11 a After the flip
process **b** first region,
c second region

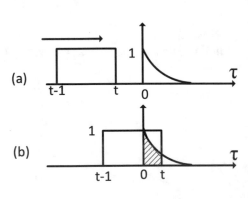

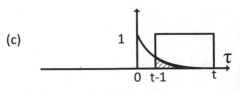

Region 2 From Fig. 3.11c t ≥ 1.

It is tempting here to think that the upper limit should be infinity, since the exponential goes to infinity, but the pulse confines the area to being between t and t − 1.

The convolution integral becomes

$$y(t) = \int_0^t u(t - \tau)g(\tau)d\tau$$
$$= \int_{\tau=0}^{\tau=t} e^{-\tau}d\tau$$
$$1 - e^{-t}$$

The same result as was shown previously.

3.3 Heaviside Piecewise Laplace Transform Method

Oliver Heaviside in the late nineteenth century invented a method that could enable engineers to take the Laplace Transform (LT) of signals that have different properties as time changes. The method can be applied to similar problems as solved by convolution by parts. For example, the LT of a unit step is known to be 1/s, but what of a pulse of a given duration? A pulse is an aperiodic signal and its LT does not appear on any table of LTs. Heaviside's method was to consider a waveform made up of component parts. Consider the three waveforms shown in Fig. 3.12a–c.

In order to take the LT of these waveforms we first need to solve a simpler problem. This problem is the LT of a delayed signal.

Fig. 3.12 a Square pulse, **b** triangular pulse, **c** arbitrary waveform

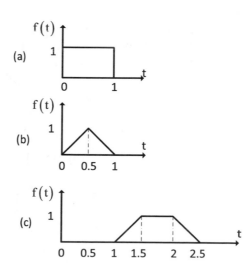

Laplace Transform of a signal f(t) *delayed by* τ.

We have already solved this problem in Chap. 2. By ordinary integration we can show that

$$\mathcal{L}\{f(t-\tau)\} = e^{-s\tau}F(s) \tag{3.13}$$

The $e^{-s\tau}$ term represents the LT of a pure time-delay. In honour of Heaviside, the Heaviside unit step function is defined as

$$\mathcal{L}\{u(t-\tau)\} = e^{-s\tau}/s \tag{3.14}$$

and u is used as a marker to tell us when a signal begins in time. A step at time zero is simply u(t). Therefore, to take the LT of Fig. 3.12a we note that the waveform consists of a step at time zero and a negative step at time 1 s. we write it as

$$f(t) = u(t) - u(t-1) \tag{3.15}$$

Its LT follows as

$$\mathcal{L}\{u(t) - u(t-1)\} = 1/s - e^{-s}/s \tag{3.16}$$

For Fig. 3.13b, it consists of a ramp of gradient 2 of duration 0.5 s and a negative ramp starting at time t = 0.5 going down and hitting zero at time 1 s. However, we cannot simply just subtract a negative going ramp of slope 2. For example, if we write

$$f(t) = 2tu(t) - 2(t - 0.5)u(t - 0.5) \tag{3.17}$$

Fig. 3.13 a, b Two incomplete representations of the triangular pulse

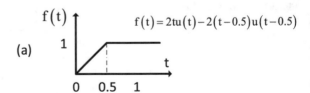

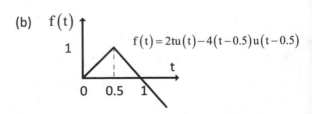

Fig. 3.14 Adding a positive slop to cancel the negative-going slope to have a resultant gradient of zero

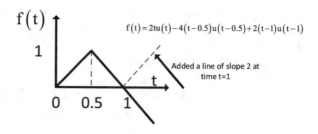

$$f(t) = 2tu(t) - 4(t-0.5)u(t-0.5) + 2(t-1)u(t-1)$$

Added a line of slope 2 at time t=1

Note that $u(t - 0.5)$ is being used as an operator here or marker to tell us that the expression on the RHS of (3.15) starts at time $t = 0.5$ s. Whereas $u(t)$ tells us that the first part starts at time zero. This will then represent a ramp as required at time zero but at time $t = 0.5$ s the two ramps cancel, and the line will go to zero gradient to time infinity. This is shown in Fig. 3.13a.

Taking away double the gradient at time $t = 0.5$ s gives us

$$f(t) = 2tu(t) - 4(t - 1)u(t - 0.5) \tag{3.18}$$

As shown in Fig. 3.13b.

This is still not a triangular pulse and therefore a positive going slope of magnitude 2 must be added at time $t = 1$ s which will cancel out the line that goes to minus infinity and let it have slope zero. Finally, the complete description is given in the time-domain by (Fig. 3.14)

$$f(t) = 2tu(t) - 4(t - 0.5)u(t - 0.5) + 2(t - 1)u(t - 1) \tag{3.19}$$

Notice that for each term in (3.19) the u operator has the same argument in brackets as the time expression before it. t, $(t - 0.5)$ and $(t - 1)$ for this example. This method will only work easily if the signals are represented in this way. Once the Heaviside equation is written then LTs can be taken directly.

In words, the triangular pulse is described as follows:

A ramp of magnitude 2 at time t = 0, minus a ramp of magnitude 4 at time t = 0.5, plus a ramp of magnitude 2 at time t = 1.

The LT of (3.19) is then readily found to be

$$F(s) = \mathcal{L}\{2tu(t) - 4(t - 0.5)u(t - 0.5) + 2(t - 1)u(t - 1)\}$$
$$= \frac{2}{s^2} - \frac{4e^{-0.5s}}{s^2} + \frac{2e^{-s}}{s^2} \tag{3.20}$$

Now consider Fig. 3.12c. In words this can be represented as.

A ramp of magnitude 2 at time t = 1 minus a ramp of magnitude 2 at time t = 1.5 minus a ramp of magnitude -2 at time t = 2 plus a ramp of magnitude 2 at time t = 2.5.

Mathematically this becomes

$$f(t) = 2(t-1)u(t-1) - 2(t-1.5)u(t-1.5)$$
$$- 2(t-2)u(t-2) + 2(t-2.5)u(t-2.5) \qquad (3.21)$$

Sometimes r(t), a shortcut for a unit slope ramp (that is $r(t) = t$ at time $t - 0$) is used instead of using u and the equation can be written instead as

$$f(t) = 2r(t-1) - 2r(t-1.5) - 2r(t-2) + 2r(t-2.5) \qquad (3.22)$$

This is more in engineering applications however than in pure mathematics. The LT is found as

$$F(s) = \mathcal{L}\{2(t-1)u(t-1) - 2(t-1.5)u(t-1.5)$$
$$- 2(t-2)u(t-2) + 2(t-2.5)u(t-2.5)\}$$
$$= \frac{2e^{-s}}{s^2} - \frac{2e^{-1.5s}}{s^2} - \frac{2e^{-2s}}{s^2} + \frac{2e^{-2.5s}}{s^2} \qquad (3.23)$$

3.3.1 Example of Heaviside Laplace Method for an RC Circuit

Consider the same problem illustrated in Fig. 3.7. A pulse of unit amplitude, duration 1 s applied to an RC network where the output is taken as the capacitor voltage.

Write the pulse in the time-domain using Heaviside operator notation.

$$V_{in}(t) = u(t) - u(t-1) \qquad (3.24)$$

See Fig. 3.15.

The transfer function for $T = 1$ s is $G(s) = \frac{1}{s+1}$. Take LTs of $V_{in}(t) = u(t) - u(t-1)$ and obtain $V_{in}(s) = \frac{1-e^{-s}}{s}$. Multiplying gives the output signal LT

Fig. 3.15 Pulse applied to an RC network

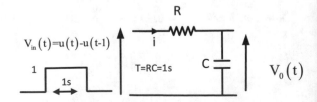

$$V_o(s) = \left(\frac{1 - e^{-s}}{s(s+1)}\right) \tag{3.25}$$

Expanding using partial fractions yields

$$V_o(s) = \frac{1}{s} - \frac{1}{s+1} - e^{-s}\left(\frac{1}{s} - \frac{1}{s+1}\right) \tag{3.26}$$

Take inverse LTs $V_o(t) = \mathcal{L}^{-1}\{\frac{1}{s} - \frac{1}{s+1} - e^{-s}(\frac{1}{s} - \frac{1}{s+1})\}$

$$V_o(s) = u(t) - e^{-t}u(t) - \left[u(t-1) - e^{-(t-1)}u(t-1)\right] \tag{3.27}$$

Interpreting (3.27) means that $u(t) - e^{-t}u(t)$ occurs from time $t = 0$ onwards and at time $t = 1$ or above $-\left[u(t-1) - e^{-(t-1)}u(t-1)\right]$ occurs. The u terms act as markers in time telling us when to start.

For time $0 \leq t \leq 1$

$$V_o(t) = 1 - e^{-t} \tag{3.28}$$

For time $t \geq 1$ the first part is still running, and we add the second part giving

$$\begin{aligned} V_o(t) &= 1 - e^{-t} - \left[1 - e^{-(t-1)}\right] \\ &= e^{-(t-1)} - e^{-t} \\ &= e^{-t}[e - 1] \end{aligned} \tag{3.29}$$

This is the same result as obtained by convolution by parts.

Now consider the same example but with a ramp waveform that goes to zero at time $t = 1$ s a shown in Fig. 3.16.

The triangular pulse has slope unity and has duration 1 s. To describe it using the Heaviside method we can say in words.

The waveform has a slope of unity which rises for 1 s from t = 0. Then we subtract a slope of minus unity which makes it go horizontal at t = 1 s. We final subtract a step of magnitude 1 at time t = 1 s to make the waveform go to zero.

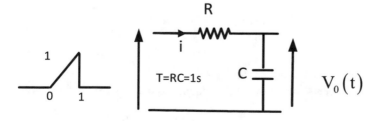

Fig. 3.16 Ramp that returns to zero at time $t = 0$ as an input to an RC network

$$V_{in}(t) = tu(t) - (t-1)u(t-1) - u(t-1) \tag{3.30}$$

Its LT is then $V_{in}(s) = \mathcal{L}\{tu(t) - (t-1)u(t-1) - u(t-1)\}$

$$
\begin{aligned}
V_{in}(s) &= \frac{1}{s^2} - \frac{e^{-s}}{s^2} - \frac{e^{-s}}{s} \\
&= \frac{1}{s^2} - e^{-s}\left[\frac{1}{s^2} + \frac{1}{s}\right] \\
&= \frac{1}{s^2} - e^{-s}\left[\frac{s+1}{s^2}\right]
\end{aligned}
\tag{3.31}
$$

Applying $V_{in}(s)$ to the transfer function $G(s) = \frac{1}{s+1}$ gives two parts

$$V_o(s) = \frac{1}{s^2(s+1)} - \frac{e^{-s}}{s^2} \tag{3.32}$$

$\frac{1}{s^2(s+1)}$ represents the LT of a signal from time $t = 0$ and $-\frac{e^{-s}}{s^2}$ engages when $t \geq 1$ in the time-domain.

The LHS of (3.32) involves $\frac{1}{s^2(s+1)}$ which we must expand using partial fractions since its transform does not appear in the tables.

$$\frac{1}{s^2(s+1)} = \frac{A}{s} + \frac{B}{s^2} + \frac{C}{s+1} \tag{3.33}$$

Multiplying both sides of (3.33) by $s^2(s+1)$

$$1 = As(s+1) + B(s+1) + Cs^2 \tag{3.34}$$

Compare coefficients of s.

s^0 $1 = B$.

s^1 $0 = A + B$ so $A = $ -1.

s^2 $0 = A + C$ giving $C = $ -1.
   Therefore

$$\frac{1}{s^2(s+1)} = \frac{1}{s^2} - \frac{1}{s} + \frac{1}{s+1} \tag{3.35}$$

and each term now appears in the tables.
   Now consider a time before and including time $t = 1$ s in the time-domain:

$$V_o(t) = \mathcal{L}^{-1}\left\{\frac{1}{s^2} - \frac{1}{s} + \frac{1}{s+1}\right\}$$

$$= tu(t) - u(t) + e^{-t}u(t) \tag{3.36}$$

The Heaviside operator is just a marker which tells us at what time things occur. Without the Heaviside operator (3.36) becomes

$$V_o(t) = t - 1 + e^{-t}, t \geq 0$$

and at time $t = 1$ s, the end of this first region, the output voltage has a value $V_o(1) = e^{-1} = 0.368$.

When time reaches or is greater than 1 s we get the second region being activated in the equation of the output. The first part still exists too, however.

The second part on its own in the time domain is found from the RHS of (3.32)

$$V_o(t) = \mathcal{L}^{-1}\left\{\frac{e^{-s}}{s^2}\right\}$$
$$= (t-1)u(t-1) \tag{3.37}$$

The $u(t-1)$ letting us know the partition between the first and second region in time beginning at $t = 1$ s. We must include the first part too in the solution (this follows from (3.32)) giving

$$V_o(t) = t - 1 + e^{-t} - (t-1)$$
$$= e^{-t}, t \geq 1$$

At time $t = 1$ s the output voltage $V_o(1) = e^{-1} = 0.368$ which corresponds to the end of the previous region. The solution is shown in Fig. 3.17.

Fig. 3.17 Response of RC network to a ramp of fixed duration

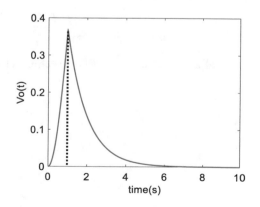

3.4   Laplace Transform of Convolution

Write the convolution integral as

$$y(t) = \int_{\tau=0}^{\tau=t} g(t-\tau)h(\tau)d\tau \tag{3.38}$$

An important result is to take the Laplace Transform of (3.38) and by writing the limits as plus infinity

$$\mathcal{L}\{y(t)\} = \int_{t=0}^{t=\infty} \int_{\tau=0}^{\tau=\infty} g(t-\tau)h(\tau)d\tau e^{-st}dt \tag{3.39}$$

This follows because for causality $g(t-\tau) = 0, \tau > t$. We then re-arrange the order of integration which in this case is straight-forward.

$$\mathcal{L}\{y(t)\} = \int_{\tau=0}^{\tau=\infty} \int_{\tau=0}^{\tau=\infty} g(t-\tau)h(\tau)e^{-st}dtd\tau \tag{3.40}$$

By substituting $x = t - \tau, dx = dt$ obtain

$$\mathcal{L}\{y(t)\} = \int_{x=0}^{x=\infty} \int_{\tau=0}^{\tau=\infty} g(x)h(\tau)e^{-s(x+\tau)}dxd\tau$$

$$= \int_{x=0}^{x=\infty} g(x)e^{-sx}dx \int_{\tau=0}^{\tau=\infty} h(\tau)e^{-s\tau}d\tau \tag{3.41}$$

The Laplace Transform of convolution is then expressed as

$$Y(s) = G(s)H(s) \tag{3.42}$$

It follows that from (3.31): *Convolution in the time-domain is the same as Multiplication in the s-domain.* This is a common result used by engineers for many applications. (though the applications usually involve a similar transform called the Fourier transform). It applies to signals as well as systems. For example, for two systems in cascade we simply multiply their transfer functions to get the total transfer function. The order of multiplication doesn't matter either (in theory), though there are always physical restraints on which system can come first. Or we can apply a signal to a system and find the response also by multiplication of their LTs the time-domain signal is then found by using inverse LTs.

3.5 Stability of LTI Systems

If a system has an input that is bounded, then it is said to be stable if the output is also bounded. (BIBO stability). By bounded we mean a signal that does not grow to infinity as time progresses. For example, a signal $f(t) = 1$, is considered bounded. Another example of a bounded signal is $f(t) = e^{-2t}$ which exponentially dies away to zero as time progresses. However, an unbounded signal is $f(t) = e^{2t}$ which grows to infinity.

Systems can be unstable if the output grows to infinity, stable if the output stays within some bound, or critically stable whereby the output oscillates at a certain frequency. Since LTI systems can be represented in transfer function form. We can define BIBO stability as any system whose poles have negative real parts.

3.5.1 Introduction to Poles and Zeros of Transfer Functions

A transfer function can be written in general form as the ratio of two polynomials.

$$G(s) = \frac{b_0 + b_1 s + b_2 s^2 + \cdots + b_m s^m}{a_0 + a_1 s + a_2 s^2 + \cdots + a_n s^n}$$
$$= \frac{N(s)}{D(s)} \tag{3.43}$$

For physical realizability the orders of the numerator must be less than or equal to the denominator $m \leq n$. An nth order polynomial can be factorized into n roots (some may of course be repeated). We likewise apply this for an mth order polynomial and (3.43) can then be written in factored form

$$G(s) = \frac{(s - z_1)(s - z_2) \cdots (s - z_m)}{(s - p_1)(s - p_2) \cdots (s - p_n)} \tag{3.44}$$

We now define the poles of the system as the values of s where

$$D(s) = 0 \tag{3.45}$$

and the values of s where

$$N(s) = 0 \tag{3.46}$$

For example, the transfer function

$$G(s) = \frac{s+5}{s^2 + 3s + 2}$$
$$= \frac{s+5}{(s+1)(s+2)}$$

Clearly $N(s) = s + 5, D(s) = (s + 1)(s + 2)$.

When each are equated to zero we find the poles are at $-1, -2$ and the single zero is at -5.

Since both poles have negative real parts (they have no imaginary parts at all in fact), the system is said to be stable. The value of the zero does not matter when defining stability though it does play a role in the type (or shape) of response.

As another example

$$G(s) = \frac{s-5}{s^2 + s + 0.5}$$
$$= \frac{s-5}{(s+0.5+j0.5)(s+0.5-j0.5)}$$

The system has a zero at $s = 0.5$ and two complex conjugate poles at $s = -0.5 \pm 0.5j$. Since the poles have negative real parts, the system is stable.

The 3rd order system

$$G(s) = \frac{s+1}{(s+2)(s^2 - 5s + 6)}$$
$$= \frac{s-5}{(s-3)(s-2)}$$

Has a zero at $s = 5$ and two poles ate $s = 3$ and $s = 2$. The system is therefore unstable.

The reason for instability can be seen quite easily for systems with simple (not complex) poles. For example, if we apply a unit step $u(s) = 1/s$ to the above transfer function, the output is

$$y(s) = \frac{s - 5}{(s - 3)(s - 2)} u(s)$$

$$= \frac{s - 5}{s(s - 3)(s - 2)}$$

Expand using partial fractions and get

$$y(s) = \frac{s - 5}{s(s - 3)(s - 2)}$$

$$= \frac{A}{s} + \frac{B}{s - 3} + \frac{C}{s - 2}$$

It matters little what the constants are for this example. Take inverse LTs and obtain

$$y(t) = A + Be^{3t} + Ce^{2t}$$

We immediately see that there are positive growing exponentials which rise to infinity in the solution. If the poles had been negative, these exponentials would have been decaying away rather than growing. The same applies for complex poles except the decay happens as exponentially decaying sine or cosine waves. For example signal of the form $y(t) = Ae^{-2t} \cos(\omega t)$ would die out whereas $y(t) = Ae^{2t} \cos(\omega t)$ would grow to infinity. An illustration of the s-plane and the type of step response that would be obtained for stable systems is shown in Fig. 3.18.

Poles are usually described using the x graphical symbol and mathematically as $s = \sigma + j\omega_d$, where σ is the real part which must have negative sign for stability. The imaginary part $\omega_d = 2\pi f_d$ if non-zero is the frequency of oscillation. Poles which

Fig. 3.18 S-Plane and region of stability. Step responses shown for various pole locations

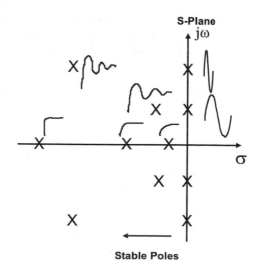

lie on the imaginary axis and which have no real part are oscillatory. The higher up the imaginary axis, the higher the frequency of oscillation. The rather unfitting name "critically stable" is used in the literature to describe such systems. From a practical point of view there is nothing stable about such a system, and unless the aim of the design is to create a sinusoidal oscillator, poles are rarely if ever designed to be there. Complex poles that lie to the left of the S-Plane are exponentially decaying poles. The further to the left means the ringing in the step response dies out faster than closer to the imaginary axis. Zeros are denoted graphically with a large O and not shown in Fig. 3.18. Their position on the S-Plane does not affect stability.

3.5.2 Final Value Theorem in Laplace Transforms

The final value theorem (FVT) is a useful tool to find the steady-state value in the time-domain of a system, without going to the trouble of inverse Laplace Transforming. It is used among other disciplines, in control engineering to find the steady-state value of the output of a stable system. If a system is unstable the FVT does not apply. For a transfer function G(s) and input signal u(s), the output is given as $y(s) = G(s)u(s)$. It's steady-state (or final value) value is found from

$$y(t)|_{\lim_{t \to \infty}} = sy(s)|_{\lim_{s \to 0}} \tag{3.47}$$

For example, For a system $G(s) = \frac{1}{s^2+3s+5}$ with a step input $u(s) = \frac{1}{s}$, then

$$y(s) = \frac{1}{s(s^2 + 3s + 5)} \tag{3.48}$$

According to the FVT, to find the steady-state value of y(t), we do not have to do any partial fraction expansion and inverse Laplace at all. This simple shortcut means that we simply multiply y(s) by s and put s = 0. Therefore

$$y(t)|_{t \to \infty} = sx\frac{1}{s(s^2 + 3s + 5)}\bigg|_{s \to 0} \tag{3.49}$$

$$= 1/5$$

Reference

1. A. Amarder, (ed.), *Analog and Digital Signal Processing* (Brooks/Cole Publishing Company, USA)

Chapter 4
Frequency-Domain Properties of Signals

4.1 Fourier Series for Periodic Waveforms

One of the most notable scientific works was made in 1822 by a French Mathematician and Scientist who probably knew very little about signals. He was working in the area of heat transfer [1] at the time when he made his great mathematical discovery. The first of these is named after him, the Fourier series. Expressed in engineering terms, it is a method whereby periodic signals only can be expressed as an infinite series of cosine and sinewaves of integer multiple frequencies of a base or fundamental frequency. A sinewave is the purest of waveforms and only has itself represented with an amplitude and frequency. Its phase of course is also important but not relevant at this stage since phase is a relative measure of the relationship of say one sinewave to another in time. Here we are interested in frequency only. A periodic signal defined from T/2 to −T/2 is one where

$$f(t + T) = f(t) \tag{4.1}$$

where T is the period or repetition rate of the signal. The signal frequency is given by $f_o = \frac{1}{T}$ Hz or $\omega_o = \frac{2\pi}{T}$ rad/s. For any such signal, Fourier showed that it can be represented with the following infinite series expansion.

$$f(t) = a_o + \sum_{k=1}^{\infty} a_k \cos(k\omega_o t) + b_k \sin(k\omega_o t) \tag{4.2}$$

where a_o is the average or dc component of the signal found from

$$a_o = \frac{1}{T} \int_{-T/2}^{T/2} f(t)dt \tag{4.3}$$

© The Author(s), under exclusive license to Springer Nature Switzerland AG 2022
T. J. Moir, *Rudiments of Signal Processing and Systems*,
https://doi.org/10.1007/978-3-030-76947-5_4

and the other coefficients of sin and cos are found from

$$a_k = \frac{2}{T} \int_{-T/2}^{T/2} f(t) \cos(k\omega_0 t) dt \qquad\qquad (4.4)$$

$$b_k = \frac{2}{T} \int_{-T/2}^{T/2} f(t) \sin(k\omega_0 t) dt \qquad\qquad (4.5)$$

Other ways to write the limits on integrals such as $a_k = \frac{2}{T} \int_{-T/2}^{T/2} f(t) \cos(k\omega_0 t) dt$
immediately follow since the waveform is periodic and it matters little where the
start and end point is other than that a whole period is considered. For example,
$a_k = \frac{2}{T} \int_{0}^{T} f(t) \cos(k\omega_0 t) dt$ is equally valid, or over a half a period (known as the range)
in special symmetric cases only, we can use $a_k = \frac{4}{T} \int_{0}^{T/2} f(t) \cos(k\omega_0 t) dt$ instead. In
this last case the integral is doubled.

When $k = 1$, (4.2) involves only the waveforms $\cos(\omega_0 t)$ or $\sin(\omega_0 t)$. This frequency
ω_0 is referred to as the *fundamental frequency* and is the same as the frequency of
the periodic waveform itself. When $k = 1, 2, 3$ etc. then $2\omega_0, 3\omega_0, 4\omega_0$ the frequencies are referred to as second, third, fourth harmonics and so on. These frequencies play certain roles in the quality of signals. For example, an audio signal that
has third harmonic present is known as having third harmonic distortion. This is
very displeasing to the human ear. Surprisingly, second harmonic or even-harmonic
distortion is more bearable to our hearing.

In order to be valid, the waveform must satisfy the *Dirichlet conditions*. Dirichlet
was a mathematician who made the early work of Fourier more rigorous. These
conditions, although important, are not of great value in engineering other than
to state that the waveform and its first derivative must be periodic and piecewise
continuous in the region of time $(-T/2, T/2)$. Most of the commonly met electrical
engineering waveforms satisfy these conditions, though mathematicians have as a
matter of rigour constructed special counterexamples which do not.

The reason that the Fourier series equations work at all is because it uses cos and
sin which are orthogonal (90° out of phase) to one another. Other basis functions are
also possible which give different series.

Equations (4.3), (4.4) and (4.5) are quite general, but a great deal of simplification
is found if the signal is either even or odd. For *even* signals the b_k terms are always
zero whereas for *odd* signals the a_k terms are always zero. For signals that are neither
even nor odd then both a_k and b_k terms must be calculated.

4.1.1 Example 1. A Square Wave with Even Symmetry

Consider the waveform with period T shown in Fig. 4.1.

Since $f(t) = f(-t)$, the signal is *even*. This means we need not bother calculating the b_k terms in (4.5). The signal is periodic with period T, hence $\omega_o = 2\pi/T$.

The average or dc term is

$$a_o = \frac{1}{T} \int_{-T/2}^{T/2} f(t)dt$$

$$= \frac{1}{T} \int_{-\tau/2}^{\tau/2} dt$$

$$= \frac{\tau}{T} \tag{4.6}$$

The a_k terms are

$$a_k = \frac{2}{T} \int_{-\tau/2}^{\tau/2} \cos(k\omega_o t)dt$$

$$= \frac{2}{kT\omega_o} \sin(k\omega_o t)\big|_{t=-\tau/2}^{t=\tau/2}$$

$$= \frac{4}{kT\omega_o} \sin\left(k\omega_o \frac{\tau}{2}\right) \tag{4.7}$$

But $\omega_o = 2\pi/T$. Substitute into (4.7) and get

$$a_k = \frac{2}{k\pi} \sin\left(k\pi \frac{\iota}{T}\right) \tag{4.8}$$

Put some numbers in to get an engineering feel for the problem. Let $T = 4$ s and $\tau = 2$ s. (equal mark-to-space ratio or 50% duty-cycle) Then $\omega_o = \frac{2\pi}{T} = \frac{\pi}{2}$ rad/s.

Fig. 4.1 Square wave

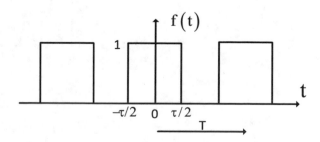

The dc term $a_o = 1/2$ and for the other a_k coefficients, $a_k = \frac{2}{k\pi} \sin\left(\frac{k\pi}{2}\right)$. Now substitute $k = 1, 2, 3 \ldots$

$$a_1 = \frac{2}{\pi} \sin\left(\frac{\pi}{2}\right) = \frac{2}{\pi}, \, a_2 = \frac{1}{\pi} \sin(\pi) = 0,$$

$$a_3 = \frac{2}{3\pi} \sin\left(\frac{3\pi}{2}\right) = -\frac{2}{3\pi}, \, a_4 = \frac{1}{2\pi} \sin(2\pi) = 0,$$

$$a_5 = \frac{2}{5\pi} \sin\left(\frac{5\pi}{2}\right) = \frac{2}{5\pi}.$$

By using (4.2) we get the Fourier series:

$$f(t) = a_o + \sum_{k=1}^{\infty} a_k \cos(k\omega_o t) + b_k \sin(k\omega_o t)$$

$$= \frac{1}{2} + \frac{2}{\pi} \cos\left(\frac{\pi}{2}t\right) - \frac{2}{3\pi} \cos\left(\frac{3\pi}{2}t\right) + \frac{2}{5\pi} \cos\left(\frac{5\pi}{2}t\right) + \ldots$$

Note that only odd harmonics exist for this case of a square wave with equal mark-to-space ratio. This is called a perfect or ideal square wave.

For these first 5 terms of the series we can plot f(t) as shown in Fig. 4.2.

This is of course a very crude approximation since there are only 5 terms. Plotting to 100 terms in Fig. 4.3 gives something more akin to a square wave.

The ringing at time in Fig. 4.3 was first analysis in detail by a mathematician called Gibbs and is named *Gibbs* phenomena in his honour. It happens when there are sudden changes in time on the original waveform. In this case at the edge of a square wave.

Fig. 4.2 First 4 terms of the Fourier series approximating the square wave

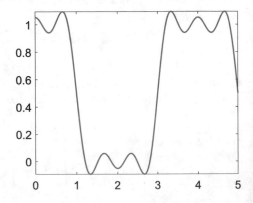

Fig. 4.3 First 100 terms of the Fourier series

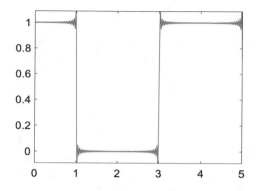

4.1.2 Half-Range Fourier Series

It is worth mentioning but not a great advantage in this example that it is possible for either an even or odd function to use only half the range to integrate. In this case instead of

$$a_k = \frac{2}{T} \int_{-\tau/2}^{\tau/2} \cos(k\omega_o t)dt$$

We half the range by using time zero as the lower limit on the integral. The integral must be doubled to make this happen, however. Evaluating

$$a_k = 2 \times \frac{2}{T} \int_{0}^{\tau/2} \cos(k\omega_o t)dt$$

Instead, gives us the same result as above. The symmetry of the waveform has been exploited. In some cases, this simplifies the problem significantly.

4.1.3 Example 2. A Square Wave Being Neither Even nor Odd Symmetry

Consider a square wave shown in Fig. 4.4. The problem is to find the 3rd and fifth harmonic. It is neither an even nor odd signal and therefore both the a and b coefficients must be solved for.

The waveform is periodic with period $T = 3$ s. The dc value can be found from integration or merely inspection. Finding the area of the positive and negative half-cycles gives $2-3 = --1$. Divide by the period gives the dc value at -1/3. The fundamental frequency $\omega_o = \frac{2\pi}{3}$ rad/s.

Fig. 4.4 Square wave

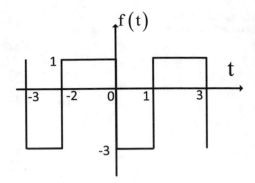

By integration

$$a_0 = \frac{1}{3} \int\limits_{0}^{1} -3 dt + \frac{1}{3} \int\limits_{-2}^{0} dt$$

$$= \frac{1}{3}[-3t]_0^1 + \frac{1}{3}[t]_{-2}^0$$

$$= \frac{1}{3}[-3 + 2] = -\frac{1}{3}$$

$$a_k = \frac{2}{T} \int\limits_{-T/2}^{T/2} f(t) \cos(k\omega_0 t) dt, \, k = 1, 2, \ldots$$

$$= -\frac{2}{3} \int\limits_{0}^{1} 3 \cos(k\omega_0 t) dt + \frac{2}{3} \int\limits_{-2}^{0} \cos(k\omega_0 t) dt$$

$$-\frac{2}{k\omega_0} \sin(k\omega_0 t) \Big|_0^1 + \frac{2}{3k\omega_0} \sin(k\omega_0 t) \Big|_{-2}^0$$

Solving the above integrals yields

$$a_k = \frac{3}{k\pi} \left[\frac{1}{3} \sin\left(\frac{4k\pi}{3}\right) - \sin\left(\frac{2\pi k}{3}\right) \right]$$

Similarly

$$b_k = \frac{2}{T} \int\limits_{-T/2}^{T/2} f(t) \sin(k\omega_0 t) dt, \, k = 1, 2, \ldots$$

Fig. 4.5 Second example
Fourier series summed for
100 coefficients of the series

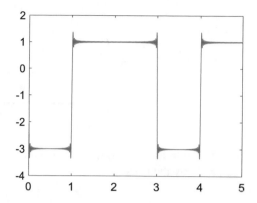

$$= -\frac{2}{3} \int_0^1 3 \sin(k\omega_o t)dt + \frac{2}{3} \int_{-2}^0 \sin(k\omega_o t)dt$$

$$= \frac{2}{k\omega_o} \cos(k\omega_o t)\Big|_0^1 - \frac{2}{3k\omega_o} \cos(k\omega_o t)\Big|_{-2}^0$$

Solving gives

$$b_k = \frac{2}{k\omega_0}[\cos(k\omega_0) - 1] - \frac{2}{3k\omega_0}[1 - \cos(2k\omega_0)]$$

$$= \frac{3}{k\pi}\left[\frac{1}{3}\cos\left(\frac{4k\pi}{3}\right) + \cos\left(\frac{2k\pi}{3}\right)\right] - \frac{4}{k\pi}$$

The Fourier series is

$$f(t) = -\frac{1}{3} + \sum_{k=1}^{\infty} a_k \cos\left(k\frac{2\pi}{3}t\right) + b_k \sin\left(k\frac{2\pi}{3}t\right)$$

The first 100 terms are plotted in Fig. 4.5 and shows the general shape of a square wave with the correct duration and amplitudes.

4.1.4 Example of an Even Triangular Waveform

Consider the waveform shown in Fig. 4.6. The waveform is periodic with period T = 2 s. It has fundamental frequency $\omega_o = \frac{2\pi}{T} = \pi$ rad/s.

The average or dc value can be found by inspection and is the area under a triangle divided by the period T = 2. This gives us ½. Verifying through integration we use

Fig. 4.6 Triangular
waveform

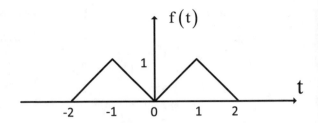

the expression for $f(t) = t$, which is valid from time 0 to time 1 and double this value
and divide by the period.

$$a_0 = \frac{2}{T} \int_0^1 t\,dt$$

$$= \frac{1}{2}[t^2]_0^1$$

$$= 1/2$$

Here we have exploited the symmetry and integrated over the half-range only but
doubled the area since the area in negative time is the same as positive time.

The waveform is even and therefore there are only cosine terms to be calculated.
They are found from

$$a_k = \frac{2}{T} \int_{-T/2}^{T/2} t\cos(k\omega_o t)dt, \ k = 1, 2, \ldots$$

Integrating over the half-range and doubling the value gives

$$a_k = \frac{4}{T} \int_0^1 t\cos(k\omega_o t)dt$$

$$= 2\left(\frac{k\omega_o \sin(k\omega_o) + \cos(k\omega_o) - 1}{k^2\omega_o^2}\right)$$

The above is found by integration by parts. The details need not concern us since
this is a standard integral problem. But $\omega_o = \pi$ therefore

$$a_k = 2\left(\frac{k\omega_o \sin(k\omega_o) + \cos(k\omega_o) - 1}{k^2\omega_o^2}\right)$$

$$= 2\left(\frac{k\pi \sin(k\pi) + \cos(k\pi) - 1}{k^2\pi^2}\right)$$

Fig. 4.7 First 5 terms of the Fourier series approximates a triangular waveform

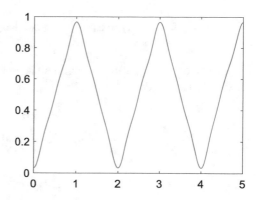

$$= 2\left(\frac{(-1)^k - 1}{k^2 \pi^2}\right)$$

where we have used the fact that for k = 1, 2, ... The values of $\sin(k\pi) = 0$ and $\cos(k\pi) = (-1)^k$.

The Fourier series is now

$$f(t) = a_o + \sum_{k=1}^{\infty} a_k \cos(k\omega_o t)$$

$$= \frac{1}{2} + 2 \sum_{k=1}^{\infty} \left(\frac{(-1)^k - 1}{k^2 \pi^2}\right) \cos(k\pi t)$$

The first 5 terms are summed, and the result is shown in Fig. 4.7

4.1.5 Fourier Series of Rectified Cosine Waveform

A fully rectified cosine is shown in Fig. 4.8. It is even and periodic with period T. We can write its Fourier series as

Fig. 4.8 Fully rectified sinewave

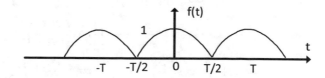

$$f(t) = a_o + \sum_{k=1}^{\infty} a_k \cos(k\omega_o t) + b_k \sin(k\omega_o t)$$

The dc or average value is found since in the region $-T/2$ to $T/2$ we have $f(t) = \cos(\frac{\pi}{T}t)$ and as usual define $\omega_o = \frac{2\pi}{T}$.

$$a_o = \frac{2}{T} \int_0^{T/2} \cos\left(\frac{\pi t}{T}\right) dt$$

$$\frac{2}{\pi}\left[\sin\left(\frac{\pi t}{T}\right)\right]_0^{T/2}$$

$$= \frac{2}{\pi}$$

The period T is arbitrary here and not initially assigned a numerical value.

This is an even waveform, so we only need cosine terms over the half-range.

$$a_k = \frac{4}{T} \int_0^{T/2} f(t) \cos(k\omega_o t) dt, \ k = 1, 2, \ldots$$

$$= \frac{4}{T} \int_0^{T/2} \cos\left(\frac{\pi}{T}t\right) \cos\left(\frac{2\pi k}{T}t\right) dt$$

$$\frac{2}{\pi(4k^2 - 1)}\left[(2k - 1)\sin\left(\frac{2\pi k + \pi}{2}\right) + (2k + 1)\sin\left(\frac{2\pi k - \pi}{2}\right)\right]$$

The above integral was computed using computer algebra. The importance in engineering is not so much the ability to solve tough integrals, but to set the problem up so that it can be solved using available tools. It can be solved manually using integration by parts.

The Fourier series is then

$$f(t) = 2/\pi + \sum_{k=1}^{\infty} \frac{2}{\pi(4k^2 - 1)}\left[(2k - 1)\sin\left(\frac{2\pi k + \pi}{2}\right)\right.$$

$$\left. + (2k + 1)\sin\left(\frac{2\pi k - \pi}{2}\right)\right]\cos\left(\frac{2\pi k t}{T}\right)$$

$$= \frac{2}{\pi} + \frac{4}{3\pi}\cos\left(\frac{2\pi t}{T}\right) - \frac{4}{15\pi}\cos\left(\frac{4\pi t}{T}\right) + \frac{4}{35\pi}\cos\left(\frac{6\pi t}{T}\right) + \cdots$$

Fig. 4.9 Fully rectified
cosine wave, first 10 terms
Fourier series

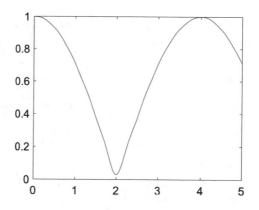

Define the period as $T = 4$ s and the first 10 terms of the series are plotted in
Fig. 4.9.

4.2   Complex Fourier Series

A more compact form of the Fourier series is known as the complex Fourier series.
Recall some basic mathematical expressions due to Euler.

$$e^{j\theta} = \cos(\theta) + j\sin(\theta) \tag{4.9}$$

$$e^{-j\theta} = \cos(\theta) - j\sin(\theta) \tag{4.10}$$

By adding and subtracting these two equations above we get

$$\cos(\theta) = \frac{e^{j\theta} + e^{-j\theta}}{2} \tag{4.11}$$

and

$$\sin(\theta) = \frac{e^{j\theta} - e^{-j\theta}}{2j} \tag{4.12}$$

Now the ordinary Fourier series is

$$f(t) = a_o + \sum_{k=1}^{\infty} a_k \cos(k\omega_o t) + b_k \sin(k\omega_o t)$$

Write

$$c_k = \frac{1}{2}\left(a_k - jb_k\right) \qquad (4.13)$$

$$c_0 = a_0 \qquad (4.14)$$

$$c_{-k} = \frac{1}{2}\left(a_k + jb_k\right) \qquad (4.15)$$

Then we write the Fourier series as

$$f(t) = \sum_{k=-\infty}^{\infty} c_k e^{jk\omega_o t} \qquad (4.16)$$

Known as the *complex Fourier series.*
The coefficients c are found from

$$c_0 = \frac{1}{T} \int_{-T/2}^{T/2} f(t)dt \qquad (4.17)$$

$$c_k = \frac{1}{T} \int_{-T/2}^{T/2} f(t)e^{-jk\omega_o t}dt \qquad (4.18)$$

4.2.1 Example: Square Wave in Complex Fourier Form

Consider the same example as shown in Fig. 4.1, this time we will apply the complex method instead.

The waveform is periodic with period T.

$$c_0 = a_0 = \frac{1}{T} \int_{-T/2}^{T/2} f(t)dt$$

$$= \frac{1}{T} \int_{-\tau/2}^{\tau/2} dt$$

$$= \frac{\tau}{T}$$

The fundamental frequency is $\omega_o = 2\pi/T$.
Now the other c terms are found from

$$c_k = \frac{1}{T} \int\limits_{-\tau/2}^{\tau/2} e^{-jk\omega_o t} dt$$

$$= -\frac{1}{Tjk\omega_o} \left[e^{-jk\omega_o t} \right]_{-\tau/2}^{\tau/2}$$

$$= -\frac{1}{Tjk\omega_o} \left[e^{-jk\omega_o \tau/2} - e^{jk\omega_o \tau/2} \right]$$

$$= \frac{1}{Tjk\omega_o} \left[e^{jk\omega_o \tau/2} - e^{-jk\omega_o \tau/2} \right]$$

$$= \frac{2}{Tk\omega_o} \frac{\left[e^{jk\omega_o \tau/2} - e^{-jk\omega_o \tau/2} \right]}{2j}$$

$$= \frac{1}{\pi k} \sin(\pi k \tau/T)$$

Let $T = 4$ s and $\tau = 2$ s for 50% duty-cycle.

$$c_k = \frac{1}{\pi k} \sin(\pi k/2) = c_{-k}$$

$$c_1 = \frac{1}{\pi} \sin(\pi/2) = \frac{1}{\pi}, c_2 = \frac{1}{2\pi} \sin(\pi) = 0,$$

$$c_3 = \frac{1}{3\pi} \sin(\pi 3/2) = -\frac{1}{3\pi}, c_4 = \frac{1}{4\pi} \sin(2\pi) = 0,$$

$$c_5 = \frac{1}{5\pi} \sin(\pi 5/2) = \frac{1}{5\pi}$$

The negative-time values of c_k have the same magnitude as the positive ones. That is $|c_k| = |c_{-k}|$. This is like a mirror image if they are plotted versus k as the independent axis. Use $k\omega_o$ as the independent axis and each magnitude of c represents the amplitude of a harmonic. This is known as the line-spectrum of the periodic signal. This tells us what the frequency content of the signal is.

Observe how the even harmonics are zero and how the odd ones die off in amplitude as frequency increases. Note that in Fig. 4.10 the concept of negative frequencies now has been introduced by this example. In mathematical terms it is like two phasers, each of 50% the size of the real-world signal, one phasor goes clockwise, and one goes anti-clockwise. Negative frequencies do not exist in the real world, however.

Fig. 4.10 Line magnitude spectrum of the square wave. Independent axis is frequency

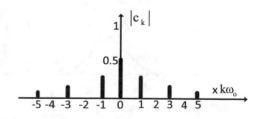

4.2.2 Complex Fourier Series of a Triangular Waveform

The triangular waveform shown in Fig. 4.11 is symmetric (even) and has period T = 2 s. The fundamental frequency is $\omega_o = \frac{2\pi}{T} = \pi$ rad/s.

In the region $0 < t < 1$ the equation of the negative-slop line is.

$f(t) = -(t - 1)$. The dc value is found inspection by calculating the area over one period and dividing by the period. This gives ½. Verifying by integration

$$c_0 = \frac{2}{T} \int_0^{T/2} f(t)dt$$

$$= -\frac{2}{2} \int_0^1 (t - 1)dt$$

$$= -\left[\frac{t^2}{2} - t\right]_0^1$$

$$= -\left[-\frac{1}{2}\right] = \frac{1}{2}$$

The c terms are found over the half range

$$c_k = -\frac{2}{T} \int_0^{T/2} (t - 1)e^{-jk\omega_o t}dt$$

$$= -\int_0^1 (t - 1)e^{-jk\omega_o t}dt$$

$$= \frac{e^{-jk\omega_o}\left((jk\omega_o - 1)e^{jk\omega_o} + 1\right)}{j^2 k^2 \omega_o^2}$$

$$= -\frac{\left(e^{-jk\omega_o} + jk\omega_o - 1\right)}{k^2 \omega_o^2}$$

$$= \frac{\left(1 - e^{-jk\omega_o} - jk\omega_o\right)}{k^2 \omega_o^2}$$

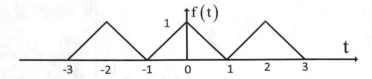

Fig. 4.11 Triangular waveform

Fig. 4.12 Single-sided line spectrum of triangular waveform

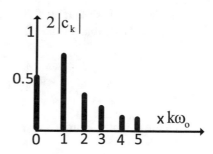

Substituting k values and $\omega_o = \pi$

$$c_1 = \frac{\left(1 - e^{-j\omega_o} - j\omega_o\right)}{\omega_o^2} = \frac{2}{\pi^2} - j\frac{1}{\pi}$$

$$|c_1| = \left|\frac{2}{\pi^2} - j\frac{1}{\pi}\right| = 0.3773$$

$$c_2 = \frac{\left(1 - e^{-2j\pi} - 2j\pi\right)}{4\pi^2} = -j\frac{1}{2\pi}$$

$$|c_2| = \left|-j\frac{1}{2\pi}\right| = 1/2\pi$$

And similarly

$$|c_2| = 0.1592, \ |c_3| = 0.1085, \ |c_4| = 0.0796, \ |c_5| = 0.0642$$

In the previous example we plotted the "two-sided" spectrum. That is, we showed negative and positive frequencies. From a theory point of view this is ok, but it makes more practical senses to show positive frequencies only since they are the real frequencies. Negative frequencies turn up all the time in signal processing but do not exist in the real world. Usually we double the magnitudes of the positive c terms instead and plot these. We leave the dc value unchanged, however Fig. 4.12.

4.2.3 Complex Fourier Series of an Impulse Train

A train of impulses is an infinite length generation of impulses separated by time T seconds. This may at first sight does not appear to be a very practical thing to study, but it is an important result used later in sampling theory. An impulse train is given by

$$f(t) = \sum_{k=1}^{\infty} \delta(t - kT)$$

Each impulse is shifted by T from the previous. This is a periodic signal and the FT coefficients are found from

$$c_k = \frac{1}{T} \int_{-T/2}^{T/2} \delta(t) e^{-jk\omega_o t} dt = \frac{1}{T}$$

And the complex Fourier series becomes

$$f(t) = \sum_{k=-\infty}^{\infty} c_k e^{jk\omega_o t}$$

$$= \frac{1}{T} \sum_{k=-\infty}^{\infty} e^{jk\omega_o t}$$

Which is an infinite sum of complex sinusoids. The integral was solved easily because the delta function is only valid at time $t = 0$.

4.2.4 More on Square Waves

Returning to example 4.2.1, we showed that the c coefficients are given by

$$c_o = \frac{1}{2} \tag{4.19}$$

$$c_k = \frac{2}{Tk\omega_o} \frac{\left[e^{jk\omega_o \tau/2} - e^{-jk\omega_o \tau/2} \right]}{2j} \tag{4.20}$$

Equation (4.20) can be written as

$$c_k = \frac{2}{Tk\omega_o} \sin\left(\frac{k\omega_o \tau}{2} \right)$$

$$= \frac{\tau}{T} \frac{\sin\left(\frac{k\omega_o \tau}{2} \right)}{\left(\frac{k\omega_o \tau}{2} \right)} \tag{4.21}$$

Equation (4.21) is of the form $c_o = \frac{\sin x}{x}$ where $x = \frac{k\omega_o \tau}{2}$. This is the well-known sinc(x) function scaled by a constant $c_o = \frac{\tau}{T}$.

In Fig. 4.13 for the basic sinc(x) function the zero crossings with the horizontal axis is at π, 2π, 3π etc. Applying this to our Eq. (4.21), it must cross at

$$\frac{k\omega_o \tau}{2} = \pi, 2\pi, 3\pi \ldots \tag{4.22}$$

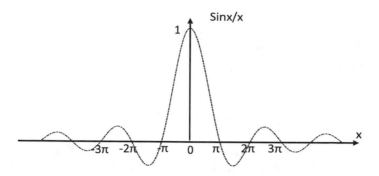

Fig. 4.13 Function $\frac{\sin x}{x}$

Giving

$$k\omega_o = \frac{2\pi}{\tau}, \frac{4\pi}{\tau}, \frac{6\pi}{\tau} \ldots \tag{4.23}$$

Furthermore, every line in the spectrum is spaced apart by $\omega_o = \frac{2\pi}{T}$. It stands to reason that the lines in the spectrum must be closer together as T increases and τ controls the zero crossing. This is illustrated in Fig. 4.14 for the square wave.

The envelope is not solid and shown as a broken line. These are line spectra only. If the absolute value is taken, then the negative part of the graph is flipped in the $k\omega_o$ axis to make it positive. Each line is separated by $\frac{2\pi}{T}$, and so if T gets very large the lines will appear to be very close together since $\omega_o \rightarrow 0$. Likewise, if the width τ is reduced, the zero-crossings widen. This is the basic idea behind the *Fourier Transform*, which unlike the Fourier series can be applied to *aperiodic* signals. Note also that since

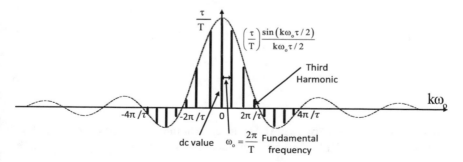

Fig. 4.14 Line spectrum for a general square wave. The lines are encased within the form of a sinc(x) function. The envelope is not solid however and only shown as a framework

$$c_k = \frac{\tau}{T} \frac{\sin\left(\frac{k\omega_0\tau}{2}\right)}{\left(\frac{k\omega_0\tau}{2}\right)}$$

Then when $k = 2$ for the second harmonic

$$c_2 = \frac{\tau}{T} \frac{\sin\left(\frac{2\pi\tau}{T}\right)}{\left(\frac{2\pi\tau}{T}\right)}$$

When $\frac{2\pi\tau}{T} = \pi$ then $c_2 = 0$. This occurs only when $\frac{\tau}{T} = \frac{1}{2}$ or 50% duty cycle. The same applies for all even harmonics. They will be zero when the square wave has 50% duty cycle. It is important to note that because this example waveform is symmetric and even, that the Fourier coefficients are real. More than often the coefficients are complex, and the magnitude line spectrum is plotted (absolute value).

4.3 The Fourier Transform

The Fourier transform is a natural extension of Fourier series. It applies to signals that are not periodic. For example, a single pulse is *aperiodic* and the Fourier transform (FT) can be applied to this waveform. It is one of the most widely used tools in signal processing and science to this day. It applies to subjects as diverse as astronomy, physics, electronics and mechanical engineering. The FT enables us to view waveforms from the domain of frequency rather than time. We have already achieved this of course with the Fourier series, but it is limited to periodic waveforms. For a time-domain signal f(t), its FT is defined as

$$\mathcal{F}\{f(t)\} = F(\omega)$$

$$= \int_{-\infty}^{\infty} f(t)e^{-j\omega t}\,dt \qquad (4.24)$$

For the FT to exist, the signal f(t) must be absolutely integrable.

$$\int_{-\infty}^{\infty} |f(t)|\,dt < \infty \qquad (4.25)$$

The FT also can be expressed in Hz rather than rad/s frequency. The inverse FT is the opposite of the FT and returns a signal already in the frequency domain to the time domain.

$$f(t) = \mathcal{F}^{-1}\{F(\omega)\}$$

Fig. 4.15 Single pulse of
width τ

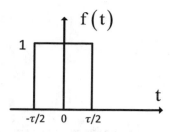

$$= \frac{1}{2\pi} \int_{-\infty}^{\infty} F(\omega)e^{j\omega t}d\omega \qquad (4.26)$$

The FT is used to examine the frequency properties of signals and systems. It is not a method for finding any form of transient analysis like the Laplace Transform. It is for steady-state analysis only. To see the importance of the FT, it is best to start with a commonly met aperiodic signal, a single pulse of width τ as shown in Fig. 4.15.

$$\mathcal{F}\{f(t)\} = \int_{-\infty}^{\infty} f(t)e^{-j\omega t}dt$$

$$= \int_{-\tau/2}^{\tau/2} e^{-j\omega t}dt \qquad (4.27)$$

Solving (4.27) manually

$$\int_{-\tau/2}^{\tau/2} e^{-j\omega t}dt = -\frac{1}{j\omega}\left[e^{-j\omega t}\right]_{-\tau/2}^{\tau/2}$$

$$= -\frac{1}{j\omega}\left[e^{-j\omega\tau/2} - e^{j\omega\tau/2}\right]$$

$$= \frac{1}{j\omega}\left[e^{j\omega\tau/2} - e^{-j\omega\tau/2}\right] \qquad (4.28)$$

To proceed further we use the Euler identity $\sin(\theta) = \frac{e^{j\theta}-e^{-j\theta}}{2j}$

$$\frac{1}{j\omega}\left[e^{j\omega\tau/2} - e^{-j\omega\tau/2}\right] = \frac{2}{\omega}\frac{\left[e^{j\omega\tau/2} - e^{-j\omega\tau/2}\right]}{2j}$$

$$= \left(\frac{2}{\omega}\right)\sin(\omega\tau/2)$$

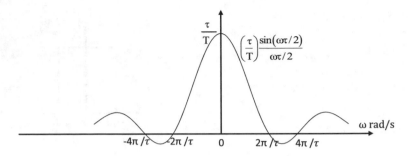

Fig. 4.16 Fourier Transform of a pulse

But this is not in a recognisable form to plot a graph. Therefore, we express in sinc(x) form as with the previous section.

$$
\left(\frac{2}{\omega}\right)\sin(\omega\tau/2) = \left(\frac{\tau}{\omega\tau/2}\right)\sin(\omega\tau/2)
$$
$$
= \tau\frac{\sin(\omega\tau/2)}{(\omega\tau/2)} \tag{4.29}
$$

This is a sinc(x) waveform with value τ at $\omega = 0$.

Unlike the periodic waveform there is of course no period in the waveform since it does not repeat. Only the width of the pulse and the frequency features in the solution Fig. 4.16.

This is an important result for such a simple waveform. The reason is as follows. Much can be gained by studying this waveform in the frequency domain. For example, as the width of the pulse reduces in width in the time domain, in the frequency domain the waveform gets wider occupying more bandwidth. The opposite happens with a wide pulse, the frequency-domain version gets narrower in frequency.

A two-way arrow is shown in Fig. 4.17 which tells us that the converse is also true. Widening in the frequency domain shortens in the time domain and vice-versa. In the limit, it can be deduced that an impulse occupies most of the frequency available. Therefore, sudden spikes in the time domain cause interference across the spectrum in say wireless receivers.

4.3.1   The Ideal Filter

The first application of the FT is now given as some analysis on whether it is possible to implement an ideal brick-wall filter. If we assume such a filter exists and has bandwidth ω_c, then we can calculate the theoretical time-domain expression. Figure 4.18 shows the brick-wall filter which has upper and lower cutoff frequencies of $-\omega_c/2$ and $\omega_c/2$ respectively. It is a pulse in the frequency domain.

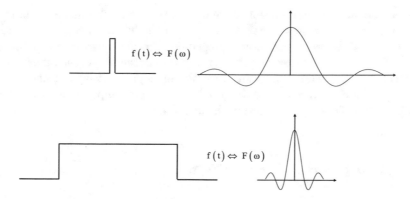

Fig. 4.17 Widening and shortening a pulse and its effect in the frequency domain

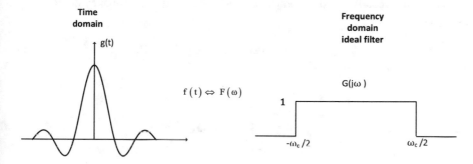

Fig. 4.18 Ideal brick-wall filter in the frequency domain

Now find the inverse FT of the ideal filter. This will give us the impulse response of the ideal filter g(t), a time domain expression.

$$g(t) = \mathcal{F}^{-1}\{F(\omega)\}$$

$$= \frac{1}{2\pi} \int_{-\infty}^{\infty} F(\omega)e^{j\omega t}d\omega$$

$$= \frac{1}{2\pi} \int_{-\omega_c/2}^{\omega_c/2} e^{j\omega t}d\omega$$

$$= \frac{\sin(\omega_c t/2)}{\pi t}$$

$$= \left(\frac{\omega_c}{2\pi}\right)\frac{\sin(\omega_c t/2)}{\omega_c t/2} \tag{4.30}$$

The finer details of this need not concern us at the point. The important thing to notice is that this is again a sinc function (this time in the time domain) and as such must have positive and negative time properties. This is enough to tell us that any filter with such an impulse response g(t) must be noncausal. As such it cannot be constructed in hardware or software. However, by adding a delay of some amount (shifting to the right into positive time) and truncating the impulse response we can approximate it. This is best achieved with digital filters.

4.3.2 Transforms of Common Signals

4.3.2.1 Inverse FT of an Impulse in the Frequency Domain

Consider the inverse FT of an impulse which occurs in the frequency domain at a frequency ω_0. Let

$$F(\omega) = \delta(\omega - \omega_0) \tag{4.31}$$

$$
\begin{aligned}
f(t) &= \mathcal{F}^{-1}\{F(\omega)\} \\
&= \frac{1}{2\pi} \int_{-\infty}^{\infty} \delta(\omega - \omega_0) e^{j\omega t} d\omega \\
&= \frac{1}{2\pi} e^{j\omega_0 t}
\end{aligned}
\tag{4.32}
$$

Therefore, we can write

$$\mathcal{F}\{e^{j\omega_0 t}\} = 2\pi\delta(\omega - \omega_0) \tag{4.33}$$

The FT of a complex sinusoid is an impulse in the frequency domain.

We can use this to find the FT of a sinewave.

4.3.2.2 FT of a Sinewave

From the Euler form

$$\sin(\omega t) = \frac{e^{j\omega t} - e^{-j\omega t}}{2j}$$

Using the result of (4.33)

Fig. 4.19 Fourier Transform of cosine or sine waveforms. Magnitude only shown

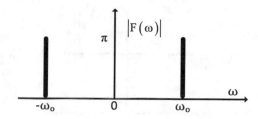

$$\mathcal{F}\{\sin(\omega t)\} = \frac{\pi}{j}[\delta(\omega - \omega_o) - \delta(\omega + \omega_o)]$$

$$= j\pi[\delta(\omega + \omega_o) - \delta(\omega - \omega_o)] \tag{4.34}$$

This is two impulses in the frequency domain at $\pm\omega_o$. We note that expressed in Hz instead of rad/s the result is slightly different:

$$\mathcal{F}\{\sin(\omega t)\} = \frac{1}{2j}[\delta(f - f_o) - \delta(f + f_o)] \tag{4.35}$$

The j term does not affect the amplitude, only the phase.
Similarly, for a cosine $\cos(\omega t) = \frac{e^{j\omega t} + e^{-j\omega t}}{2}$ and

$$\mathcal{F}\{\cos(\omega t)\} = \pi[\delta(\omega - \omega_o) + \delta(\omega + \omega_o)] \tag{4.36}$$

There are no j terms here and hence zero phase-shift Fig. 4.19.

4.3.2.3 FT of a Constant

From (4.33)

$$\mathcal{F}\{e^{j\omega_o t}\} = 2\pi\delta(\omega - \omega_o)$$

Substitute $\omega_o = 0$ and obtain

$$\mathcal{F}\{1\} = 2\pi\delta(\omega) \tag{4.37}$$

In Fig. 4.20 we see that dc is compressed into an impulse in the frequency domain.

4.3.2.4 FT of an Impulse in the Time Domain

$$\mathcal{F}\{\delta(t)\} = \int_{-\infty}^{\infty} \delta(t)e^{-j\omega t}dt = 1 \tag{4.38}$$

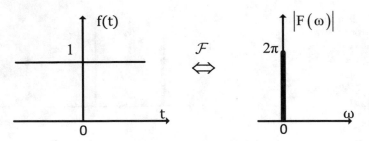

Fig. 4.20 Fourier transform of a constant (dc)

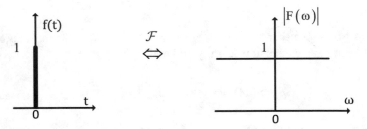

Fig. 4.21 Fourier transform of an impulse in the time domain is a constant

The FT of an impulse is a constant across the whole spectrum in the frequency domain.

This is the converse of Fig. 4.20. The result is shown in Fig. 4.21.

4.3.2.5　FT of a Delayed Signal

For a signal f(t) with FT F(ω), if delayed by T to give f(t-T), then its FT is found from

$$\mathcal{F}\{f(t - T)\} = \int_{-\infty}^{\infty} f(t - T)e^{-j\omega t}dt \tag{4.39}$$

Substitute $x = t-T$, $dx = dt$ (limits are the same) and we quickly get

$$\mathcal{F}\{f(t - T)\} = \int_{-\infty}^{\infty} f(x)e^{-j\omega(x+T)}dx$$

$$= e^{-j\omega T} \int_{-\infty}^{\infty} f(x)e^{-j\omega x}dx$$

$$= e^{-j\omega T}F(\omega) \tag{4.40}$$

Therefore we just take the FT of the original signal f(t) and get $F(\omega)$, then multiply by the FT of a pure time-delay $e^{-j\omega T}$.

4.3.2.6   FT of an Exponential Decay $f(t) = e^{-at}$

$$\mathcal{F}\{f(t)\} = \int_{-\infty}^{\infty} e^{-at}e^{-j\omega t}dt$$

$$= \int_{-\infty}^{\infty} e^{-(j\omega+a)t}dt$$

$$= \frac{1}{j\omega + a} \tag{4.41}$$

4.3.3   Some Properties of the FT

Some common properties are now shown which come in useful for many engineering and science problems.

4.3.3.1   Property: Frequency Shifting Theorem

Multiply a signal f(t) by a complex exponential to give $f(t)e^{j\omega_o t}$ and take its FT

$$\mathcal{F}\{f(t)e^{j\omega_o t}\} = \int_{-\infty}^{\infty} e^{j\omega_o t}e^{-j\omega t}dt$$

$$= \int_{-\infty}^{\infty} e^{-(j\omega-\omega_o)t}dt$$

$$= F(\omega - \omega_o) \tag{4.42}$$

This is a shift upwards in frequency of the spectrum by ω_o rad/s. we can shift downwards by multiplying by $e^{-j\omega_o t}$ to get $F(\omega + \omega_o)$. This is used in wireless communication.

4.3.3.2   Property: Modulation Theorem

A variation on the frequency shifting theorem. If we multiply an arbitrary signal f(t) by a cosine (or sine for practical cases as phase doesn't really come into play here)

$$\mathcal{F}\{f(t)\cos(\omega_0 t)\} = \int_{-\infty}^{\infty} \cos(\omega_0 t)e^{-j\omega t}dt \qquad (4.43)$$

But recall that $\cos(\omega_0 t) = \left(\frac{e^{j\omega_0 t}+e^{-\omega_0 t}}{2}\right)$. Therefore, using the frequency shifting theorem

$$\mathcal{F}\{f(t)\cos(\omega_0 t)\} = \frac{1}{2}[F(\omega + \omega_0) + F(\omega - \omega_0)] \qquad (4.44)$$

This creates two sidebands, a spectrum shift above ω to $\omega - \omega_0$ and one below to $\omega + \omega_0$. The spectra are the same, only centred at different frequencies and halved in magnitude. In radio receivers, this idea has existed since the days of one of the founding fathers of FM radio, Edwin Armstrong, an American inventor. More recently it has been found that Frenchman Lucien Lèvy had an invention predating Armstrong's. He used the method to get better filtering in a radio by introducing the idea of an *intermediate frequency* (IF) in a radio. Incoming signals were multiplied by a local oscillator (which changes with the frequency of the selected radio channel) to ensure that the shifted spectrum is always centred on a certain frequency known as the IF frequency. Usually this has been 10.7 MHz for FM receivers but the same idea is used in AM radio for 455 kHz. The invention is one usually named the superheterodyne receiver. It gives far better frequency selectivity than a basic radio receiver without the principle (the tuned radio frequency (TRF) receiver which predated the superheterodyne).

4.3.3.3   Scaling Theorem

The purpose of this is theorem is to see the effects in the frequency domain when a time domain signal is stretched or shrunk in time. For a signal f(t), let a scaling factor be α which is any non-zero real number. Then we need to find the FT of $f\left(\frac{t}{\alpha}\right)$.

$$\mathcal{F}\left\{f\left(\frac{t}{\alpha}\right)\right\} = \int_{-\infty}^{\infty} f\left(\frac{t}{\alpha}\right)e^{-j\omega t}dt \qquad (4.45)$$

Let $x = \frac{t}{\alpha}$ giving $dt = \alpha dt$. Then

$$\mathcal{F}\left\{f\left(\frac{t}{\alpha}\right)\right\} = \int\limits_{-\infty}^{\infty} f(x)e^{-j(\alpha\omega)x}\alpha dx$$

$$= |\alpha|F(\omega\alpha) \tag{4.46}$$

The absolute value is required to prevent problems when α is negative. Much can be learned from (4.46) in terms of the relationship between time and frequency domains. In words, it says that if a signal is lengthened in time duration by a factor 2 ($\alpha = 2$), then in frequency terms it will be halved in bandwidth. Conversely if a signal is shortened in time by half $\alpha = 0.5$, its frequency content will be doubled in bandwidth. This is commonly found when old vinyl records are sped up or reduced in speed, the change in frequency content is easily audible. Speeding up the record (shortening its length in time) increases high frequencies and reducing the speed (increasing its length) lowers the frequencies.

4.3.3.4 Linearity of the FT

Following the basic rules for linearity of scaling and additivity, if we scale two signals and add then we get

$$f(t) = \alpha x_1(t) + \beta x_2(t) \tag{4.47}$$

Then

$$\mathcal{F}\{\alpha x_1(t) + \beta x_2(t)\} = F(\omega)$$

where

$$F(\omega) = \alpha X_1(\omega) + \beta X_2(\omega) \tag{4.48}$$

The FT of a sum of signals is the same as the sum of the individual FTs. Multiplying by constants in the time domain gives rise to the same scaling in the frequency domain.

4.3.3.5 Convolution Theorem

The convolution theorem is a widely used theorem in signal processing. It has already been shown for Laplace Transforms. If the convolution of two signals u(t) and g(t) give rise to f(t) then we say f(t) = u(t)*g(t) where

$$f(t) = \int\limits_{-\infty}^{\infty} g(t - \tau)u(\tau)d\tau \tag{4.49}$$

The FT is found from

$$\mathcal{F}\{f(t)\} = \int_{-\infty}^{\infty} \int_{-\infty}^{\infty} g(t - \tau)u(\tau)d\tau e^{-j\omega t}dt \tag{4.50}$$

Substituting $x = t - \tau$ so that $x + \tau = t$, $dt = dx$, we can easily find that

$$\mathcal{F}\{g(t) * u(t)\} = F(\omega)$$

$$= \int_{-\infty}^{\infty} g(x)e^{-j\omega x}dx \int_{-\infty}^{\infty} u(\tau)e^{-j\omega \tau}d\tau$$

$$= G(\omega)U(\omega) \tag{4.51}$$

In words this means that: *the FT of convolution in the time domain is the same as multiplication in the frequency domain.*

This applies to systems too. Therefore, we multiply the frequency responses of two systems in cascade to get the total frequency response.

4.3.3.6   FT of a Product in the Time Domain

Multiplying two signals in the time domain, it can be shown that this is equivalent to convolution in the frequency domain.

$$\mathcal{F}^{-1}\{G(\omega) * U(\omega)\} = \frac{1}{2\pi} \int_{-\infty}^{\infty} G(\beta)U(\omega - \beta)d\beta$$

$$= g(t)u(t) \tag{4.52}$$

In words this means that: *the inverse FT of convolution in the frequency domain is the same as multiplication in the time domain.*

4.3.4   Table of Common Fourier Transforms

A Fourier transform table of the most common functions used in signal processing is shown in Table 4.1 below.

4.4 Parseval's Theorem

Parseval's theorem is also known as Rayleigh's theorem and Plancherel's theorem. Parseval appears to have been earliest in 1799. It states that the energy in a signal as expressed in the time domain is the same energy you get if you calculate it in the frequency domain.

$$\int_{-\infty}^{\infty} |f(t)|^2 dt = \frac{1}{2\pi} \int_{-\infty}^{\infty} |F(\omega)|^2 d\omega \qquad (4.53)$$

The energy in periodic signals however is infinite if averaged over an infinite amount of time. Therefore they are referred to as power signals instead. For example, a sinewave has infinite energy (over an infinite time) but finite power as defined mathematically. Whereas an energy signal like a pulse (or any aperiodic signal) has finite energy (since it goes to zero) but zero power if averaged over an infinite amount of time.

Therefore, the following definition is used for periodic signals from Fourier series which is based on power instead:

$$\frac{2}{T} \int_{-T/2}^{T/2} f(t)^2 dt = a_o^2/2 + \sum_{k=1}^{\infty} a_k^2 + b_k^2 \qquad (4.54)$$

4.4.1 Example Using Parseval's Theorem

Consider a signal $f(t) = c^{-at}$, which is an energy signal since it is not periodic. Then $|f(t)|^2 = e^{-2at}$. Integrating this within (4.53)

$$\int_{-\infty}^{\infty} |f(t)|^2 dt = \int_{-\infty}^{\infty} e^{-2at} dt$$

$$= \frac{1}{2a} \qquad (4.55)$$

This is the energy in the time domain. In the frequency domain we need the FT of e^{-at} and this is readily found to be $\mathcal{F}\{e^{-at}\} = \frac{1}{j\omega+a}$ (see Table 4.1).

Now $\left|\frac{1}{j\omega+a}\right|^2 = \frac{1}{\omega^2+a^2}$.

From (4.53)

Table 4.1 Table of Fourier transforms

Time-Domain f(t)	Fourier Domain F(ω)
$\delta(t)$	1
e^{-at}	$\frac{1}{j\omega+a}$
$e^{j\omega_o t}$	$2\pi\delta(\omega - \omega_o)$
$\cos(\omega_o t)$	$\pi[\delta(\omega - \omega_o) + \delta(\omega + \omega_o)]$
$\sin(\omega_o t)$	$j\pi[\delta(\omega + \omega_o) - \delta(\omega - \omega_o)]$
$\text{rect}(\frac{t}{\tau})$	$\tau\,\text{sinc}(\frac{\omega\tau}{2})$
$e^{-a\vert t\vert}$	$\frac{2a}{a^2+\omega^2}$
$\sum\limits_{k=1}^{\infty} \delta(t - kT)$	$\frac{1}{T} \sum\limits_{k=-\infty}^{\infty} e^{jk\omega_o t}$
$f(t - \tau)$	$e^{-j\omega\tau}F(\omega)$

$$\frac{1}{2\pi} \int_{-\infty}^{\infty} |F(\omega)|^2 d\omega = \frac{1}{2\pi} \int_{-\infty}^{\infty} \frac{d\omega}{\omega^2 + a^2}$$

$$= \frac{1}{2\pi a} \tan^{-1}\left(\frac{\omega}{a}\right)_{\omega=-\infty}^{\omega=\infty}$$

$$= \frac{1}{2a} \tag{4.56}$$

Hence energy in the time and frequency domains are the same.

Reference

1. J. Fourier, *The analytic theory of heat. Reprint and translation of the original 1822 French edition: Théorie analytique de la chaleur* Dover Pheonix, 2003

Chapter 5
Sampling of Signals and Discrete Mathematical Methods

5.1 The Sampling Theorem

The mathematics of sampling was first discovered by Whittaker [1] in 1915. As a mathematician, he did not realise the engineering implications of the theory however and it was left to two others, namely a Russian Kotelnikov [2] in 1933 and independently by an American Claude Shannon [3] in 1948. The theory considers ideal sampling, that is samples that are obtained instantaneously by a unit impulse. This is the basic sampling theory that underpins modern digital signal processing, communications theory and control-systems. Sampling in hardware is done with an analogue to digital convertor (ADC). The way the ADC acquires data is not the same as the theory. There are no impulses used during conversion. The theory is highly accurate however and has had most of the latter part of the twentieth century to be tested in practical design. Mathematically it makes sense to model the sampling process as a series of impulses (or impulse train) multiplied by the signal itself. This is illustrated in Fig. 5.1.

Let the impulse train be an infinite series of pulses separated by the sampling interval T_s. The sampling frequency is defined as the reciprocal $f_s = \frac{1}{T_s}$ Hz or $\omega_s = \frac{2\pi}{T_s}$ rad/s.

$$\delta_T = \sum_{k=0}^{\infty} \delta(t - kT_s) \tag{5.1}$$

Multiplying the impulse train by a signal f(t) gives the sampled signal $f^*(kT_s)$.

$$f^*(kT_s) = f(t) \sum_{k=0}^{\infty} \delta(t - kT_s) \tag{5.2}$$

The object is to first find what is happening in the frequency domain. This is best done using the Fourier transform (FT). We use the fact that convolution in

© The Author(s), under exclusive license to Springer Nature Switzerland AG 2022
T. J. Moir, *Rudiments of Signal Processing and Systems*,
https://doi.org/10.1007/978-3-030-76947-5_5

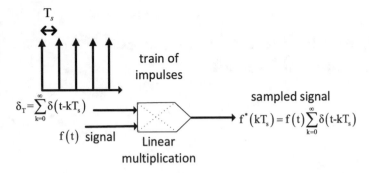

Fig. 5.1 Block diagram of ideal sampling process

the frequency domain is the same as the FT of the product in the time domain. Equation (5.2) is a product so we need the individual FTs so we can convolve them in the Fourier domain. The FT of f(t) is $F(\omega)$ and the complex Fourier series of an impulse train has already been found in Chap. 4.

$$\mathcal{F}\{e^{j\omega_s t}\} = 2\pi\delta(\omega - \omega_s) \tag{5.3}$$

Use the result from FT tables that $\delta_T = \sum\limits_{k=0}^{\infty} \delta(t - kT_s)$. Then take FT of (5.3) thus

$$R(\omega) = \frac{1}{T_s} \int\limits_{-\infty}^{\infty} \sum\limits_{k=-\infty}^{\infty} e^{jk\omega_o t} e^{-j\omega t} dt$$

$$= \frac{2\pi}{T_s} \sum\limits_{k=-\infty}^{\infty} \delta(t - kT_s) \tag{5.4}$$

Convolving in the frequency domain (this procedure is not too different than the time-domain by introducing dummy frequency shift variable β) thus

$$\mathcal{F}[r(t)f(t)] = \frac{1}{2\pi} \int\limits_{-\infty}^{\infty} R(\beta)F(\omega - \beta)d\beta$$

$$= \frac{1}{2\pi} \int\limits_{-\infty}^{\infty} \frac{2\pi}{T_s} \sum\limits_{k=-\infty}^{\infty} \delta(\beta - k\omega_s)F(\omega - \beta)d\beta$$

$$= \frac{1}{T_s} \sum\limits_{k=-\infty}^{\infty} F(\omega - k\omega_s) \tag{5.5}$$

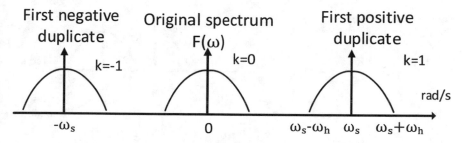

Fig. 5.2 Spectra of a sampling process. Shows first duplicate, positive and negative spectra only

Equation (5.5) is the mathematical description of the spectra of the sampled signal. Reading the equation indicates an infinity number of duplicated spectra (of the original signal). The spacing of these spectra depends on the sampling frequency. This is best viewed graphically. The shape of the original spectrum of the signal is of no great importance other than the beginning and end frequency. Assume that the signal spectrum has minimum and maximum frequencies of $\pm\omega_h$ rad/s and *zero energy outside of those frequencies*. Now view Fig. 5.2 which shows the first of these set of infinite duplicate spectra.

If the original spectrum is to be recovered, then it can be done using a low pass filter. The steepness of this filter will depend how close the first (k = 1) and original (k = 0) spectra are apart. They can be touching or even overlapping if we are not careful. The lowest frequency of the first positive duplicate must be a minimum of just greater than the sampling frequency (the highest edge of the first spectrum). Then

$$\omega_h < \omega_s - \omega_h$$

or

$$\omega_s > 2\omega_h \tag{5.6}$$

This is the Sampling Theory as popularised by Shannon. His colleague Nyquist also played a role and it is often named after both and even all four authors. The Whittaker, Kotelnikov, Nyquist-Shannon sampling theory. Equation (5.6) says that the sampling frequency must be greater than twice the maximum frequency of the original signal. However, this is only because the original spectrum has been shown from dc upwards. There are many spectra that do not start at dc. The proper definition is that for a bandlimited signal:

Sampling must take place at a frequency at least twice the bandwidth of the signal in order for the original signal to be exactly reconstructed from its samples.

The frequency $\omega_s/2$ is usually known as the *Nyquist frequency* or half-sampling frequency. We note this for future reference.

Fig. 5.3 Illustration of aliasing in the frequency domain

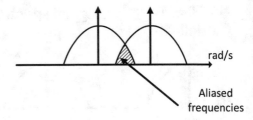

If the sampling theorem is not satisfied, there will be overlap of the adjacent spectra and the original spectrum cannot be recovered. This overlap of frequencies is known as Aliasing (Fig. 5.3).

Reconstruction of the samples

To reconstruct the original time domain signal, the sampled signal must pass through a bandpass filter. Let us assume the impulse response of this filter is g(t). We can also represent the sampled signal from (5.2) as

$$f^*(kT_s) = \sum_{k=-\infty}^{\infty} f(kT_s)\delta(t - kT_s) \tag{5.7}$$

With adjustment of the lower limit on the summation to minus infinity for mathematical convenience. If this is then passed through a bandpass filter, we get the reconstructed signal via convolution

$$f_r(t) = \sum_{k=-\infty}^{\infty} f(kT_s)\delta(t - kT_s) * g(t) \tag{5.8}$$

where $f_r(t)$ is the reconstructed signal in the time domain. For an ideal filter as covered in Sect. 4.3.1, its impulse response will be a sinc function and will have a brick wall cutoff at the Nyquist frequency $\frac{\omega_s}{2}$ rad/s. The impulse response will therefore be from Eq. (4.30) with $\omega_c = \omega_s = \frac{2\pi}{T_s}$

$$g(t) = \left(\frac{\omega_s}{2\pi}\right)\frac{\sin(\omega_s t/2)}{\omega_s t/2}$$
$$= \left(\frac{1}{T_s}\right)\frac{\sin(\omega_s t/2)}{\omega_s t/2} \tag{5.9}$$

Multiply by (5.9) by T_s to get unity at $t = 0$ for proper scaling

$$g(t) = \frac{\sin(\omega_s t/2)}{\omega_s t/2} = T_s \sin\left(\frac{\omega_s}{2}t\right)/\pi t \tag{5.10}$$

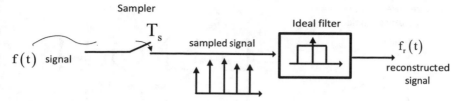

Fig. 5.4 Ideal sampling and reconstruction

Then

$$f_r(t) = \sum_{k=-\infty}^{\infty} f(kT_s)\delta(t - kT_s) * T_s \sin\left(\frac{\omega_s}{2}t\right)/\pi t \qquad (5.11)$$

The impulses shift the sin term by $t - kT_s$. So, the answer becomes

$$f_r(t) = T_s \sum_{k=-\infty}^{\infty} f(kT_s)\frac{\sin\left(\frac{\omega_s}{2}(t - kT_s)\right)}{\pi(t - kT_s)} \qquad (5.12)$$

This is known as *ideal bandlimited interpolation*. The sinc(x) functions add up to smooth-out and reconstruct the original signal. The ideal sampling and reconstruction process is illustrated in Fig. 5.4.

Of course, there is no such thing as ideal sampling, but the theory fits the practice quite closely. Practically the digital to analogue convertor (DAC), has a sinc type function built in by virtue of the *zero-order hold*. The holds the last value constant before the next sample appears to be outputted. It gives a staircase type waveform at the output.

5.2 Zero-Order Hold (ZOH)

The ZOH is hardware within a DAC that keeps a sample constant until the next sample arrives to be sent out. This should not be confused with a *sample and hold* which is at the input of an ADC. The ZOH is at the DAC and produces the staircase waveform as shown in Fig. 5.5.

The staircase waveform has harmonics which are usually filtered using a lowpass filter set around half-sampling frequency. This smooths out the edges. The impulse response of the ZOH is also shown in Fig. 5.5 and has an impulse response which is unity and returns to zero at the sampling interval. Expressing mathematically using Heaviside notation

$$h(t) = u(t) - u(t - T_s) \qquad (5.13)$$

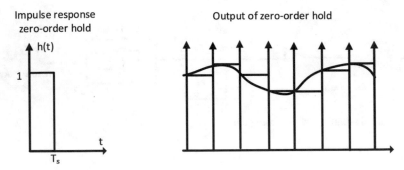

Fig. 5.5 ZOH output and impulse response of ZOH

Taking Laplace Transforms

$$H(s) = \frac{1 - e^{-sT_s}}{s} \tag{5.14}$$

We could also have taken Fourier Transforms, but substituting $s = j\omega$ into (5.14) gives the same result.

$$
\begin{aligned}
H(j\omega) &= \frac{1 - e^{-j\omega T_s}}{j\omega} \\
&= \frac{e^{-j\omega T_s/2}\left(e^{j\omega T_s/2} - e^{-j\omega T_s/2}\right)}{j\omega} \\
&= \frac{2}{\omega} e^{-j\omega T_s/2} \frac{\left(e^{j\omega T_s/2} - e^{-j\omega T_s/2}\right)}{2j} \\
&= \frac{T_s}{\omega T_s/2} e^{-j\omega T_s/2} \sin\left(\frac{\omega T_s}{2}\right) \\
&= e^{-j\omega T_s/2} T_s \operatorname{sinc}\left(\frac{\omega T_s}{2}\right)
\end{aligned}
$$

Then the magnitude of this becomes (for a frequency–response plot)

$$|H(j\omega)| = \left| T_s \operatorname{sinc}\left(\frac{\omega T_s}{2}\right) \right| \tag{5.15}$$

which is a sinc(x) function as required by ideal sampling. This also has a nasty phase-shift due to the $e^{-j\omega T_s/2}$ delay term. The phase-shift is linear and given by $-\omega T_s/2$. In signal processing applications it is of little consequence, but when feedback is put around a sampled system (a digital controller) it will de-stabilise the closed-loop system. Therefore, for feedback control using digital controllers, usually much higher sampling frequencies are used so that this delay is minimised. The frequency

Fig. 5.6 The Zero-order
hold frequency–response

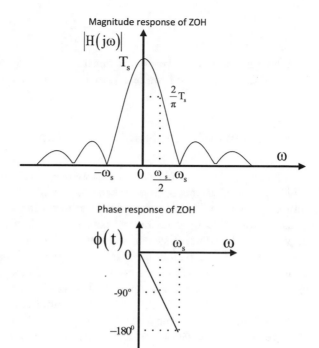

and phase-response are shown in Fig. 5.6. Note that there is minus 90° phase-shift
at half-sampling (the Nyquist frequency). This follows because from the time-delay
$e^{-j\omega T_s/2}$,the phase-shift $\phi(t)$ is

$$\phi(t) = -\omega T_s/2$$

$$= -\left(\frac{\omega}{\omega_s}\right)\pi \tag{5.16}$$

Then when $\omega = \omega_s/2$, $\phi(t) = -\pi/2$. Likewise, were we to sample ten times faster
than the maximum frequency of interest our phase-shift at that frequency would be −
18° or in general −180/n where n is the sampling factor (usually 10 to 20 for control
systems).

By substituting for $T_s = 2\pi/\omega_s$ into the sinc function we get an alternative
expression for magnitude

$$|H(j\omega)| = \left|T_s \operatorname{sinc}\left(\frac{\omega\pi}{\omega_s}\right)\right| \tag{5.17}$$

We find the magnitude at half-sampling frequency as

$$|H(j\omega)|_{\omega=\omega_s/2} = |T_s \operatorname{sinc}(\pi/2)|$$

$$= \frac{2}{\pi} T_s \tag{5.18}$$

Which is approximately -4 dB down from T_s, its highest point at dc. This is known as ZOH droop and can be compensated for if required.

5.3 Aliasing of Signals and Prevention

Despite the sampling theory stating that we need to sample at least twice the bandwidth, this does not mean we have to work at such a low sampling rate. Depending on the application vital features of the waveform in the form of higher harmonics could be lost. Usually a far higher sampling rate is used but this depends on the application. Most control-systems operate from dc and need to be sampled at a much higher rate than say a bandlimited signal. A signal could have bandwidth 50 kHz and centred at 100 MHz. This does not mean we need to sample at twice of 100 MHz + 50 kHz. Instead we can sample at 100 kHz minimum which is twice the bandwidth of the signal. This is not violating the sampling theory since it is bandwidth that is important and not the maximum frequency. When the lowest frequency is dc then we have no choice but to sample higher than twice the maximum frequency. Now consider a problem of sampling a sinewave whose frequency is higher than the Nyquist frequency (hence violating the sampling theory). This is shown in Fig. 5.7. The Nyquist frequency (or half-sampling frequency) is like an upper bound on the digital sampled world. Nothing can pass through!

For example, if the sampling frequency is 20 kHz and a signal appears at 12 kHz (here 10 kHz is the Nyquist frequency—like a maximum barrier for signals) then it gets folded back by the same amount it is over the Nyquist "barrier". In this case the signal is 2 kHz greater than 10 kHz, so it gets folded back to 8 kHz. This is also

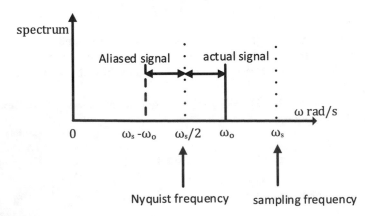

Fig. 5.7 When a frequency occurs higher than the Nyquist frequency, and with no filter, it is aliased back like a mirror

20–12 kHz. It may seem obvious that all we need do is prevent this happening. In real life this could be some form of noise that we don't know about. It is necessary therefore to put a lowpass filter in to prevent any signals getting past the Nyquist barrier. Ideally it should be a brick wall filter but not such filter can be made since it is physically unrealisable. Therefore, some figures need to be used to design a filter that is practical enough to do the job. Usually if a noise term (in this case an aliased signal) is at least 40 dB below the important information it can be considered to not cause any significant harm. This is a design specification for a lowpass filter.

The lowpass filter or anti-aliasing filter as it is commonly known is analogue and made up of operational amplifier active filters. This provides buffering of impedance whilst at the same time some amplification of the signal can take place. If the slope of the filter is too steep the order of the filter will be high, and much circuitry will be needed. It is a trade-off between simple circuitry and high enough bandwidth. We can use a low order filter by placing it at a low frequency cut-off. But this will reduce the available bandwidth of the signal or system. In some applications such as when negative feedback is used, the analogue anti-aliasing filter will introduce undesired phase-shift of its own and de-stabilised a control-loop. The filter needs to be high enough (in cut-off frequency) so as not to reduce bandwidth or introduce phase problems whilst still able to attenuate the undesired signals above the Nyquist frequency.

5.3.1 Anti-Aliasing Filtering

A lowpass filter is required to band limit the signal to be less than the Nyquist frequency. In the ideal case we considered a brick wall filter, but this cannot be achieved in practice. There are several real-world options available. All LTI filters with only poles and no zeros, must roll off at $-20n$ dB/decade asymptotically, where n is the order of the filter. The manner in which they attenuate in the stopband before they reach high frequencies however will make some filter types of more importance than others. Refer to Fig. 5.8. The Chebyshev filter has a sharper attenuation than most but suffers from ripple in the passband whereas the Butterworth filter has a flat passband but has not so severe attenuation in the stopband. The elliptic filter has the best attenuation of all but has ripple in both the passband and stopband. Finally, the Bessel filter has the worst attenuation but has good phase characteristics which are important for some applications. We therefore commonly use the Butterworth filter.

The Butterworth filter is an old idea, going back to 1930 [4] and vacuum tube technology, but it is still very popular today with operational amplifier designs. The implementation (or realisation) must be analogue since the filtering takes place in the analogue domain before sampling has yet happened. We only consider here lowpass Butterworth filters though if the signal is narrowband in nature a bandpass filter may be needed instead. If we choose the Butterworth method, the magnitude-squared lowpass *prototype* is given by

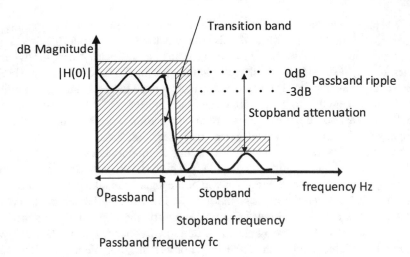

Fig. 5.8 Generic filter type frequency response

$$|G(j\omega)|^2 = \frac{1}{1 + \left(\frac{\omega}{\omega_c}\right)^{2n}} \tag{5.19}$$

where n is the filter order, ω_c is the passband edge frequency and ω is frequency in rad/s. The formula (5.19) is replaceable by an equivalent one in Hz

$$|G(jf)|^2 = \frac{1}{1 + \left(\frac{f}{f_c}\right)^{2n}} \tag{5.20}$$

Equations (5.19) or (5.20) do require a little explanation if not met before since they are not of the usual transfer function form. This is because in the field of filter design the magnitude-squared is defined in terms of special polynomials. We define the passband as being at unity (called the prototype filter) and the stopband frequency to attenuate relative to this as a ratio $\frac{f}{f_c}$. When the filter is found, it is a normalised one and needs to be transformed to the desired passband frequency at a later step. Furthermore, it is magnitude squared that is defined. For example, for an ordinary first-order filter, say $G(j\omega) = \frac{1}{1+j\omega T}$, its magnitude-squared is found to be $|G(j\omega)|^2 = \frac{1}{1+(\omega T)^2}$ and this can also be factorised and written as $|G(j\omega)|^2 = G(j\omega)G(-j\omega)$. This is the product of two transfer functions, one causal with its poles in the left-half plane and one anticausal with poles in the right-half plane like a mirror-image. There are therefore 2n poles for a filter of order n instead of the usual n poles. Since half of these poles are unstable the two mirror transfer functions need to be separated. This is called *spectral factorization*, but in filter design there are more direct ways to perform this which are detailed in the literature.

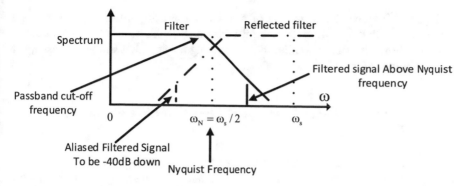

Fig. 5.9 Effect of an anti-aliasing filter

The attenuation in the passband is -3 dB though it is possible with slight modification to get less. This is known as passband ripple, but in the Butterworth case it is more of a gradual droop to reach the passband edge frequency. The *attenuation in the stopband* is given from (5.20) in dB as

$$A_s = 10Log_{10}\left(1 + \left(\frac{f}{f_c}\right)^{2n}\right) \tag{5.21}$$

and is defined to be positive (even though the gain will go negative as measured in dB).

We can re-arrange (5.21) to get any variable we like. For example, for a given integer order n, we can find what the passband frequency f_c Hz is, to attenuate a frequency at f Hz with attenuation A_s dB:

$$f_c = \frac{f}{\left[10^{0.1A_s} - 1\right]^{\frac{1}{2n}}} \tag{5.22}$$

For the -3 dB ripple case a set of Butterworth polynomials is readily available to give the prototype lowpass filter for a given order n. The passband frequency is then used to transform the prototype into the actual Laplace transfer function [5]. From there it can be physically realised in hardware.

For a given desired frequency f Hz which we wish to attenuate by A_s dB and passband frequency f_c Hz, we can also find the filter order by taking the nearest integer value of:

$$n = \frac{Log_{10}\left[10^{0.1A_s} - 1\right]}{2Log_{10}\left(\frac{f}{f_c}\right)} \tag{5.23}$$

Figure 5.9 shows the basic idea of how an anti-aliasing filter works. There is a signal outside half-sampling (the Nyquist frequency) which gets aliased back at the same amplitude if there is no filter. By attenuating the signal, it still gets aliased, but the amplitude is far less.

5.3.1.1 Illustrative Example. Design of a Butterworth Anti-Aliasing Filter

A signal is sampled at 20 kHz but a noise term appears at 12 kHz of amplitude 100 mV. Design a Butterworth filter to attenuate the aliased spectrum by 40 dB whilst maintaining a good passband frequency and ensuring the order of the filter is not excessively large.

Solution

We need to decide on a passband frequency. Sampling at 20 kHz gives us an available bandwidth of 10 kHz, so it would be nice to be as close as possible to 10 kHz.

First try

The easiest design is to say that a second order filter will have an attenuation of -40 dB/decade and to put the passband frequency one decade down from the unwanted signal at 12 kHz. If the passband frequency is designed as $f_c = 1.2\,\text{kHz}$ then 12 kHz will be attenuated by 40 dB and the aliased signal at $20-12 = 8$ kHz will be 100 mV/100 = 1 mV in size. By using this design, we have restricted our passband frequency down to 1.2 kHz.

Second try

Suppose instead we set out to get a passband frequency of $f_c = 6\,\text{kHz}$. We can find the order of filter necessary to attenuate by $A_s = 40\,\text{dB}$ from (5.23). We use $\frac{f}{f_c} = 12/6 = 2$ and obtain (rounded up to the nearest integer)

$$n = \frac{\text{Log}_{10}\left[10^4 - 1\right]}{2\text{Log}_{10}(2)} \simeq \frac{\text{Log}_{10}\left[10^4\right]}{2\text{Log}_{10}(2)} = 2/0.3 = 7$$

However, this will result in a lot of circuitry and extra noise.

Third and best approach

As a compromise, let us use a 4th order filter and find what passband we can achieve for 40 dB attenuation. Using (5.22) with $n = 4$ and $A_s = 40\,\text{dB}$ we find the passband edge frequency to be:

$$f_c = \frac{f}{\left[10^{0.1A_s} - 1\right]^{\frac{1}{2n}}} \simeq \frac{12}{\left(10^4\right)^{1/8}} = \frac{12}{\sqrt{10}} = 3.8\,\text{kHz}.$$

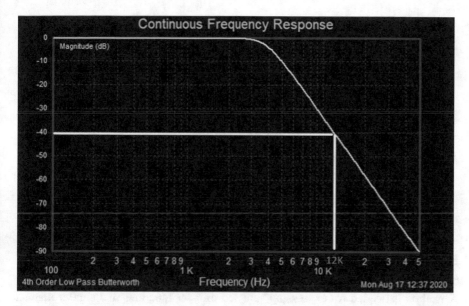

Fig. 5.10 Frequency response of 4th order Butterworth for anti-aliasing filter example (*Filter Solutions by Nuhertz*)

Settling on this available passband frequency we can then go ahead and realise the filter. Nowadays it is not necessary to do the physical realisation part by hand as this can be tedious with higher order filters. Software is readily available to perform such tasks.

Here we use *Filter Solutions* by *Nuhertz* software to find the transfer function and hardware. Specifying the order as 4 and passband frequency as 3.8 kHz the software gives as the following frequency response of the Butterworth filter (Fig. 5.10).

The attenuation is −40 dB at 12 kHz as desired. There is no ripple in the passband since we have chosen a Butterworth solution. The transfer function is found using the software as

$$H(s) = \frac{3.25 \times 10^{17}}{s^4 + 6.239 \times 10^4 s^3 + 1.946 \times 10^9 s^2 + 3.557 \times 10^{13} s + 3.25 \times 10^{17}}$$

One possible realisation is given using the Sallen and Key method in Fig. 5.11. This method was first used in 1955 [6] using vacuum-tubes but was later transferred to operational amplifiers when they became available.

The software enables us to use many other topologies, for example multiple feedback filters, state-variable filters and even passive filters.

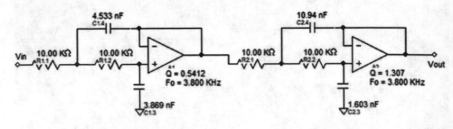

4th Order Low Pass Butterworth

Pass Band Frequency = 3.800 KHz

Mon Aug 17 12:45 2020

Fig. 5.11 Sallen and Key topology realisation of 4th order Butterworth anti-aliasing filter

5.4 The Z-Transform

A method is needed for expressing sampled transfer functions in a similar manner as continuous-time ones. We begin by taking Laplace Transforms.

By taking Laplace Transforms of the sampled signal $f^*(kT_s)$ (that is the output of a multiplier times an impulse train in (5.7))

$$\mathcal{L}\{f^*(kT_s)\} = \int_{-\infty}^{\infty} \sum_{k=-\infty}^{\infty} f(kT_s)\delta(t - kT_s)e^{-st}dt \qquad (5.24)$$

The above integral is zero everywhere except when $t = kT_s$ and (5.24) becomes

$$F^*(s) = \sum_{k=-\infty}^{\infty} f(kT_s)e^{-skT_s} \qquad (5.25)$$

For mathematical convenience it is convention to drop the kT_s in $f(kT_s)$ since it is just multiples of the sampling interval and replace it with $f(k)$ instead. We therefore have

$$F^*(s) = \sum_{k=-\infty}^{\infty} f(k)e^{-skT_s} \qquad (5.26)$$

We then define $z = e^{sT_s}$ in (5.26) as the z-transform operator and define the z-transform itself as

$$F(z) = \sum_{k=-\infty}^{\infty} f(k)z^{-k} \qquad (5.27)$$

We can also write in shorthand notation using $\mathcal{Z}$ in a similar manner as Laplace Transforms.

$$\mathcal{Z}\{f(k)\} = F(z) \qquad (5.28)$$

We note that (5.27) is written as a sum from minus to plus infinity. This is known as the *Bilateral* (*or two-sided*) *Z-Transform*. If we restrict ourselves to positive time only for the time being, we can define the *Unilateral Z-Transform* provided $f(k) = 0, k < 0$ as

$$F(z) = \sum_{k=0}^{\infty} f(k)z^{-k} \qquad (5.29)$$

The signal $f(t)$ is said to be in continuous time whereas $f(k)$ is said to be in discrete time. A continuous time signal is a function of time t and for a discrete time signal we only have a collection of numbers at discrete multiples of the sampling interval. That is $f(0), f(1), f(2), \ldots$ at sampling times $0, T_s, 2T_s, 3T_s, \ldots$ etc. Since $z = e^{sT_s}$ and the Laplace Transform operator s is already complex, it follows that z is complex too. For Laplace we talk of the *s-plane* and for z we use the *z-plane*.

5.4.1 Z-Transforms of Common Signals

Discrete Step signal

A discrete step signal (Fig. 5.12) is given by 1, 1, 1, 1, ... We can write it as

$$f(k) = 1, k = 0, 1, 2, \ldots$$

we can take its Z-transform by substituting $f(k) = 1$ into (5.29)

$$F(z) = \sum_{k=0}^{\infty} z^{-k}$$
$$= 1 + z^{-1} + z^{-2} + z^{-3} + \cdots \qquad (5.30)$$

This is a geometric progression. More generally a geometric progression or sequence has the form for n terms as the summation

$$S_n = a + ar + ar^2 + ar^3 + \cdots + ar^{n-1} \qquad (5.31)$$

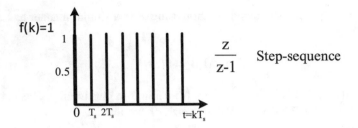

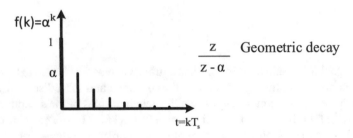

Fig. 5.12 Discrete step and Geometric decay

where r is called the growth factor which must have magnitude less than 1, or $|r| < 1$ and a is the first term in the sequence. By multiplying (5.31) by r and subtracting we obtain eventually

$$S_n = a\left(\frac{1 - r^n}{1 - r}\right) \tag{5.32}$$

Now provided $|r| < 1$, then $r^n \rightarrow 0, n \rightarrow \infty$. The sum to infinity becomes

$$S_\infty = \frac{a}{1 - r} \tag{5.33}$$

Equation (5.30) now becomes in closed form

$$F(z) = \frac{1}{1 - z^{-1}}$$
$$= \frac{z}{z - 1} \tag{5.34}$$

Which is the z-transform of a unit step. In the time domain, to avoid confusion with step signals that go backwards in time in a later section, we use a discrete version of the Heaviside operator to denote that the step starts at time zero. A unit step at time zero become $u(k)$. Similarly, a step at some other time (in this case some other integer sample instance) m is $u(k - m)$. A step that goes backwards in time from say time $-m$ is given the notation $u(-k - m)$ and we shall use this later.

Also, to avoid confusion when we consider anti-causal signals, we must define the region of convergence (ROC) of the geometric sequence. We have from (5.30) that to converge $\left|z^{-1}\right| < 1$ so that $|z| > 1$ in the z-plane. This is a region described in the z-plane (which is complex) given by values outside of a circle of unit radius.

Discrete Geometric decay

Now consider a second example, similar to a decaying exponential in continuous time.

$$f(k) = \alpha^k u(k), |\alpha| < 1 \tag{5.35}$$

Here the discrete Heaviside step operator u(k) is being used to tell us that k goes forwards in time from time 0. The Heaviside operator is not necessary, and neither is the ROC if we assume all signals and systems are causal. For practical problems this is clearly the case, but nevertheless in mathematical filter design the negative time anti-causal information in an impulse response often appears in a solution as well as the causal positive time part. The two need to be separated to create a causal filter.

Take Z-transforms of (5.35)

$$\begin{aligned} F(z) &= \sum_{k=0}^{\infty} f(k) z^{-k} \\ &= \sum_{k=0}^{\infty} \alpha^k z^{-k} \\ &= 1 + \alpha z^{-1} + \alpha^2 z^{-2} + \ldots \end{aligned} \tag{5.36}$$

This is the sum to infinity of a geometric sequence and converge must happen provided $\left|\alpha z^{-1}\right| < 1$. The ROC is therefore $|z| > \alpha$. This is convergence outside of a circle radius α. Using (5.33) we get the sum to infinity giving the Z-transfer function:

$$\begin{aligned} F(z) &= \frac{1}{1 - \alpha z^{-1}} \\ &= \frac{z}{z - \alpha} \end{aligned} \tag{5.37}$$

Unit Impulse

The unit impulse at time k = 1 is defined as the Kronecker delta

$$f(k) = \delta(k) \tag{5.38}$$

Take Z-Transforms

$$F(z) = \sum_{k=0}^{\infty} f(k)z^{-k}$$

$$= \sum_{k=0}^{\infty} \delta(k)z^{-k}$$

$$= \delta(0)$$

$$= 1 \tag{5.39}$$

This is not even a function of z, and the ROC is all of the z-plane.

Delayed-Impulse by m samples

$$f(k) = \delta(k - m) \tag{5.40}$$

The Z-Transform is

$$F(z) = \sum_{k=0}^{\infty} f(k)z^{-k}$$

$$= \sum_{k=0}^{\infty} \delta(k - m)z^{-k} \tag{5.41}$$

The above summation with the delta signal must be zero everywhere except at time k = m. Therefore

$$\mathcal{Z}\{\delta(k - m)\} = z^{-m} \tag{5.42}$$

At this stage we can then use the delta signal to describe discrete values at different times.

For example, to describe a signal of a sequence of numbers which occur at specific times. Suppose f(k) is 1 at time k = 0, 3 at time k = 4 and 5 at time k = 6. Then we can write

$$f(k) = \delta(k) + 3\delta(k - 4) + 5\delta(k - 6) \tag{5.43}$$

The Z-transform of (5.43) is $\mathcal{Z}\{\delta(k) + 3\delta(k - 4) + 5\delta(k - 6)\} = 1 + 3z^{-4} + 5z^{-6}$.

Delayed Signal

Consider a discrete signal f(k) with Z-Transform F(z), delayed by m samples giving f(k − m). Its Z-Transform becomes

$$\mathcal{Z}\{f(k - m)\} = \sum_{k=0}^{\infty} f(k - m)z^{-k} \tag{5.44}$$

Substitute $n = k - m$ so that $k = n + m$. Then

$$\sum_{k=0}^{\infty} f(k - m)z^{-k} = \sum_{k=0}^{\infty} f(n)z^{-(k+m)}$$

$$= z^{-m} \sum_{k=0}^{\infty} f(n)z^{-n}$$

$$= z^{-m}F(z) \tag{5.45}$$

We can then use z^{-m} as an operator to shift a discrete signal at time k to time k − m. Thus

$$z^{-m}f(k) = f(k - m) \tag{5.46}$$

Equation (5.46) is a bit of an abuse of notation since time and Z-domain are being mixed, but nevertheless this notation is common in much of the electrical engineering literature. A more precise definition is to replace z^{-m} with an operator q^{-m} but this amounts to the same end result.

Z-Transform of a pulse of width N − 1

A discrete pulse has samples 0, 1, 2, ..., N − 1 which is N samples total. If the amplitude is one, then there are N samples each of amplitude unity.

$$F(z) = \sum_{k=0}^{N-1} z^{-k}$$

$$= 1 + z^{-1} + z^{-2} + \cdots + z^{-(N-1)} \tag{5.47}$$

From (5.31) and (5.32) the summation becomes

$$F(z) = \frac{1 - z^{-N}}{1 - z^{-1}} \tag{5.48}$$

This can also be deduced a different way by using the Heaviside method. Equation (5.47) is a sequence of ones that start at time $k = 1$ up to and including time $k = N - 1$. If we then define a discrete step at time zero and subtract a second step of equal magnitude at time N (we cannot subtract at time N − 1 since this would take that value to zero) we get

$$f(k) = u(k) - u(k - N)$$

Take Z-transforms, the second part is just a delayed step by N samples:

$$F(z) = \frac{1}{1 - z^{-1}} - \frac{z^{-N}}{1 - z^{-1}}$$

which is (5.48).

5.4.2 Table of Unilateral Z-Transforms

The Z-Transfer function applies to discrete signals and systems. Some signals and systems (as are the case in continuous time Laplace) have the same Z-transform (Table 5.1). For example, a discrete step signal and a discrete integrator have the same Z-transform.

Z-Transfer functions have poles and zeros in the same manner as continuous time Laplace. For instance, the transfer function $G(z) = \frac{(z-0.1)}{(z-0.9)(z-0.5)}$ has a zero at $z = 0.1$ and two poles at 0.9 and 0.5 respectively. Unlike the Laplace Transform, the Z-transform can have transfer functions with more zeros than poles or even no poles at all. The system $G(z) = (z - 0.2)(z - 0.95)$ only has zeros.

Tables are also available for commonly met signals, their Laplace Transform and the equivalent Z-Transform as shown (Table 5.2).

Table 5.1 A table of commonly met Z-Transforms for a unit sample-interval

Sampled time-domain sequence	Z transform
$\delta(k)$ unit Impulse at $k = 0$	1
$\delta(k - m)$ delayed impulse at $k = m$	z^{-m}
$u(k)$, unit step function	$\frac{z}{(z-1)}$
$ku(k)$, unit ramp function	$\frac{z}{(z-1)^2}$
$\alpha^k u(k)$, $|\alpha| < 1$ geometric decay	$\frac{z}{z-\alpha}$
$k\alpha^{k-1}u(k)$, $|\alpha| < 1$	$\frac{z}{(z-\alpha)^2}$
$(-\alpha)^k u(k)$, $|\alpha| < 1$	$\frac{z}{z+\alpha}$
$f(k - m)$, signal $f(k)$ delayed by m steps	$z^{-m}F(z)$
$\cos(\Omega k)u(k)$	$\frac{z^2-z\cos(\Omega k)}{z^2-2z\cos(\Omega k)+1}$
$\sin(\Omega k)u(k)$	$\frac{z\sin(\Omega k)}{z^2-2z\cos(\Omega k)+1}$
$\alpha^k \cos(\Omega k)u(k)$	$\frac{z^2-z\alpha\cos(\Omega k)}{z^2-2\alpha z\cos(\Omega k)+\alpha^2}$
$\alpha^k \sin(\Omega k)u(k)$	$\frac{z\alpha\sin(\Omega k)}{z^2-2\alpha z\cos(\Omega k)+\alpha^2}$

Table 5.2 Continuous time, laplace and Z-transform equivalent

Continuous time	Laplace transform	Z-transform
$u(t)$	$\frac{1}{s}$	$\frac{z}{z-1}$
e^{-at}	$\frac{1}{s+a}$	$\frac{z}{z-e^{-aT}}$
$\delta(t-kT)$	e^{-kTs}	e^{-k}
t	$\frac{1}{s^2}$	$\frac{Tz}{(z-1)^2}$
te^{-at}	$\frac{1}{(s+a)^2}$	$\frac{Tze^{-aT}}{(z-e^{-aT})^2}$
$\sin(\omega t)$	$\frac{\omega}{s^2+\omega^2}$	$\frac{z\sin(\omega T)}{z^2-2z\cos(\omega T)+1}$
$\cos(\omega t)$	$\frac{s}{s^2+\omega^2}$	$\frac{z(z-\cos(\omega T))}{z^2-2z\cos(\omega T)+1}$
$e^{-at}\sin(\omega t)$	$\frac{\omega}{(s+a)^2+\omega^2}$	$\frac{e^{-aT}z\sin(\omega T)}{z^2-2e^{-aT}z\cos(\omega T)+e^{-2aT}}$
$e^{-at}\cos(\omega t)$	$\frac{s+1}{(s+a)^2+\omega^2}$	$\frac{z^2-e^{-aT}z\cos(\omega T)}{z^2-2e^{-aT}z\cos(\omega T)+e^{-2aT}}$

5.4.3 Inverse Z-Transforms

There are at least 3 ways to find the inverse Z-Transform. The first method is to use the standard tables and partial fraction expansion, the second method is to use long-division and the third is integration in the complex z-plane and Cauchy residue theory. To understand the methods, it is best to consider some examples.

Example 1 Find the inverse Z-Transform of $F(z)$ where

$$F(z) = \frac{0.5z^2}{(z-0.5)(z-1)}$$

(a)   Using Partial Fractions

First divide F(z) by z. This is because the tables have an entry $\frac{z}{z-\alpha}$ but not $\frac{1}{z-\alpha}$, and we multiply across afterwards by the z we divided by to re-introduce it.

$$F(z)/z = \frac{0.5z}{(z-0.5)(z-1)} = \frac{A}{z-0.5} + \frac{B}{z-1}$$

Multiplying out

$$0.5z = A(z-1) + B(z-0.5)$$

From which by comparing coefficients of z gives $A = -0.5$, $B = 1$. Then multiply back by z to give

$$F(z) = \frac{-0.5z}{z-0.5} + \frac{z}{z-1}$$

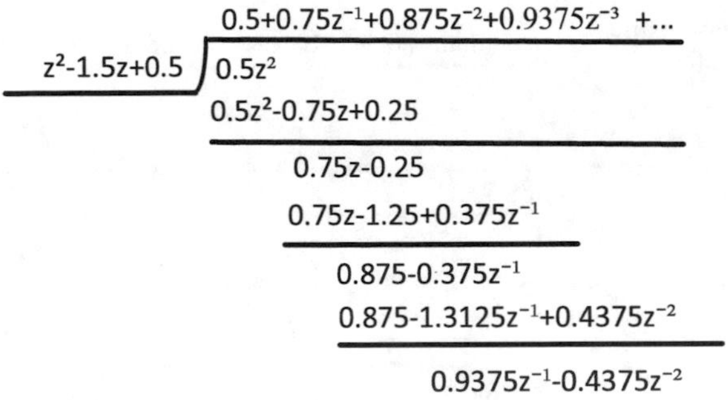

$$\begin{array}{r} 0.5+0.75z^{-1}+0.875z^{-2}+0.9375z^{-3}\ +... \\ \hline z^2-1.5z+0.5\ \big)\ 0.5z^2 \\ 0.5z^2-0.75z+0.25 \\ \hline 0.75z-0.25 \\ 0.75z-1.25+0.375z^{-1} \\ \hline 0.875-0.375z^{-1} \\ 0.875-1.3125z^{-1}+0.4375z^{-2} \\ \hline 0.9375z^{-1}-0.4375z^{-2} \end{array}$$

Fig. 5.13 Long division method to find the inverse Z-Transform

Using tables

$$\mathcal{Z}^{-1}\{F(z)\} = u(k) - 0.5(0.5)^k u(k)$$

As k gets larger the steady-state value ends up at 1.

(b)   Using long division

This method is perhaps more time-consuming and doesn't result in a closed-form solution either. When dividing numerator by denominator we ensure that each is written in descending powers of z. This makes the division in terms of negative powers of z and not positive. Negative powers of z are needed for causality. The long division is shown in Fig. 5.13.

The solution to the division is

$$F(z) = 0.5 + 0.75z^{-1} + 0.875z^{-2} + 0.9375z^{-3} + \cdots$$

This is interpreted as the sequence (or power-series)

$$f(k) = 0.5\delta(k) + 0.75\delta(k-1) + 0.875\delta(k-3) + 0.9375\delta(k-4) + \cdots$$

Or written as a series of numbers

$$f(k) = \{0.5, 0.75, 0.875, 0.9375, \ldots\}$$

We can see it has not got a closed-form solution but is an infinite series. Provided the poles of the transfer function have magnitude less than or equal to one then this power-series will be convergent.

(c)   Residue Method

The third method stems from the eighteenth century work of mathematician Augustin-Louis Cauchy. For a Z-transfer function $F(z)$, its inverse Z-Transform is found from the following integration in the complex z-plane.

$$f(k) = \frac{1}{2\pi j} \oint_C F(z)z^k \frac{dz}{z} \tag{5.49}$$

where C is a counter clockwise closed contour on the z-plane encircling the origin. Fortunately, a direct integration is not necessary, and it can be simplified by knowing that the solution to (5.49) is

$$f(k) = \frac{1}{2\pi j} \oint_C F(z)z^k \frac{dz}{z}$$

$$= \sum_{poles} \text{residues}$$

The sum of residues at the poles *within the contour* (the unit circle) of the transfer function. For a simple pole at $z = p$, the residue is found from

$$\text{Res}(F, p) = (z - p)F(z)z^{k-1}\big|_{z \to p} \tag{5.50}$$

For a pole of order m, the residue is found from

$$\text{Res}(F, p) = \frac{1}{(m-1)!} \frac{d^{m-1}}{dz^{m-1}} (z - p)^m F(z)z^{k-1}\bigg|_{z \to p} \tag{5.51}$$

Our example has two simple poles within the contour. p1 at $z = 1$ and p2 at $z = 0.5$.

Residue p1 at z = 1

$$\text{Res}(F, p1) = (z - p1)F(z)z^{k-1}\big|_{z \to p1}$$

$$= (z - 1)\frac{0.5z^{k+1}}{(z - 0.5)(z - 1)}\bigg|_{z \to 1} = 1$$

Residue p2 at z = 0.5

$$\text{Res}(F, p2) = (z - p2)F(z)z^{k-1}\big|_{z \to p1}$$

$$= (z - 0.5)\frac{0.5z^{k+1}}{(z - 0.5)(z - 1)}\bigg|_{z \to 0.5} = -(0.5)^{k+1}$$

Summing the residues gives

$$f(k) = 1 - (0.5)^{k+1}, k = 0, 1, 2, \ldots$$

Which can be written as

$$f(k) = u(k) - (0.5)(0.5)^k u(k)$$

Which is the same solution as found using partial fractions.

Example 2 Find the inverse Z-Transform of F(z) where

$$F(z) = \frac{z^3}{(z - 0.5)^2 (z - 0.1)}$$

Use Partial Fractions and the residue method. (Long division is rather tedious and does not give a nice closed-form solution in discrete time index k).

(a)   Partial fraction method.

Divide F(z) by z and get

$$F(z)/z = \frac{z^2}{(z - 0.5)^2 (z - 0.1)}$$

$$= \frac{A}{(z - 0.5)} + \frac{B}{(z - 0.5)^2} + \frac{C}{(z - 0.1)}$$

Multiply across and get

$$z^2 = A(z - 0.5)(z - 0.1) + B(z - 0.1) + C(z - 0.5)^2$$

Compare coefficients of z on both sides of the equation and we get three equations and three unknowns.

$$1 = A + C$$
$$0 = -0.6A + B - C$$
$$0 = 0.05A - 0.1B + 0.25C$$

We can solve these simultaneously, but this is a bit tedious. A matrix format can be used instead

$$\begin{bmatrix} 1 & 0 & 1 \\ -0.6 & 1 & -1 \\ 0.05 & -0.1 & 0.25 \end{bmatrix} \begin{bmatrix} A \\ B \\ C \end{bmatrix} = \begin{bmatrix} 1 \\ 0 \\ 0 \end{bmatrix}$$

To find the unknow vector (A, B, C) we can either invert the matrix or solve using Cramer's rule. Either method involves a 3×3 matrix. Nowadays we prefer to solve using numerical methods using software such as MATLAB. We find the solution is $A = 0.9375$, $B = 0.625$, $C = 0.0625$. Multiply back by z and get

$$F(z) = \frac{0.9375z}{(z - 0.5)} + \frac{0.625z}{(z - 0.5)^2} + \frac{0.0625z}{(z - 0.1)}$$

Using the table of Z-transforms we find

$$\mathcal{Z}^{-1}\left\{\frac{0.9375z}{(z - 0.5)} + \frac{0.625z}{(z - 0.5)^2} + \frac{0.0625z}{(z - 0.1)}\right\}$$
$$= 0.9375(0.5)^k + 0.625k(0.5)^{k-1} + 0.0625(0.1)^k$$
$$= (0.5)^k[0.9375 + 1.25k]u(k) + 0.0625(0.1)^k u(k)$$

This dies out to zero as k approaches infinity.

(b)   Using Cauchy's residue method

$$f(k) = \frac{1}{2\pi j} \oint_C F(z)z^k \frac{dz}{z}$$

$$= \frac{1}{2\pi j} \oint_C \frac{z^3}{(z - 0.5)^2(z - 0.1)} z^k \frac{dz}{z}$$

$$= \sum_{\text{poles}} \text{residues}$$

There are two poles within the unit circle. Pole 1 is at $z = 0.1$ and is a simple pole. Pole 2 is at $z = 0.5$ and of order 2.

For Pole 1: $z = 0.1$. The residue is for a simple pole

$$\text{Res}(F, p1) = (z - p1)F(z)z^{k-1}\big|_{z \to p1}$$

$$= (z - 0.1)\frac{z^{k+2}}{(z - 0.5)^2(z - 0.1)}\bigg|_{z \to 0.1} = 0.0625(0.1)^k$$

For Pole 2. The residue is for the pole at $z = 0.5$ of order 2.

$$\text{Res}(F, p2) = \frac{d}{dz}(z - p2)^2 F(z)z^{k-1}\bigg|_{z \to p2}$$

$$= \frac{d}{dz}\frac{z^{k+2}}{(z - 0.1)}\bigg|_{z \to 0.5}$$

$$= \frac{(k+2)(z-0.1)z^{k+1} - z^{k+2}}{(z-0.1)^2}\Bigg|_{z\to 0.5}$$

$$= (0.5)^k[1.25k + 0.9375]$$

Adding the residues gives

$$f(k) = (0.5)^k[0.9375 + 1.25k]u(k) + 0.0625(0.1)^k u(k)$$

Which is the same result as partial fractions. This method is easier for this example as it only requires a simple differentiation compared to solving three linear simultaneous equations.

5.4.4 The Bilateral Z-Transform

Recall from (5.27)

$$F(z) = \sum_{k=-\infty}^{\infty} f(k)z^{-k}$$

Which is the Bilateral Z-Transform. The summation here is taken from minus infinity which means it can deal with causal as well as anticausal signals and systems. Figure 5.14 shows three possible signals or impulse responses of systems. Causal is a so-called right-sided sequence, anticausal is left-sided and noncausal is both left and right.

A sequence with both positive and negative powers of z is called a Laurent sequence and is given by

$$L(z) = \cdots + b_3 z^3 + b_2 z^2 + b_1 z + a_0 + a_1 z^{-1} + a_2 z^{-2} + a_3 z^{-3} + \cdots \tag{5.52}$$

If we take a first order Z-transfer function

$$G(z) = \frac{z}{z-a}, \ |a| < 1 \tag{5.53}$$

Then by long division we can write the MacLaurin series

$$G(z) = 1 + az^{-1} + a^2 z^{-2} + a^3 z^{-3} + \cdots \tag{5.54}$$

which converges provided $\left|az^{-1}\right| < 1$ or alternatively $|z| > a$.
Its inverse Z-Transform is from tables

Fig. 5.14 Causal, anticausal and noncausal definitions

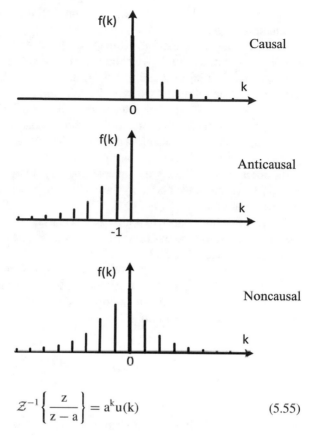

$$\mathcal{Z}^{-1}\left\{\frac{z}{z-a}\right\} = a^k u(k) \qquad (5.55)$$

Which due to the discrete Heaviside operator u(k), is valid for $k = 0, 1, 2, \ldots$
A problem arises however if we re-arrange (5.53)

$$G(z) = \frac{z}{z-a}$$

$$= -\frac{z}{a-z} = -\left(\frac{z}{a}\right)\frac{1}{(1-z/a)}$$

$$= -\left(\frac{z}{a}\right)\left(1 + z/a + z^2/a^2 + z^3/a^3 + \cdots\right)$$

$$- z/a - z^2/a^2 - z^3/a^3 - \cdots \qquad (5.56)$$

This is a power series in positive powers of z and represents a sequence going backwards in time. It can be written in the time domain as

$$\mathcal{Z}^{-1}\left\{\frac{z}{z-a}\right\} = -a^k u(-k-1) \qquad (5.57)$$

Which is valid for $k = -1, -2, -3, \ldots$ only. The Heaviside operator tells us this information in (5.57).

Therefore, the inverse Z-transform $\frac{z}{z-a}$ can have two possible solutions, one causal and one anticausal. This is why the region of convergence (ROC) is quoted. For (5.53) the ROC is $|z| > a$ and for (5.57) it is $|z| < a$. If we are dealing only with causal signals and systems we can of course relax this convention. In some areas of signal processing, optimal filter design gives a solution that is noncausal. The causal part must then be separated from the anticausal part. It is of interest that a system that is causal and stable will be unstable if time is reversed. For example, in (5.56) (the anticausal sequence or system) the powers of z go up as the reciprocal of the coefficient a. Clearly the inverse powers get larger if $|a| < 1$ and for powers of a in (5.54) (the causal system) the coefficients when raised to higher powers get progressively smaller.

It is of some importance to note that for a causal system the power series includes a term at time zero whereas for anticausal systems the power series starts at time k $= -1$ *and excludes time zero.*

As an alternative to using the Heaviside operator, a more elegant approach is used in the scientific literature. We can define the Z-transfer function of a causal system in negative powers of z only and a noncausal system in positive powers of z. This is best achieved using polynomials. Let a general causal Z-transfer function be defined as

$$G(z^{-1}) = \frac{b(z^{-1})}{a(z^{-1})} \tag{5.58}$$

where the $b(z^{-1})$ polynomial of degree m is found from

$$b(z^{-1}) = b_1 z^{-1} + b_2 z^{-2} + \cdots + b_m z^{-m} \tag{5.59}$$

and the $a(z^{-1})$ polynomial of degree n is found from

$$a(z^{-1}) = 1 + a_1 z^{-1} + a_2 z^{-2} + \cdots + a_n z^{-n} \tag{5.60}$$

Equation (5.60) is quite general since any term which is not unity in the zeroth coefficient can be divided into the numerator and denominator in (5.58) resulting always with $a(0) = 1$. We can define (5.58) being stable, which would mean that all the n roots of the polynomial $a(z^{-1})$ having magnitude less than unity when solved in terms of z. Simply by replacing z^{-1} with z, the anticausal system can then be defined as

$$G(z) = \frac{b(z)}{a(z)} \tag{5.61}$$

Often the arguments for z and its inverse are dropped and $G = G(z^{-1})$, $G^* = G(z)$ used instead. This also applies to the polynomials. The G^* (star as superscript) is often called the conjugate since $z = e^{j\theta} = \cos(\theta) + j\sin(\theta)$ is complex and $z^{-1} = e^{-j\theta} = \cos(\theta) - j\sin(\theta)$ follows. Hence z and z^{-1} are complex conjugates of one another.

All roots of a(z) have magnitude greater than unity provided all roots of $a(z^{-1})$ have magnitude less than unity. As a simple example consider the stable and causal system

$$G(z^{-1}) = G = \frac{b}{a}$$

$$= \frac{0.9z^{-1}}{1 - 0.5z^{-1}}$$

Which has a root in the denominator at $z = 0.5$ (stable) and its anticausal counterpart

$$G^* = G(z) = \frac{b^*}{a^*}$$

$$= \frac{0.9z}{1 - 0.5z}$$

has a root in the denominator at $z = 1/0.5$ (unstable for positive time index). If time runs backwords however the anticausal system will be stable. Using this notation means that time $k = 0$ is excluded from both models and time always starts at $k = 1$ or -1 depending on causality. This makes practical sense because a discrete-time system cannot respond instantaneously anyway, and a minimum delay of one step must always be present. It is easy to verify that G^* has no $k = 0$ component in its inverse transform (impulse response) by using Cauchy's residue method. For a system $G^* = \frac{z}{1-az}$ with $|a| < 1$ it has only one pole and it is outside the unit circle. The inverse transform is found from

$$g(k) = \frac{1}{2\pi j} \oint_{|z|=1} G^*(z)z^{k-1}dz$$

$$\frac{1}{2\pi j} \oint_{|z|=1} \frac{z}{1 - az}z^{k-1}dz$$

This is the sum of residues at the poles encircled by the unit circle. *For positive k there are no poles within the unit circle and the integral is zero.* For $k = 0$, likewise there are no poles within the unit circle and the integral is zero. For negative k however, for example $k = -1$ we have one pole at $z = 0$ which is within the unit circle and hence must have a residue given by

$$g(-1) = \frac{1}{2\pi j} \oint_{|z|=1} \frac{1}{1-az} \frac{dz}{z}$$

$$= 1$$

As k becomes more negative, we get multiple poles at $z = 0$ and we find $g(-1) = a$, $g(-2) = a^2$ and so on.

If we don't put the z^{-1} term in the definition for the first term of the b polynomial it means that its anticausal counterpart will have a $k = 0$ term and there is no symmetry for the notation since having a term at $k = 0$ means it is not entirely anticausal. If notation like this is not used, then the Heaviside operator must be used for each individual case.

5.4.5   Discrete-Time Final Value Theorem

In analogy with the continuous time Laplace case, the final value theorem (FVT) can come in useful for certain problems, particularly finding the steady-state value for a step input to a system.

For a *stable* (bounded) sequence,

$$f(k)|_{\underset{k\to\infty}{\text{Lim}}} = (z-1)\,F(z)|_{\underset{z\to1}{\text{Lim}}} \tag{5.62}$$

Example: Final value Theorem Consider a causal second order system $F(z) = \frac{z^2}{(z-0.5)(z-0.8)}$ with a unit step applied to it. Find the final value of the output.

The system is $F(z) = \frac{z^2}{(z-0.5)(z-0.8)}$ and the Z-transform of a unit step is $u(z) = \frac{z}{(z-1)}$.

Let the Z-transform of the output be y(z) where

$$y(z) = F(z)u(z)$$

$$= \frac{z^3}{(z-0.5)(z-0.8)(z-1)}$$

Partial Fraction method

First find the final value the hard way, using partial fractions. Let

$$y(z)/z = \frac{z^2}{(z-0.5)(z-0.8)(z-1)}$$

$$= \frac{A}{(z-0.5)} + \frac{B}{(z-0.8)} + \frac{C}{(z-1)}$$

Solving for the constants yields A $= 1.67$ B $= -10.667$, C $= 10$.

Then

$$y(z) = \frac{1.67z}{(z-0.5)} - \frac{10.67z}{(z-0.8)} + \frac{10}{(z-1)}$$

Using the table of transforms we find the inverse (omitting the Heaviside operator as this is a causal system anyway) as

$$y(k) = 10 + 1.67(0.5)^k - 10.67(0.8)^k$$

And as k gets large the last two terms die off to zero giving

$$y(k)|_{k\to\infty} = 10$$

Using the Final value theorem

$$y(k)|_{\underset{k\to\infty}{Lim}} = (z-1)\,y(z)|_{\underset{z\to1}{Lim}}$$

$$= (z-1)\frac{z^3}{(z-0.5)(z-0.8)(z-1)}\Big|_{z\to1}$$

$$= \frac{1}{0.5\times0.2} = 10$$

A much faster method. If the system is unstable the method fails since the output will go to infinity, there being no steady-state value of the output.

5.5 Finite Difference Equations

In continuous time we have differential equations describing the physics of real-world processes and this in turn can be converted into Laplace Transforms in the form of a transfer function. In discrete time differential equation do not exist. Instead *finite difference equations* take their place. The difference equation (for short) is found by using the Z-Transform as an operator. We use the fact that for some signal say y(k) which means a signal at time sample k, it can be shifted backwords or forwards in time by the z operator. Thus

$$z^{-1}y(k) = y(k-1) \tag{5.63}$$

This is an abuse of notation of course because z is a frequency quantity operating on a time signal, but nevertheless it has become standard in most digital signal processing literature. We define z^{-1} as the backwards shift operator and similarly z as the forward shift operator.

$$zy(k) = y(k + 1) \tag{5.64}$$

This can then be directly applied to a Z-Transform. For example, define a stable transfer function.

$$G(z) = \frac{z - 0.5}{z^2 - z + 0.5} \tag{5.65}$$

It has two complex poles within the unit circle and is therefore stable.
By definition, the transfer function relates some output y to an input u.

$$y(z) = \frac{z - 0.5}{z^2 - z + 0.5} u(z) \tag{5.66}$$

where y(z) is the Z-Transform of the output signal and u(z) is the Z-Transform of the input driving signal. Then we abuse the notation and write z as an operator (we can use another symbol for z but the result is the same!)

$$y(k) = \frac{z - 0.5}{z^2 - z + 0.5} u(k) \tag{5.67}$$

There follows

$$y(k)(z^2 - z + 0.5) = (z - 0.5)u(k) \tag{5.68}$$

Now use z as the forward shift operator

$$y(k + 2) - y(k + 1) + 0.5y(k) = u(k + 1) - 0.5u(k) \tag{5.69}$$

Write on the left-hand side of (5.69) by re-arranging it the most current value of output y(k + 2).

$$y(k + 2) = y(k + 1) - 0.5y(k) + u(k + 1) - 0.5u(k) \tag{5.70}$$

This is the finite difference equation of the system. We note that y(k + 2) is written on the left as the output and not say y(k + 1). If y(k + 1) was written on the left it would mean that the output at time k + 1 was dependent on an output at time y(k + 2), one step ahead into the future and this makes no sense. We can also write (5.67) in backwards shift form

$$y(k) = \frac{z^{-1} - 0.5z^{-2}}{1 - z^{-1} + 0.5z^{-2}} u(k) \tag{5.71}$$

This can be written as

$$y(k)\left(1 - z^{-1} + 0.5z^{-2}\right) = \left(z^{-1} - 0.5z^{-2}\right)u(k) \tag{5.72}$$

Using the backwards shift (5.63) we expand (5.72)

$$y(k) - y(k-1) + 0.5y(k-2) = u(k-1) - 0.5u(k-2)$$

Now write the output at the current time y(k) and re-arrange

$$y(k) = y(k-1) - 0.5y(k-2) + u(k-1) - 0.5u(k-2) \tag{5.73}$$

It makes sense to have y(k) as the output since the right-hand side of (5.73) has past values of output and input. We cannot calculate the present output value based on future values of output. It is easier to deal with the backwards shift method but really there is no difference and either method can be used. In fact, by subtracting 2 from each k in (5.73) we get (5.70).

If we represent filters or controllers as Z-transfer functions, we can quickly convert them to a difference equations and from there they can be implemented in software. The details are shown later but from say (5.73) we can substitute values of k = 0, 1, 2, 3, ... and find y(0), y(1), y(2), y(3) etc. and continue forever.

For k = 0 we get from (5.73)

$$y(0) = y(-1) - 0.5y(-2) + u(-1) - 0.5u(-2)$$

Now since y(−1), y(−2), u(−1), u(−2) have not yet occurred (they are in negative time), we consider them to be zero and the output zero.

For time k = 1 we get from (5.74)

$$y(1) = y(0) - 0.5y(-1) + u(0) - 0.5u(-1)$$

If an input occurs at time k = 0, then u(0) will have some value. The other terms are zero. Then we have

For time k = 2

$$y(2) = y(1) - 0.5y(0) + u(1) - 0.5u(0)$$

But we already know the values of y(1) and y(0) from the previous calculations so we find y(2). By this time we have a new sample at the input, namely u(1), the input at sample instant k = 1. We continue forever and the system responds for whatever input is applied. The uncertainty in the first few values of output is the transient period of the waveform when starting up. The more past values in a difference equation the more terms that initially have negative time values and must be assigned zero values.

5.5.1 Steady-State Output of a Difference Equation

A useful thing to know is how to find the steady-state output for a particular constant input. This can also be done using the final value theorem. If the input to a difference equation is held constant (say $u(k) = u(k-1) = u(k-2)$ etc. $= u$), then in steady-state we must have $y(k) = y(k-1) = y(k-2)$ etc. $=$ some constant y. Suppose we have (5.73)

$$y(k) = y(k-1) - 0.5y(k-2) + u(k-1) - 0.5u(k-2)$$

For a unit step input $u(k) = 1$ for all positive integer values of k. Let the steady-state output be $y(k) = y(k-1) = y(k-2) = y$. Then we get

$$y = y - 0.5y + 1 - 0.5$$

resulting in $0.5y = 0.5$ or $y = 1$. This only works for *stable difference equations* *since* unstable ones have unbounded outputs. This value of output is also known as the *equilibrium point*. There can only be one equilibrium point for a LTI system.

References

1. E.T. Whittaker, On the functions which are represented by the expansions of the interpolation theory. Proc. Roy. Soc. Edinburgh **35**, 181–194 (1915)
2. V.A. Kotelnikov, On the carrying capacity of the "ether" and wire in telecomminucations, in *Material dfor the First All-Union Conference on Questions of Communications*, vol. Izd. (Red. Upr. Svyazi RKKA, Moscow, 1933)
3. C.E. Shannon, A mathematical theory of communications. Bell Syst. Tech. J. **27**, 379–423 (1948)
4. S. Butterworth, On the theory of filter amplifiers. Engineer **7**, 536–541 (1930)
5. A. Amarder, *Analog and Digital Signal Processing*, 2nd ed. (Brooks/Cole Publishing Company, USA, 1999)
6. R.P. Sallen, E.L. Key, A practical method of designing RC active filters. IRE Trans. Circ. Theor. **2**(1), 74–85 (1955)

Chapter 6
Properties of Discrete-Time Systems and Signals

6.1 Stability of Discrete-Time Systems

In Chap. 5 we show that the Z transfer function of a causal LTI discrete-time system can be written in polynomial form:

$$G(z^{-1}) = \frac{b(z^{-1})}{a(z^{-1})} \tag{6.1}$$

where the polynomials $a(z^{-1})$ and $b(z^{-1})$ are of order n and m respectively. As with continuous-time systems, we need more detail on the stability properties of such systems. The key is to examine the relationship between the s and z domains via the relationship already derived in sampling theory, namely

$$z = e^{sT_s} \tag{6.2}$$

In (6.2) z is the Z-transform operator, s is the Laplace operator and T_s is the sampling interval. We can write the Laplace operator in complex form

$$s = \sigma + j\omega \tag{6.3}$$

and substitute into (6.2) giving

$$z = e^{(\sigma + j\omega)T_s}$$
$$= e^{\sigma T_s} e^{j\omega T_s} \tag{6.4}$$

The region of stability for continuous-time system is when poles of the system lie in the left-half s-plane. That is a region when a pole satisfies

$$\text{real}(\sigma) < 0 \tag{6.5}$$

© The Author(s), under exclusive license to Springer Nature Switzerland AG 2022
T. J. Moir, *Rudiments of Signal Processing and Systems*,
https://doi.org/10.1007/978-3-030-76947-5_6

137

Since from (6.2), z must be a complex number like s, we can write it as

$$z = u + jv \tag{6.6}$$

Taking the magnitude of z in (6.4) gives us the locus that satisfies stability.

$$|z| = \left| e^{(\sigma + j\omega)T_s} \right|$$
$$= \left| e^{\sigma T_s} \right| \left| e^{j\omega T_s} \right| \tag{6.7}$$

For any complex number of the form $e^{j\omega T_s}$, its magnitude must be unity since $e^{j\omega T_s} = \cos(\omega T_s) + j\sin(\omega T_s)$ and taking the sum of squares leads to one. Therefore

$$|z| = \left| e^{\sigma T_s} \right| \tag{6.8}$$

For stability σ must be negative. On the boundary of stability (the $s = j\omega$ axis in the s plane) $\sigma = 0$ so that

$$|z| < 1 \tag{6.9}$$

From (6.9), substituting the complex form of z from (6.6)

$$|z| = \sqrt{u^2 + v^2} < 1 \tag{6.10}$$

Equation (6.10) represents the inner region of a circle on the (u, v) complex Z plane. Clearly the $s = j\omega$ axis in the s plane corresponds to a circle of unit radius in z. This is illustrated in Fig. 6.1. We can substitute frequency in Hz and the theory remains unchanged. Note that only a strip in the s-plane maps into the interior of the

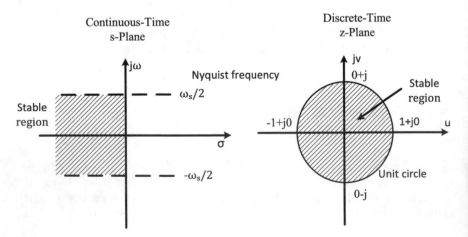

Fig. 6.1 Mapping from stable region in s-plane to stable region in z-plane

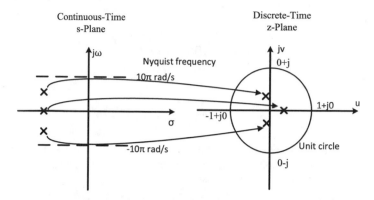

Fig. 6.2 Mapping of three stable poles in s to z-plane

circle. This is because the sampling theory means we can only reach frequencies up to less than the Nyquist frequency which is $\omega = \omega_s/2$ (half-sampling frequency).

To illustrate this further. Suppose we have a sampling frequency of $f_s = 10\,\text{Hz}$ giving a Nyquist frequency $f_N = 5\,\text{Hz}$ and a sampling interval of $T_s = 0.1\,\text{s}$. The strip in the s-plane will go from the positive and negative Nyquist frequency of 10π to -10π rad/s or 5 to -5 Hz.

Let a negative real pole in continuous-time be at $s = -10 + j0$. Using $z = e^{(\sigma + j\omega)T_s}$ we get the z-domain equivalent as $z = e^{(-10+j0)T_s} = e^{-10\times0.1} = 0.3679$. This has magnitude less than unity.

Now consider a pair of stable complex poles at $s = -10 + j20$ and $s = -10 - j20$. This complex frequency when $\omega = 20$ rad/s is less than the Nyquist frequency 10π rad/s and so no aliasing occurs. Take the pole with positive imaginary part: $s = -10 + j20$ and it maps into $z = e^{(-10+j20)T_s} = e^{(-10+j20)0.1} = e^{(-1+2j)}$.

Now $e^{(-1+2j)} = e^{-1}(\cos(2) + j\sin(2)) = -0.1531 + 0.334j$. Taking the magnitude of $-0.1531 + 0.334j$ we find it is $|-0.1531 + 0.334j| = \sqrt{-0.1531^2 + 0.334^2} = 0.3674 < 1$ which is well within the unit circle and clearly stable. Its complex conjugate partner at $s = -10 - j20$ must also have the same magnitude and be stable. This is illustrated in Fig. 6.2.

In summary, for a discrete-LTI system described by (6.1), the n roots of $a(z^{-1})$ (poles) must all lie within the unit circle of the z-plane. The m zeros of the system are given by the roots of $b(z^{-1})$ do not directly affect stability. This can change if negative feedback is applied around the system, however.

6.2 Impulse Response and Convolution

The impulse response of a discrete-time system is the output of that system when it is driven by a unit impulse. Convolution is defined as a summation rather than as an integral in continuous time. For two discrete signal g(k) and h(k) their convolution is defined as y(k) where

$$y(k) = g(k) * h(k)$$
$$= h(k) * g(k)$$
$$= \sum_{i=0}^{k} g(k - i)h(i) \tag{6.11}$$

The above summation can have lower and upper limits of minus and plus infinity where necessary when calculating certain theoretical results. The summation allows us to find the output of a system if we know its input and impulse response. It is easy to implement on software but can be computationally intensive if the impulse responses has many terms. The impulse response can then be found by defining as an input $h(k) = \delta(k)$. Substitute into (6.11)

$$y(k) = \sum_{i=0}^{k} g(k - i)\delta(i) \tag{6.12}$$

The summation is zero everywhere except when $i = 0$. Then

$$y(k) = g(k) \tag{6.13}$$

But this can also be interpreted as the inverse Z-Transform of a system $G(z^{-1})$. Therefore

$$g(k) = \mathcal{Z}^{-1}\{G(z^{-1})\} \tag{6.14}$$

This also follows with analogy to the continuous-time case where the impulse response is the inverse Laplace Transform of the Laplace transfer function.

6.2.1 Examples of Discrete-Time Convolution

6.2.1.1 Example 1. Step Response

As an example, consider the step response of the system $G(z^{-1}) = \frac{1}{1-0.5z^{-1}}$. The input is $H(z^{-1}) = \frac{1}{1-z^{-1}}$. Let us first find the output using Z-Transforms.

Working in terms of z instead of its inverse we find the output is

$$y(z) = \frac{z^2}{(z - 1)(z - 0.5)} = \frac{2z}{(z - 1)} - \frac{z}{(z - 0.5)}$$

Take inverse Z-Transforms

$$y(k) = \mathcal{Z}^{-1}\left\{ \frac{2z}{(z-1)} - \frac{z}{(z-0.5)} \right\}$$
$$= 2u(k) - (0.5)^k u(k)$$

Now substitute some values of the time index $k = 0, 1, 2, 3...$

$$y(0) = 2 - 1 = 1$$
$$y(1) = 2 - 0.5 = 1.5$$
$$y(2) = 2 - 0.25 = 1.75$$
$$y(3) = 2 - 0.125 = 1.875$$

With the steady-state output approaching 2. This can be verified using the final value theorem.

Now use convolution where $h(k) = u(k) = \{1, 1, 1, 1...\}$ (unit step 1 from time 0), and the system impulse response is the inverse Z-transform:

$$g(k) = \mathcal{Z}^{-1}\{G(z^{-1})\}$$
$$= \mathcal{Z}^{-1}\left\{ \frac{z}{z-0.5} \right\}$$
$$= (0.5)^k u(k)$$
$$= \{1, 0.5, 0.25, 0.125...\}$$

Then substituting both terms into the convolution summation:

$$y(k) = \sum_{i=0}^{k} g(k-i)h(i)$$
$$= \sum_{i=0}^{k} (0.5)^{k-i}$$
$$= \sum_{i=0}^{k} (0.5)^i$$

Take the first 4 values of k and find y(k)

$$y(0) = (0.5)^0 = 1$$
$$y(1) = 0.5 + 1 = 1.5$$
$$y(2) = 0.25 + 0.5 + 1 = 1.75$$
$$y(3) = 0.125 + 0.25 + 0.5 + 1 = 1.875$$

We can see the same solution graphically in Fig. 6.3. Here g(k) and h(k) are converted to a change in independent variable g(i) and h(i). Since we need g(k − i) we need to then fold g vertically as shown. It then needs to be shifted to the right, multiplied and added sample by sample as shown for k = 0, 1, 2, 3 in Fig. 6.4.

In Fig. 6.4 at time k = 0 we multiply 01 × 1 since h(i) is always 1 at all samples and g(i) starts at 1 and goes down by a multiple of 0.5 at each sample. That is we have h(i) = {1, 1, 1, 1...}, g(i) = {1, 0.5, 0.25, 0.125...}.

Expressed in terms of g(k − i) this is g(k − i) = {...0.125, 0.25, 0.5, 1} (reverse order). So at time k = 0, y(0) = 1. At time k = 1 we have y(1) = 1 + 0.5, at time k = 2 we have y(2) = 1 + 0.5 + 0.25 and finally at time k = 3 we have y(3) = 1 + 0.5 + 0.25 + 0.125. So at any time k we have $y(k) = \sum_{i=0}^{k} (0.5)^i$. This has a closed-form solution since it is a geometric sequence summed to k terms. It can be

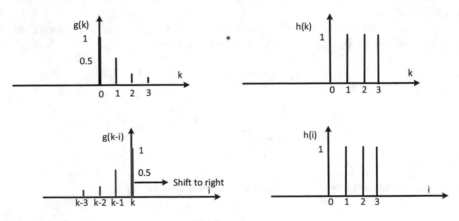

Fig. 6.3 Original impulse responses and folding of g

$$y(k) = \sum_{i=0}^{k} g(k-i)h(i)$$

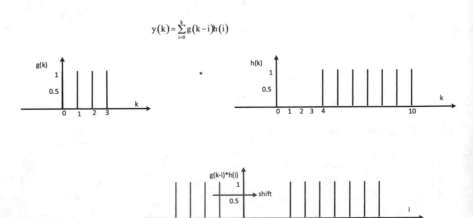

Fig. 6.4 Convolution summation formed by shifting, multiplication and summing

Fig. 6.5 Convolution by shift, multiply and addition of two sequences

g(k)={1,2,3,4}   h(k)={5,6,7,8}

k=0

```
        5 6 7 8
      4 3 2 1
      _____
          5
```

```
        5 6 7 8          k=1
      4 3 2 1
      _____
        10+6
```

k=2   5 6 7 8
      4 3 2 1
      _____
      15+12+7

```
        5 6 7 8
      4 3 2 1              k=3
      _____
      20+18+14+8
```

written as (see (5.31) and (5.32)), $y(k) = \frac{1-(0.5)^{k+1}}{1-0.5} = 2(1 - (0.5)^{k+1})$. We could of course have arrived at the same result by folding h(k) instead of g(k).

6.2.1.2   Example 2. Numerical Method

We can also glean a more mechanical method of discrete convolution from this example. We take one of the sequences and reverse it in time then overlap it with the other sample by sample. At each sample we shift and add the overlapping multiples together. This is shown in Fig. 6.5 for the first 4 shifts of two impulse sequences.

If we continue shifting beyond k = 3 in Fig. 6.5 we get k = 4, 5, 6 and then we run out of data. For two sequences of finite length p and q samples, we can show that the number of shifts will be p + q − 1 in total.

6.2.1.3   Example 3. Convolution of Two Pulses

Now consider a third example of the convolution of two finite impulse responses. Both are pulses. The first is a pulse with amplitude 1 starting at sample k = 4 and continuing up to and including time k − 10.

$$h(k) = u(k - 4) - u(k - 11)$$

The second is a pulse of amplitude 1 starting from time k = 0 and going up to and including time k = 3.

$$g(k) = u(k) - u(k - 4)$$

We write g(k) as g(k − i) and h(k) as h(i) and put on the same graph. Let the output be y(k) = g(k) * h(k). We then note that g has been reversed with respect to index i. It is slid along to the right sample by sample until the two impulse responses overlap. Before the overlap the multiplication is zero for times y(k), k = 0, 1, 2, 3.

They first overlap at time k = 4. Multiplication gives the solution at this time as y(4) = 1. The overlap continues past k = 4–5 where multiplication and adding gives

$1 + 1 = 2$. Hence $y(5) = 2$. Similarly $y(6) = 3$ and $y(7) = 4$. Then the smaller pulse has been swallowed up by the larger one and further multiplication and adding gives $y(8) = 4$, $y(9) = 4$, $y(10) = 4$. Then beyond $= 10$ the value falls: $y(11) = 3$, $y(12) = 2$, $y(13) = 1$ and $y(14)$ and beyond is zero again.

6.2.1.4   Solution via the Heaviside Operator Method

An elegant and much faster method is to use the discrete-time Heaviside operator. Write both pulses for the previous example in terms of the Heaviside step.

$$h(k) = u(k - 4) - u(k - 11)$$

and

$$g(k) = u(k) - u(k - 4)$$

Take Z-Transforms

$$H(z) = \left(\frac{z}{z - 1}\right)\left(z^{-4} - z^{-11}\right)$$

$$G(z) = \left(\frac{z}{z - 1}\right)\left(1 - z^{-4}\right)$$

Multiply to find $y(z)$

$$y(z) = \left(\frac{z}{z - 1}\right)^2 \left(z^{-4} - z^{-11}\right)\left(1 - z^{-4}\right)$$

$$= \frac{z^2\left(z^{-4} - z^{-11}\right)\left(1 - z^{-4}\right)}{(z - 1)^2}$$

$$= \frac{z^2}{(z - 1)^2}\left(z^{-4} - z^{-8} - z^{-11} + z^{-15}\right)$$

$$= \frac{z}{(z - 1)^2}\left(z^{-3} - z^{-7} - z^{-10} + z^{-14}\right)$$

Inverse Z-Transforming gives a series of ramps which come into play at various delays.

$$y(k) = (k - 3)u(k - 3) - (k - 7)u(k - 7)$$
$$- (k - 10)u(k - 10) + (k - 14)u(k - 14)$$

This is a discrete ramp of slope 1 at time $k = 3$ followed by the subtraction of a ramp at time $k = 7$ which makes the plot go horizontal. Then at time $k = 10$ the

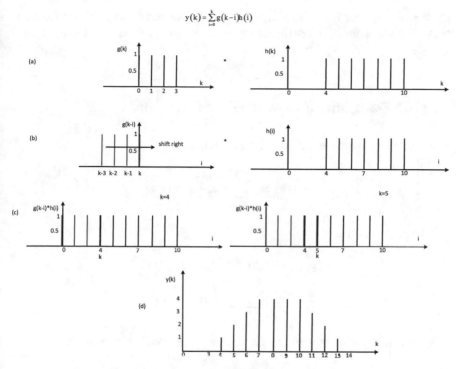

Fig. 6.6 Convolution of two sampled pulses. **a** Original impulse responses, **b** folded and ready to shift, **c** folded and overlap for k = 4 and k = 5, **d** convolution output y(k)

signal reduces with unit slope until it reaches time $k = 14$ when a positive sloping ramp cancels it out and makes the signal go flat as in Fig. 6.6d.

6.2.1.5 Polynomial Multiplication

For two polynomials P(x) and Q(x), where

$$P(x) = p_0 + p_1 x + p_2 x^2 + \ldots p_{np} x^{np}$$
$$Q(x) = q_0 + q_1 x + q_2 x^2 + \ldots q_{np} x^{nq}$$

Their product polynomial is $R(x) = r_0 + r_1 x + r_2 x^2 + \ldots r_{nr} x^{nr}$ and the coefficients are given by the convolution summation

$$r_k = \sum_{i=0}^{k} p_{k-i} q_i$$

where $nr = np + nq$ is the highest power of $R(x)$ and the number of coefficients in $R(x)$ is the total number of coefficients minus 1, or $np + 1 + nq + 1 - 1 = np + nq - 1$.

6.2.2 Z-Transform of Discrete-Time Convolution

Write the convolution summation in the general form

$$y(k) = \sum_{i=-\infty}^{\infty} g(k-i)h(i)$$

Take Z-Transforms

$$\mathcal{Z}\{y(k)\} = y(z)$$

$$= \sum_{k=-\infty}^{\infty} \sum_{i=-\infty}^{\infty} g(k-i)h(i)z^{-k}$$

Substitute a change of variable. Let $m = k - i$, so that $k = m + i$. Then

$$y(z) = \sum_{k=-\infty}^{\infty} \sum_{i=-\infty}^{\infty} g(m)h(i)z^{-(m+i)}$$

Then we can write

$$y(z) = \sum_{k=-\infty}^{\infty} g(m)z^{-m} \sum_{i=-\infty}^{\infty} h(i)z^{-i}$$

or

$$y(z) = G(z)H(z)$$

The Z-Transform of convolution is just the product of the two Z-Transfer functions. This only applies to LTI discrete-time systems. This is somewhat intuitive since the Laplace product case also applies in continuous time. Furthermore, the convolution theorem in Fourier transforms has also been shown in a previous chapter.

6.3 Frequency Response of Discrete-Time Systems

Returning to the fundamental relationship between the s and z domains

$$z = e^{sT_s} \tag{6.15}$$

Consider the continuous time $j\omega$ axis only since this represents frequency in rad/s

$$z = e^{j\omega T_s}$$

Substitute for the sampling interval $T_s = \frac{2\pi}{\omega_s}$, where ω_s is the sampling frequency in rad/s.

$$z = e^{j\omega T_s}$$
$$= e^{j2\pi\left(\frac{\omega}{\omega_s}\right)} \tag{6.16}$$

Equation (6.16) is the equation of a circle radius 1, (which we call the unit circle) and any value on the edge of the circle will represent frequency. For example, when $\omega = \omega_s/2$, the half-sampling or Nyquist frequency occurs and

$$z = e^{j\pi} \tag{6.17}$$

and this is the point $z = -1 + j0$ on the unit circle. At one quarter sampling frequency $\omega = \omega_s/4$

$$z = e^{j\pi/2} \tag{6.18}$$

Similarly, any frequency in rad/s or Hz can be represented as a fraction of the sampling frequency. This is a great convenience and is known as the normalised frequency θ, where

$$\theta = \omega T_s \tag{6.19}$$

We refer to θ as an angle in radians rather than Hz or rad/s since that is its basic units. The maximum frequency is the Nyquist frequency or $\theta = \pi/2$. Refer to Fig. 6.7 to see the idea of normalised frequency.

The main advantage of using normalised frequency (6.19) is that filters can be designed independent of sampling frequency. This means we could design a notch filter at quarter sampling frequency and by a change of sampling frequency change the actual frequency that the notch occurs. Zeros act as infinite attenuators at a complex frequency and poles are like infinite amplifiers. Consider a system with z transfer function

$$G(z) = \frac{z^2 + 1}{z^2 - z + 0.5} \tag{6.20}$$

By solving the two quadratics on numerator and denominator we can find the zeros and poles. The poles are complex at $z = 0.5 \pm j0.5$ and the two zeros are also

Fig. 6.7 Normalised
frequency on the z-plane unit
circle

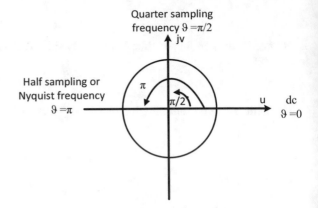

complex with values $z = \pm j$. The poles have magnitude less than one and the system
is stable. However, the zeros have magnitude one and are on the unit circle (Fig. 6.8),
which means that at whatever frequency they are at the magnitude must go to zero
at this frequency. In this case the zeros are at quarter sampling frequency.

If we substitute our transfer function (6.20) with a z value that covers all points on
the complex plane, we then have a three-dimensional view of things. It is a 3d plot
of the complex plane in its true form. Usually we only see a plan view looking down
from above. However, the magnitude is out of the plane of the usual 2d plot. Looking
at the 3d plot obtained in Fig. 6.9, the magnitude goes to infinity at the poles (they
look like wooden poles sticking up rom the ground) and the zeros are areas where
the magnitude goes to zero.

Usually we do not bother with a 3d plot and plot frequency response as a separate
plot altogether. This we can do by substituting $z = e^{j\theta}$ as a complex variable into the
z transfer function and taking magnitude of the result. We can also take the phase
angle too if necessary. Normalised frequency θ is varied from 0 to pi to give the full
range of frequency from dc to half sampling. This can be algebraically tedious task
except for all but simple systems. The usual method is to plot using computer tools
such as MATLAB. Unlike for continuous time, there are no shortcut methods that

Fig. 6.8 Pole-zeros plot on
unit circle

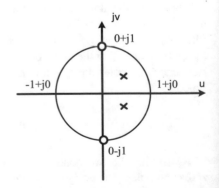

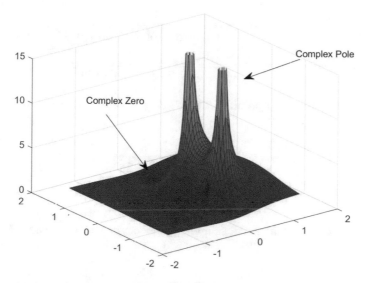

Fig. 6.9 A 3d plot of magnitude (vertical axis), and $z = u + jv$ looking from above

use asymptotic approximations to get a frequency response (or Bode-plot as is the common name). The s-domain theory that is so simple and elegant does not carry through well to the z-domain.

6.3.1 Example: Calculation of Frequency Response

$$H(z) = \frac{1}{1 - \alpha z^{-1}}, \ |\alpha| < 1 \tag{6.21}$$

Find the discrete-time frequency response of the system.
Solution:
Substitute $z^{-1} = e^{-j\theta}$ into (6.21)

$$H\left(e^{j\theta}\right) = \frac{1}{1 - \alpha e^{-j\theta}}$$

$$= \frac{1}{1 - \alpha \cos(\theta) + \alpha j \sin(\theta)} \tag{6.22}$$

Take the magnitude of (6.22)

$$\left|H\left(e^{j\theta}\right)\right| = \left| \frac{1}{1 - \alpha \cos(\theta) + \alpha j \sin(\theta)} \right|$$

$$= \frac{1}{\left|1 - \alpha \cos(\theta) + \alpha j \sin(\theta)\right|}$$

$$= \frac{1}{\sqrt{(1 - \alpha\cos(\theta))^2 + \alpha^2\sin(\theta)^2}}$$

$$= \frac{1}{\sqrt{1 + \alpha^2 - 2\alpha\cos(\theta) + \sin(\theta)^2}} \tag{6.23}$$

A few things we can glean off (6.23). It's dc gain occurs when $\theta = 0$ giving $z = 1$ and

$$|H(1)| = \frac{1}{\sqrt{1 + \alpha^2 - 2\alpha}}$$

$$= \frac{1}{1 - \alpha}$$

Its high-frequency gain at half sampling frequency can be found when $\theta = \pi$ and $z = -1$

$$|H(-1)| = \frac{1}{\sqrt{1 + \alpha^2 + 2\alpha}}$$

$$= \frac{1}{1 + \alpha}$$

Of course, this could be found directly from (6.21) as well. The frequency–response is then a plot for all gain values with θ in $0 < \theta < \pi$. The phase is found from (6.22)

$$\left| H(e^{j\theta}) \right| = \left| \frac{1}{1 - \alpha e^{-j\theta}} \right|$$

$$= \left| \frac{1}{1 - \alpha\cos(\theta) + \alpha j\sin(\theta)} \right|$$

$$= -\tan^{-1}\left(\frac{\alpha\sin(\theta)}{1 - \alpha\cos(\theta)} \right) \tag{6.24}$$

When $\theta = 0$ the phase is zero and when $\theta = \pi$ the phase is zero. Select $\alpha = 0.9$ and plot the frequency response. The result is shown in Fig. 6.10. Frequencies up to half sampling are shown only since the frequency response will repeat after this frequency and the phase will flip becoming positive instead of negative. All LTI discrete systems have *conjugate symmetry* of the frequency response.

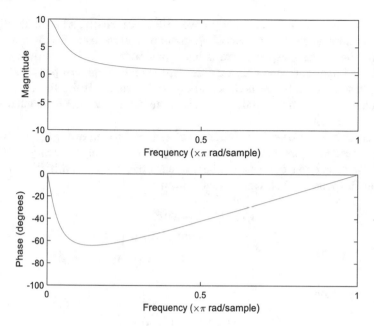

Fig. 6.10 Frequency response of discrete-time first order system up to half sampling (normalised) frequency

6.3.2   *Example: Frequency Response of a Finite-Impulse Response Filter (FIR Filter)*

Consider the system

$$H(z) = 1 + z^{-2} \tag{6.25}$$

The above system (filter) has an impulse response equal to just the coefficients of the filter. The impulse response is $h(k) = \{1, 0, 1\}$ allowing for the fact that there is no z^{-1} coefficient. It can also be written more concisely as $h(k) = \delta(k) + \delta(k-2)$. It has no obvious poles, but we can write (6.25) as

$$H(z) = 1 + z^{-2}$$
$$= \frac{z^2 + 1}{z^2} \tag{6.26}$$

Indicating that two poles are at the origin. The filter is said to have a finite impulse response (FIR) as opposed to a pole-zero one which is infinite. They are known as Infinite impulse response filters (IIR filters). For example, an IIR filter will may have an impulse response like the previous example $h(k) = \mathcal{Z}^{-1}\{\frac{1}{1-\alpha z^{-1}}\} = \alpha^k, k = 0, 1, 2, \ldots$ which theoretically has infinite length as k increases.

The FIR filter has a few advantages over IIR filters. Firstly, it has poles all at the origin and so is never stable. Secondly, filters of its type can be shown in certain cases to have a phase response that is linear (a straight-line characteristic). This is particularly important when filtering pulse-type signals. Nonlinear phase can jumble up the phase spacing of harmonics and distort the signal. It has the disadvantage that it can be much higher order to achieve the same amount of attenuation as a lower-order IIR filter.

The zeros of (6.26) are the solutions to $z^2 + 1 = 0$ giving two imaginary roots at $z = \pm j$. These roots are on the unit circle at quarter sampling frequency. Therefore, for a given sample rate there should be a notch at one quarter of this frequency. The frequency response is found theoretically from

$$\begin{aligned} H(e^{j\theta}) &= 1 + e^{-2j\theta} \\ &= e^{-j\theta}(e^{j\theta} - e^{-j\theta}) \\ &= 2e^{-j\theta}\cos(\theta) \end{aligned} \tag{6.27}$$

The magnitude is

$$\left|H(e^{j\theta})\right| = 2|\cos(\theta)| \tag{6.28}$$

A MATLAB plot is shown in Fig. 6.11.

The phase is just given by the phase of $e^{-j\theta}$, which is $-\theta$. This gives us zero phase at dc and $-\pi/2$ radians at quarter sampling frequency. Beyond quarter sampling the

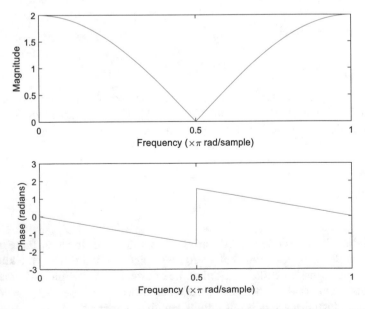

Fig. 6.11 Frequency response of a second order FIR notch filter. The phase is shown wrapped

cosine term magnitude flips to being negative and produces a jump in phase shift of another $-\pi$ radians. This would give us $-1.5\,\pi$. However, as is convention, the phase is shown flipped vertically by $+\pi$ resulting in $\pi/2$ instead. This is to prevent the scale going too far negative. This is known as phase wrapping.

6.3.3 Design of a Simple Digital Notch Filter

Suppose we have a sampling frequency of 1 kHz and need to notch out a signal at 50 Hz. Can we design a simple notch filter to fulfil the task?

For a sampling frequency $f_s = 1000$ we require a notch at 50 Hz which is 50/1000 = 1/20th sampling frequency or two complex zeros at $z = e^{j\frac{2\pi}{20}} = e^{0.1\pi}$ right on the unit circle. The first complex zero is $z = e^{0.1\pi} = \cos(0.1\pi) + j\sin(0.1\pi)$. Both are therefore $z_1, z_2 = 0.9511 \pm j0.309$ and found from a polynomial $(z - z_1)(z - z_2) = z^2 - 1.9021z + 1$. It would appear that this could be the filter, but if the frequency response is plotted it will be found that the high frequencies get amplified too much. In Fig. 6.12 the notch is present but the high frequency gain is unacceptable.

We need a filter that has a flat frequency response except for the dip due to the notch. We need two poles to bring the high frequencies down. This is achieved by having a transfer function with poles that are also complex and radially in line with the zeros. They must have magnitude less than 1 for stability. Figure 6.13 shows the pole-zero combination.

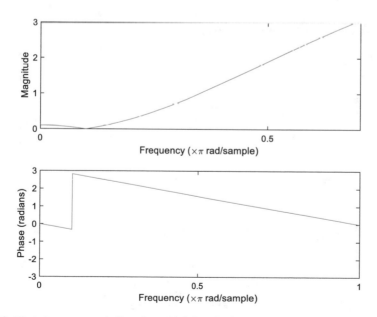

Fig. 6.12 First attempt at notch filter shows high frequencies are amplified

Fig. 6.13 Pole-zero combination for second order digital notch filter

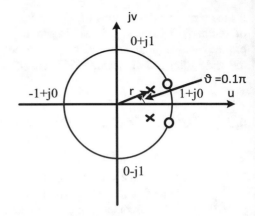

Let the poles have magnitude $r = 0.95$ with the same phase angle giving $p_1, p_2 = 0.95e^{0.1\pi} = 0.95(\cos(0.1\pi) \pm j\sin(0.1\pi)) = 0.9035 \pm 0.2936j$. The pole polynomial becomes $(z - p_1)(z - p_2) = z^2 - 1.807z + 0.9025$. The transfer function of our filter is now $H(z) = \frac{z^2 - 1.9021z + 1}{z^2 - 1.807z + 0.9025}$ which can also be expressed in backwards shift notation $H(z) = \frac{1 - 1.9021z^{-1} + z^{-2}}{1 - 1.807z^{-1} + 0.9025z^{-2}}$. The frequency response is shown in Fig. 6.14. The notch occurs at a normalised frequency 0.1π which is 50 Hz for a sampling rate of 1 kHz.

Because the transfer function has poles, the phase is not linear like the previous example. We can generalise the above filter for a notch at any normalised frequency

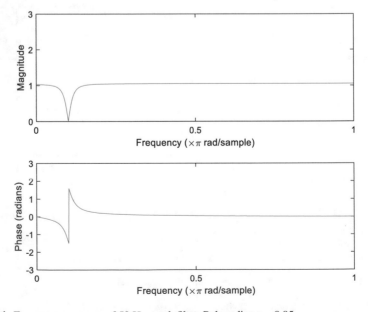

Fig. 6.14 Frequency response of 50 Hz notch filter. Pole radius $r = 0.95$

θ and pole radius $r < 1$. The transfer function is

$$H(z) = \frac{(z - \cos(\theta) + j\sin(\theta))(z - \cos(\theta) - j\sin(\theta))}{(z - r\cos(\theta) - jr\sin(\theta))(z - r\cos(\theta) + jr\sin(\theta))} \tag{6.29}$$

Simplifying we get

$$H(z) = \frac{z^2 - 2\cos(\theta) + 1}{z^2 - 2r\cos(\theta) + r^2} \tag{6.30}$$

As r gets closer to 1 the width of the notch gets narrower. For a pole radius $r = 0.99$ the frequency response is shown in Fig. 6.15.

Finally in this section, it is useful to describe a digital notch filter similar to the above called the *dc blocker*. It only needs one zero at $z = 1$ and a pole close to the zero but inside the unit circle at $z = r$. It has transfer function

$$H(z) = \frac{z - 1}{z - r}$$

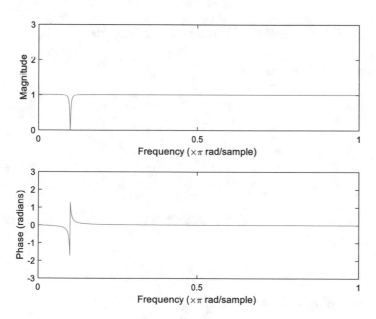

Fig. 6.15 Notch filter frequency response. Pole radius $r = 0.99$

6.4   The Discrete-Time Fourier Transform (DTFT)

The DTFT is a continuous-time measure of frequency properties of sampled signals. Just as the Fourier transform can be found from the Laplace, we can find the discrete-time simply by substitution of $z = e^{j\theta}$ into the Z-Transform. The DTFT becomes

$$F\left(e^{j\theta}\right) = \sum_{i=-\infty}^{\infty} f(k)e^{-jk\theta} \tag{6.31}$$

The DTFT is periodic with period 2π. The limits depend on the nature of the signal. If the signal is causal, then we can take the lower limit from $k = 0$. The length of the signal determines the upper limit. The inverse DTFT is given by

$$f(k) = \frac{1}{2\pi} \int_{-\pi}^{\pi} F\left(e^{j\theta}\right)e^{j\theta}d\theta \tag{6.32}$$

6.4.1   DTFT of a Delayed Impulse by m Samples

Let the time-domain signal be

$$f(k) = \delta(k - m)$$

$$F\left(e^{j\theta}\right) = \sum_{i=-\infty}^{\infty} \delta(k - m)e^{-jk\theta}$$

$$= e^{-jm\theta} \tag{6.33}$$

This is a pure time-delay with magnitude $\left|F\left(e^{j\theta}\right)\right| = \left|e^{-jm\theta}\right| = 1$ and phase $\underline{\left|F\left(e^{j\theta}\right)\right.} = -km\theta$.

6.4.2   DTFT of a Geometric Decay

$$f(k) = \alpha^k, \ |\alpha| < 1 \tag{6.34}$$

$$F\left(e^{j\theta}\right) = \sum_{i=-\infty}^{\infty} \alpha^k e^{-jk\theta}$$

$$= 1 + \alpha e^{-j\theta} + \alpha^2 e^{-2j\theta} + \alpha^3 e^{-3j\theta} + \cdots \tag{6.35}$$

Equation (6.35) is a geometric sequence with growth factor $\alpha e^{-j\theta}$ summed to infinity. It can be written in closed form as

$$F(e^{j\theta}) = \frac{1}{1 - \alpha e^{-j\theta}} \tag{6.36}$$

Taking magnitude and phase of (6.36) are found from

$$F(e^{j\theta}) = \frac{1}{1 - \alpha(\cos(\theta) - j\sin(\theta))} \tag{6.37}$$

$$|F(e^{j\theta})| = \left| \frac{1}{1 - \alpha\cos(\theta) + j\alpha\sin(\theta)} \right|$$

$$= \frac{1}{\sqrt{(1 - \alpha\cos(\theta))^2 + \alpha^2\sin^2(\theta)}}$$

$$= \frac{1}{\sqrt{1 + \alpha^2 - 2\alpha\cos(\theta)}} \tag{6.38}$$

The phase is found from (6.37)

$$\left| F(e^{j\theta}) \right| = -\tan^{-1}\left[\frac{\alpha\sin(\theta)}{1 - \alpha\cos(\theta)} \right] \tag{6.39}$$

We can plot for $\alpha = 0.92$ and this is shown in Fig. 6.16 from which we see it has a lowpass characteristic. It is shown as being periodic, but only normalised frequencies up to $\theta = \pi$ are of use.

6.4.3   DTFT of a Delayed Signal

For s signal f(k) with DTFT $F(e^{j\theta})$, the DTFT of the delayed signal by m samples $f(k - m)$ is found from $\sum_{i=-\infty}^{\infty} f(k - m)e^{-jk\theta}$. Substituting $k - m = n$ and hence $k = n + m$ we get $\sum_{i=-\infty}^{\infty} f(n)e^{-j(n+m)\theta} = e^{-jm\theta}\sum_{i=-\infty}^{\infty} f(n)e^{-jn\theta}$. This is just $e^{-jm\theta}F(e^{j\theta})$ where the $e^{-jm\theta}$ is a pure time-delay. This only affects the phase of the spectrum of the original signal and not the magnitude.

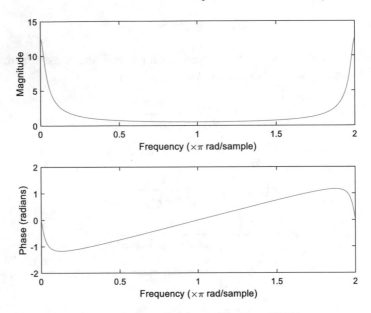

Fig. 6.16 Magnitude and phase of geometric decay as found from DTFT

6.4.4 DTFT of a Discrete-Time Pulse

Consider a sampled pulse of magnitude 1, duration N samples from time k = 0. We have f(k) = {1, 1, 1 ... .1} or f(0) = 1, f(2) = 1, ...f(N − 1) = 1. Mathematically this becomes

$$f(k) = \begin{cases} 0, & k < 0 \\ 1, & 0 \le k < N - 1 \\ 0, & k > N - 1 \end{cases} \tag{6.40}$$

Then the DTFT follows by direct substitution

$$F(e^{j\theta}) = \sum_{k=0}^{N-1} e^{-jk\theta} \tag{6.41}$$

This is the sum to N − 1 terms of a geometric sequence with growth rate $e^{-j\theta}$. The summation in closed form becomes

$$F(e^{j\theta}) = \frac{1 - e^{-j\theta N}}{1 - e^{-j\theta}} \tag{6.42}$$

Simplifying (6.42) we write it as

$$F(e^{j\theta}) = \frac{e^{-j\theta N/2}(e^{j\theta N/2} - e^{-j\theta N/2})}{e^{-j\theta/2}(e^{j\theta/2} - e^{-j\theta/2})} \tag{6.43}$$

Before continuing, recall that $e^{jx} - e^{-jx} = 2j\sin(x)$. Use this in (6.43) and get

$$F(e^{j\theta}) = e^{-j\theta\left(\frac{N-1}{2}\right)} \frac{\sin\left(\frac{\theta N}{2}\right)}{\sin\left(\frac{\theta}{2}\right)} \tag{6.44}$$

The exponential term in (6.44) only contributes to the phase and not magnitude and so we ignore it for this example. Unfortunately, this is not a sinc(x) function like we had in continuous time. It is another special function we can use called the *Dirichlet kernel*. It is defined as

$$D_N(\theta) = \frac{\sin\left(\frac{\theta N}{2}\right)}{N\sin\left(\frac{\theta}{2}\right)} \tag{6.45}$$

and also called Diric(θ, N). Then we can write (6.44) as

$$F(e^{j\theta}) = e^{-j\theta\left(\frac{N-1}{2}\right)} N D_N(\theta) \tag{6.46}$$

The nineteenth century mathematician *Peter Gustav Dirichlet* was prominent in putting Fourier's work into a more precise mathematical framework, particularly for the convergence of Fourier series.

Figure 6.17 shows the spectra for N = 4 and N = 5. The number of zero crossings is always N − 1 and the spectrum is periodic. The period of repetition varies from 2π to 4π depending on whether N is even or odd respectively. Similar properties exist as for continuous time when the pulse width narrows or widens. The spectrum widens for a narrowing pulse and vice-versa.

6.4.5   DTFT Properties

The properties of the DTFT are the same as the ordinary Fourier transform with a few changes in variables (Table. 6.1).

It is interesting to solve a problem based on Parseval's theorem. Consider the geometric decay problem in (6.34)

$$f(k) = \alpha^k, |\alpha| < 1.$$

Parseval's theorem states that the energy in the time and frequency domains are the same. Namely

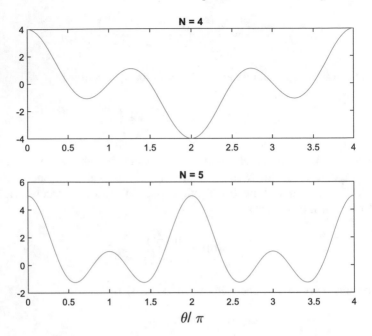

Fig. 6.17 DTFT plot of discrete-time pulse for $N = 4$ and $N = 5$ samples. Represents $ND_N(\theta)$

Table 6.1 Discrete-time Fourier transform properties

	$f(k)$	$F(\theta)$
Linearity	$a_1 f_1(k) + a_2 f_2(k)$	$a_1 F_1(\theta) + a_2 F_2(\theta)$
Time-delay	$f(k - m)$	$e^{-jm\theta} F(\theta)$
Frequency-shift	$f(k) e^{jk\theta_0}$	$F(\theta - \theta_0)$
Convolution in time-domain	$f(k) * g(k)$	$F(\theta) G(\theta)$
Convolution in the frequency-domain	$f(k) g(k)$	$F(\theta) * G(\theta)$
Parseval's theorem	$\displaystyle\sum_{k=-\infty}^{\infty} f^2(k)$	$\dfrac{1}{2\pi} \displaystyle\int_{-\pi}^{\pi} \left| F(e^{j\theta}) \right|^2 d\theta$

$$\sum_{k=-\infty}^{\infty} f^2(k) = \frac{1}{2\pi} \int_{-\pi}^{\pi} \left| F(e^{j\theta}) \right|^2 d\theta \tag{6.47}$$

Substituting in the time-domain part gives a convergent geometric sequence which is summed to infinity giving

$$\sum_{k=-\infty}^{\infty} f^2(k) = \sum_{k=0}^{\infty} \alpha^{2k}$$

$$= \frac{1}{1 - \alpha^2} \tag{6.48}$$

In the frequency-domain, we know that the DTFT is from the previous example of Eq. (6.36), $F(e^{j\theta}) = \frac{1}{1 - \alpha e^{-j\theta}}$. Then

$$\frac{1}{2\pi} \int_{-\pi}^{\pi} \left| F(e^{j\theta}) \right|^2 d\theta = \frac{1}{2\pi} \int_{-\pi}^{\pi} \left| \frac{1}{1 - \alpha e^{-j\theta}} \right|^2 d\theta$$

$$= \frac{1}{2\pi} \int_{-\pi}^{\pi} \frac{1}{1 + \alpha^2 - 2\alpha \cos(\theta)} d\theta, \ |\alpha| < 1 \tag{6.49}$$

Equation (6.49) is a tricky integral but can be solved using computer algebra to give the solution (6.48).

Chapter 7
A More Complete Picture

7.1 Link Between Various Methods

Now that we have studied continuous and discrete-time signals and systems, we need a more complete picture of how it all fits together. A summary of the theory up to this chapter is shown below in Fig. 7.1.

A lot depends in whether we are dealing with signals or with systems. For systems we can find the differential equation via mathematical modelling and then take the Laplace Transform to get the differential equation. This gives us the transfer function in the s-domain. For circuitry we can jump straight to the transfer-function by using the Laplace representation for impedances. This is for electrical systems of course but similar approaches can be made in mechanical and other systems types. From the s-domain we can move to the frequency response if we are dealing with a system and to the spectrum if it is a signal. Alternatively, we can move directly to the spectrum from the Fourier transform. This is not very practical for real-world signals however but ok for standard textbook types.

Sampling a digital signal or system leads to a similar story in the digital domain. We have difference equations instead of differential equations and the Z-transform to replace the Laplace. Similarly, the DTFT replaces the Fourier Transform. Convolution is not on the diagram because this takes place when we either pass signals through systems or when we cascade systems together. Frequency for sampled signals is usually defined by normalisation with respect to the sampling frequency, so the maximum achievable frequency is the Nyquist frequency or half-sampling which is π radians. The one thing that is missing is a bridge between the s and z domains. Given a Laplace transfer function, how do we convert it to the z-domain?

T. J. Moir, *Rudiments of Signal Processing and Systems*,
https://doi.org/10.1007/978-3-030-76947-5_7

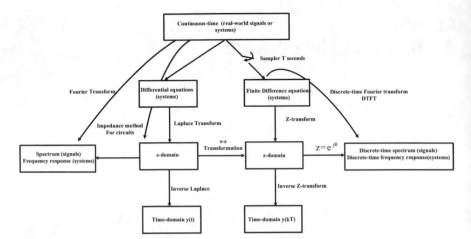

Fig. 7.1 Summary of continuous and discrete-time theory

7.2   Conversion of G(s) to G(Z)

The large wealth of literature on the design of filters and dynamic systems should not
be thrown away after realising we can sample a system and represent it in software. In
order to best exploit this however, we need a method of accurately converting transfer
function from the s to z domains. Sampling had been known in mathematics long
before it was applied to electrical engineering. The area of numerical integration for
instance necessitates the division of a graph into slices and the area of each slice taken
and summed with the others. The literature is full of a great many methods of how to
do this. If we narrow down the number of methods to a few of the most popular, we
have three main ways to do this. Rectangular integration or Euler's method is perhaps
the simplest and in turn there are two techniques used. We begin with a continuous
time integral to find the area under the curve x(t).

$$y(t) = \int_0^t x(t)dt \tag{7.1}$$

We can see that (7.1) is an integrator and we could also represent the above by
its Laplace equivalent and say that $y(s) = (1/s)x(s)$. Numerically we can do the
integration by defining as usual $t = kT_s$ and then

$$y(kT_s) = \int_0^{t=kT_s} x(kT_s)dt \tag{7.2}$$

Which can be split into two integrals

$$y(kT_s) = \int_0^{(k-1)T_s} x(kT_s)dt + \int_{(k-1)T_s}^{kT_s} x(kT_s)dt \qquad (7.3)$$

The first of these integrals we can define as

$$y(k-1) = \int_0^{(k-1)T_s} x(kT_s)dt \qquad (7.4)$$

And we get a recursion

$$y(k) = y(k-1) + \int_{(k-1)T_s}^{kT_s} x(kT_s)dt \qquad (7.5)$$

This last integral depends on which area we take. Assume we take a rectangle then we have two choices as shown in Fig. 7.2.

For the forward integration method, we have

$$y(k) = y(k-1) + x(k-1)T_s \qquad (7.6)$$

And for backwards integration we have

$$y(k) = y(k-1) + x(k)T_s \qquad (7.7)$$

Finally, we can get a better approximation by assuming the area is a Trapezium instead. See Fig. 7.3.

This gives us

$$y(k) = y(k-1) + \frac{T_s}{2}[x(k) + x(k-1)] \qquad (7.8)$$

For each case we can take Z-transforms and this in turn reveals the usefulness of each method.

If we take Z-transforms of (7.6) and (7.7) we get respectively

$$y(z) = \frac{T_s}{z-1}x(z) \qquad (7.9)$$

and

$$y(z) = \frac{zT_s}{z-1_s}x(z) \qquad (7.10)$$

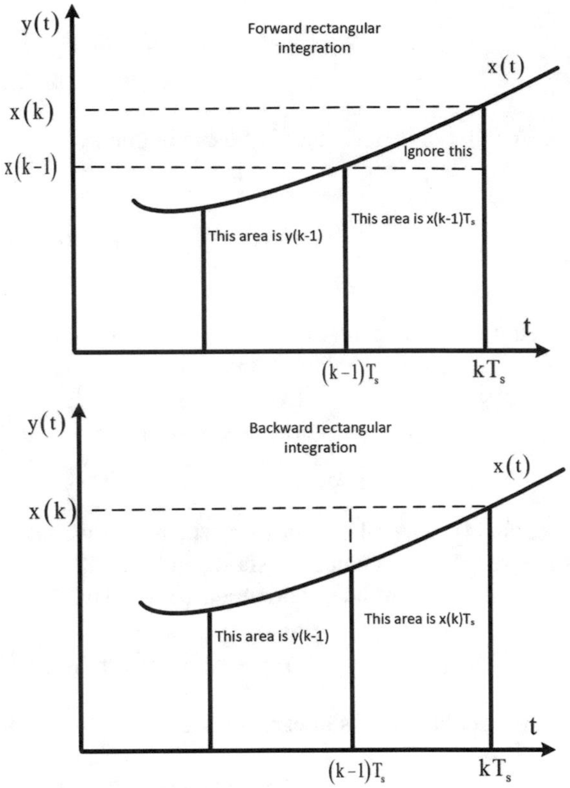

Fig. 7.2 Forward and backward rectangular integration methods

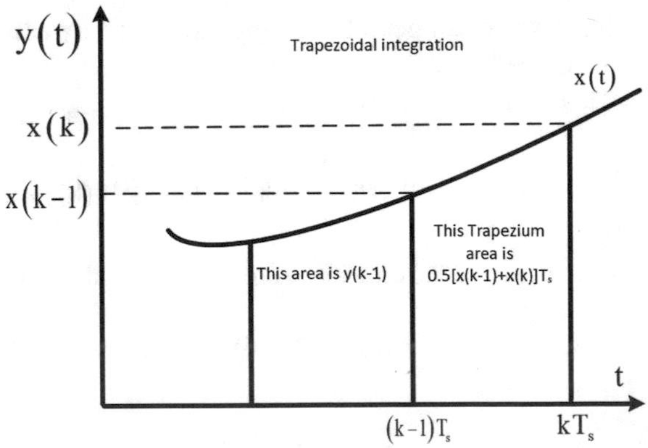

Fig. 7.3 Trapezoidal integration

From (7.8) obtain

$$y(z) = \frac{T_s}{2}\left(\frac{z+1}{z-1}\right)x(z) \tag{7.11}$$

Now Eqs. (7.9), (7.10) and (7.11) are three different approximations to integration. Each has different properties. We can obtain the same results directly from the fundamental relationship between s and z as found in sampling theory.

$$z = e^{sT_s} \tag{7.12}$$

Expanding (7.12) in a Maclaurin series gives

$$z = e^{sT_s} = 1 + sT_s + \frac{s^2 T_s^2}{2!} + \ldots \tag{7.13}$$

To a first order approximation we get

$$z \simeq 1 + sT_s$$

or

$$s = \frac{z-1}{T_s} \tag{7.14}$$

This is just the differentiator version of (7.9). Similarly, from

$$z = \frac{1}{e^{-sT_s}} \simeq \frac{1}{1 - sT_2} \tag{7.15}$$

Then

$$s = \frac{z-1}{zT_s} \tag{7.16}$$

Finally, for the trapezoidal case obtain

$$z = \frac{e^{sT_s/2}}{e^{-sT_s/2}} \simeq \frac{1 + sT_s/2}{1 - sT_s/2} \tag{7.17}$$

From which

$$s = \frac{2}{T_s}\left(\frac{z-1}{z+1}\right) \tag{7.18}$$

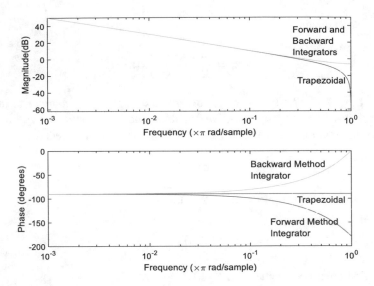

Fig. 7.4 Frequency response of all three integrators

This last approximation is often known as the Bilinear transform or Tustin's method. The three approximations are for integration $\frac{zT_s}{z-1_s}$, $\frac{T_s}{z-1_s}$ and $\frac{T_s}{2_s}\left(\frac{z+1}{z-1}\right)$. Their inverse gives differentiators. In Z-Transform tables usually the sample time is omitted and taken to be unity. We can first compare their frequency domain properties in terms of their frequency response. Each has a pole at $z = 1$. See Fig. 7.4 for the plot of magnitude and phase for each of the methods.

The forward and backward integrators have identical magnitudes but different phase. An analogue integrator has a slope of -20 dB/decade and a phase of -90 degrees at all frequencies. Both of these integrators have errors in phase as compared with continuous time. The trapezoidal method has exactly -90 degrees phase and is a better matchup. To a normalised frequency of approximately 0.02π radians, all three integrators are identical in magnitude and phase. This is quite a low frequency (one hundredth sampling frequency) and would mean very high sampling would have to occur for them to behave the same. When used as solo integrators (for example as part of a PID controller), either integrator can be used successfully and usually the simplest of these, the Euler forward method is used. The two Euler methods only differ by a one-step time-delay. For practical purposes a digital system cannot respond instantaneously and at least one step delay must happen from input to output.

For the purposes of a transformation method from s to z however, only one of the three methods should be used. The reason is illustrated in Fig. 7.5.

Fig. 7.5a shows the mapping for the forward difference integration method. It has

$$z = sT_s + 1 \tag{7.19}$$

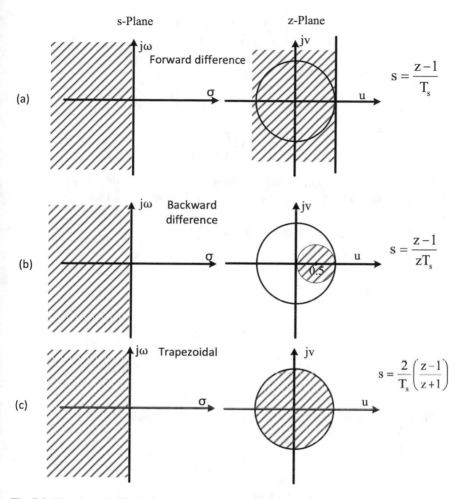

Fig. 7.5 Mapping of left half plane to z plane for **a** Forward difference **b** backward difference and **c** trapezoidal methods

since $s = \sigma + j\omega$, then when $\sigma = 0$, $z = j\omega T_s + 1$ the imaginary axis in s maps onto $z = 1$. This is a straight line at $z = 1$ on the boundary of the z-plane but covers an area outside of the plane too. When $\sigma < 0$ which indicates the left hand z-plane or stable region this maps into anything left of the line at $z = 1$. Therefore, there is a good chance that by using this method the Z-transform will end up unstable.

Fig. 7.5b shows the mapping for the backward difference method. Here we have from (7.15)

$$z = \frac{1}{1 - sT}$$

$$= \frac{1}{2} + \frac{1}{2}\left(\frac{1 + sT_s}{1 - sT_s}\right)$$

Then

$$z - \frac{1}{2} = \frac{1}{2}\left(\frac{1 + sT_s}{1 - sT_s}\right)$$

$$\left|z - \frac{1}{2}\right| = \left|\frac{1}{2}\left(\frac{1 + sT_s}{1 - sT_s}\right)\right| \tag{7.20}$$

When $s = j\omega$ the imaginary axis maps onto a circle of radius ½ centred at $z = 0.5$ on the z plane. This does not cover the whole z plane and will not give accurate results.

Finally, for the trapezoidal method of Fig. 7.5c

$$s = \frac{2}{T_s}\left(\frac{z - 1}{z + 1}\right)$$

and

$$z = \frac{1 + sT_s/2}{1 - sT_s/2}$$
$$= \frac{2 + sT_s}{2 - sT_s}$$

When $s = j\omega$, the imaginary axis maps onto the unit circle in z since

$$|z| = \left|\frac{2 + j\omega T_s}{2 - j\omega T_s}\right| = 1 \tag{7.21}$$

When $s = \sigma + j\omega$ we have the mapping

$$|z| = \left|\frac{2 + \sigma T_s + j\omega T_s}{2 - \sigma T_s - j\omega T_s}\right|$$
$$= \sqrt{\frac{(2 + \sigma T_s)^2 + \omega^2 T_s^2}{(2 - \sigma T_s)^2 + \omega^2 T_s^2}}$$

And clearly when σ is negative the magnitude of z must be less than unity (the numerator must always be smaller than the denominator since $\sigma T_s < -\sigma T_s$, if σ is negative) indicating values within the unit circle. Clearly the trapezoidal method guarantees stability and uses the entire unit circle unlike the backward difference method. This method is also known as the Bilinear transform or Tustin's method [1].

7.2.1 Example of Using the Bilinear Transform or Trapezoidal Integration

Consider a first order transfer function with a gain of 10 (20 dB).

$$G(s) = \frac{10}{1 + sT} \tag{7.22}$$

where $T = 0.159$ ms. This is a first order transfer function of a low pass filter with a cut-off (-3 dB or corner) frequency 1 kHz. Consider two sampling frequencies. The first of these will be ten times and the second three times the cut-off frequency. For $f_s = 10$ kHz the passband edge of the filter is at 1/10th sampling frequency or a normalised frequency of 0.2π rads. The sampling interval is $T_s = 0.1$ ms. The Bilinear transform is $s = \frac{2}{T_s}\left(\frac{z-1}{z+1}\right)$. Substituting this into (7.22) gives

$$
\begin{aligned}
G(z) &= \frac{10}{1 + sT}\bigg|_{s = \frac{2}{T_s}\left(\frac{z-1}{z+1}\right)} \\
&= \frac{10}{1 + \frac{2T}{T_s}\left(\frac{z-1}{z+1}\right)} \\
&= \frac{10\,T_s(z + 1)}{T_s(z + 1) + 2T(z - 1)}
\end{aligned} \tag{7.23}
$$

Re-arranging gives

$$G(z) = \frac{10T_s}{(T_s - 2T)}\left[\frac{z + 1}{1 + z\left(\frac{T_s + 2T}{T_s - 2T}\right)}\right] \tag{7.24}$$

Substituting values gives (Fig. 7.6). This is also expressed in negative powers of z with a little algebra.

$$G(z) = 2.3923\left(\frac{1 + z^{-1}}{1 - 0.5215\,z^{-1}}\right) \tag{7.25}$$

The normalised frequency is 0.2π rads when the phase is -45 degrees, and the gain is -3 dB down from 20 dB all indicating an accurate digital replica of the analogue system.

The system can be implemented in software by the difference equation obtained from (7.25), namely

$$y(k) = 0.5215y(k - 1) + 2.3923u(k) + 2.3923\,u(k - 1) \tag{7.26}$$

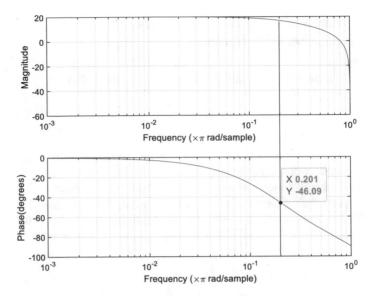

Fig. 7.6 Frequency response of first order discrete-time system when sampling frequency is 10 kHz

Now consider the same example but for a sampling frequency $f_s = 3$kHz or $T_s = 0.333$ms. Equation (7.24) becomes

$$G(z) = 5.1177\left(\frac{1 + z^{-1}}{1 + 0.0235z^{-1}}\right) \tag{7.27}$$

We can show that $G(1)$, the dc gain is approximately 10, which is the same as the dc gain for the analogue system. The frequency response is shown in Fig. 7.7.

We can see that the phase is now only -17.14 degrees and the gain is nearer 20 dB indicating an error in the cut-off frequency. The reason for this is in the next section.

7.3 Frequency Warping of the Bilinear Transform

In the previous section it was shown that for the bilinear transform to give good results the sampling frequency needed to be at least 10 times higher than the cut-off frequency of the system that is to be converted. Further analysis is needed to show the relationship between the original analogue frequency and the transformed digital frequency. Begin with the relationship from (7.18)

$$s = \frac{2}{T_s}\left(\frac{z - 1}{z + 1}\right)$$

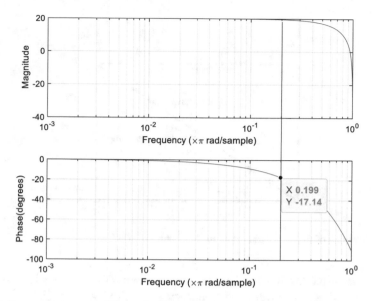

Fig. 7.7 Frequency response of first order discrete-time system when sampling frequency is 3 kHz

Substitute $z = e^{j\theta}$ and find

$$
\begin{aligned}
s &= \frac{2}{T_s}\left(\frac{e^{j\theta} - 1}{e^{j\theta} + 1}\right) \\
&= \frac{2e^{j\theta/2}}{T_s e^{j\theta/2}}\left(\frac{e^{j\theta/2} - e^{-j\theta/2}}{e^{j\theta/2} + e^{-j\theta/2}}\right) \\
&= \frac{2}{T_s}\left(\frac{2j\sin(\theta/2)}{2\cos(\theta/2)}\right) \\
&= \frac{2j}{T_s}\tan(\theta/2)
\end{aligned}
\tag{7.28}
$$

But $s = j\omega$, therefore

$$
\omega = \frac{2}{T_s}\tan\left(\frac{\theta}{2}\right)
\tag{7.29}
$$

The left-hand side is the analogue frequency ω and in the right is the normalised digital frequency θ. If this relationship is linear then the bilinear transform will give perfect results at all frequencies. Unfortunately, the tan function is only approximately linear at low values of $\theta/2$. We can see the relationship with the graph in Fig. 7.8. It can be seen that the relationship between the analogue and digital domain is nonlinear. This is commonly known as *frequency warping*.

The analogue and digital frequencies are separated by means of the nonlinear tan function. Tan is approximately linear at low frequencies (relative to sampling

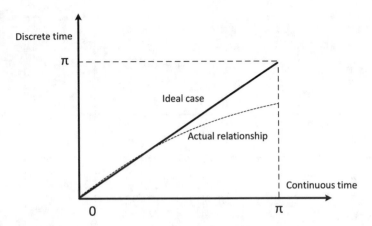

Fig. 7.8 Warping of frequency caused by the bilinear transform

rate) but warps the result if the sampling rate is about 3 times or less. To overcome this, we can simply *pre-warp* the frequency we desire by using (7.29). For $\theta = 0.2\pi$ we get $\omega = \frac{2}{T_s} \tan\left(\frac{0.2\pi}{2}\right)$ with $T_s = 0.333$ms, we get $\omega = 1.95 \times 10^3$rad/s or 310.3 Hz. We use this frequency instead of the original 1 kHz as the cut-off for our system. Our new time-constant is $T = 0.513$ ms. Our new digital filter is

$$G(z) = \frac{10T_s}{(T_s - 2T)} \left[\frac{z+1}{1 + z\left(\frac{T_s+2T}{T_s-2T}\right)} \right].$$

$$= 2.45 \frac{\left(1 + z^{-1}\right)}{\left(1 - 0.51z^{-1}\right)}$$

By substituting $z = 1$ we can find the dc gain to be approximately 10. The frequency response is shown in Fig. 7.9 and it has the desired characteristics.

This whole process is known as *frequency pre-warping*. For a lowpass filter we need only pre-warp the corner frequency, but for a bandpass filter we need to pre-warp both upper and lower band edge frequencies.

7.4 Impulse Invariance Method

This is a method of transforming from s to z domain transfer functions by using pole by pole partial fraction expansion. It doesn't work as well when there are zeros. Any LTI continuous time system can be expanded as partial fractions from its transfer function. However, if the filter is high pass then aliasing will occur and the method is not suitable. Each pole is transformed individually preserving the impulse response of each mode (or time constant). This is like the summation of n impulse responses for an nth order system. Each impulse response will be infinite in length (IIR) but it's not so much the impulse responses that we are after but their individual Z-transforms

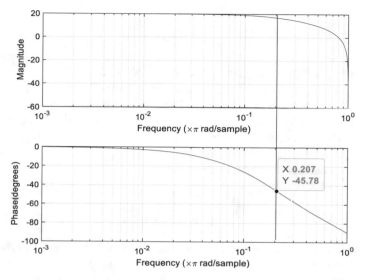

Fig. 7.9 Frequency response after pre-warping

which sum to give us the full Z-transfer function. For systems with complex poles we apply the same method by defining complex poles. Consider a system with n poles expressed as a partial fraction expansion:

$$G(s) = \sum_{i=1}^{n} \frac{a_i}{s - p_i} \tag{7.30}$$

The poles are defined with negative real part for stability. The inverse Laplace transform is then the individual impulse responses of the various modes. Note that $\mathcal{L}^{-1}\left\{\frac{1}{s|p}\right\} = e^{-pt}$ and we have a minus sign in the denominator of (7.30). Therefore

$$g(t) = \sum_{i=1}^{n} a_i e^{p_i t} \tag{7.31}$$

When g(t) is sampled with sampling interval T_s, its equivalent discrete-time form is

$$g(kT_s) = T_s \sum_{i=1}^{n} a_i e^{p_i kT_s} \tag{7.32}$$

Then we take Z-transforms and get a sum of equivalent discrete-time transfer functions.

$$G(z) = T_s \sum_{i=1}^{n} \frac{za_i}{z - e^{p_i T_s}} \tag{7.33}$$

It is worth pausing at this stage to examine (7.33) in more detail. What we have is that each continuous time pole p_i maps to $z_i = e^{p_i T_s}$ in the z-plane for $i = 1,2\ldots n$. This doesn't take account of zeros at all. The other more general point is that if the sampling interval is very small then via $z_i = e^{p_i T_s}$, the z-domain pole can become very close to unity causing possible instability problems with rounding. Hence too high a sampling interval is also a problem just as is too low!

Example: Consider a continuous time system:

$$G(s) = \frac{10^5}{(s + 100)(s + 1000)} \tag{7.34}$$

This is a second order system with two time constants which lead to a frequency response which is flat with unity gain ($|G(0)| = 1$) up to a corner frequency at $\omega = 100$ rad/s where it rolls off at -20 dB/decade up until it reaches a second corner frequency of $\omega = 1000$ rad/s a decade later and then rolls off with a slope -40 dB/decade. A suitable sampling rate might be $\omega_s = 3000$ rad/s or three times the second corner frequency of the system. This would give us a sampling interval of $T_s = 2.1$ ms. Expand (7.34) using partial fractions:

$$G(s) = \frac{111.11}{s + 100} - \frac{111.11}{s + 1000} \tag{7.35}$$

The poles are at $p_1 = -100$, $p_2 = -1000$. Using (7.33) we get

$$G(z) = T_s \left[\frac{111.111z}{z - e^{-100T_s}} - \frac{111.111z}{z - e^{-1000T_s}} \right]$$

$$= 0.0021 \left[\frac{111.111z}{z - 0.811} - \frac{111.111z}{z - 0.123} \right]$$

$$0.233 \left[\frac{z}{z - 0.811} - \frac{z}{z - 0.123} \right]$$

Adding gives

$$G(z) = \frac{0.16z}{z^2 - 0.934z + 0.099}$$

Figure 7.10a, b show respectively the frequency responses of the analogue and digital filter. They match well in magnitude but in phase there is some deviation. The two corner frequencies in continuous-time are 100 and 1000 rad/s whereas in discrete-time with the sample rate of $T_s = 2.1$ ms, these become normalised frequencies of 0.21 and 2.1 rad/sample (0.0668π and 0.668π on the scale of Fig. 7.10b).

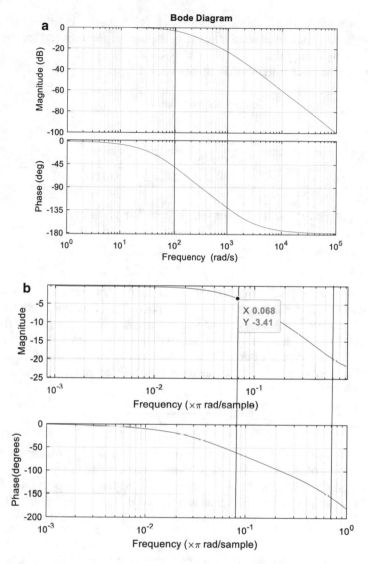

Fig. 7.10 **a** Bode plot of continuous-time transfer function example with two poles, **b** frequency response of frequency sampled discrete-time filter for two pole example

7.5 Analogue and Digital Lowpass Butterworth IIR Filters

The general diagram of how a filter works is shown by the lowpass prototype in Fig. 5.8. The Butterworth filter has no ripple in the passband and usually a gradual loss of 3 dB is designed for most filters. In other words, the filter is −3 dB down on 0 dB (or whatever the dc gain is) at the edge of the passband. We can either select a given attenuation in the stopband and find the order of the filter from this, or we can just select a filter order from the outset. For Butterworth filters a set of 3db Butterworth polynomials is readily available for design purposes and the method is relatively straight forward.

Butterworth filter polynomials up to order 4

Order n	Butterworth polynomial
1	$s + 1$
2	$s^2 + \sqrt{2}s + 1$
3	$s^3 + 2s^2 + 2s + 1$
4	$s^4 + 2.6131\,s^3 + 3.4142\,s^2 + 2.6131\,s + 1$

The prototype for a second order filter is therefore given by

$$H_p(s) = \frac{1}{s^2 + \sqrt{2}s + 1} \tag{7.36}$$

Note that a second order Butterworth polynomial is not two first orders and likewise a 4th order is not 2 second orders. The polynomials are not multiples of one another. The prototype is no good as a filter since it is like a normalised filter and only works for unity frequency passband. We must transform it depending on the filter type. For a lowpass filter we use the lowpass to lowpass transformation and substitute $s \rightarrow s/\omega_c$ into (7.36), where ω_c is the passband edge. We obtain

$$H(s) = \frac{\omega_c^2}{s^2 + \omega_c\sqrt{2}s + \omega_c^2} \tag{7.37}$$

This step is relatively easy, but for a real filter we need to realise the filter in hardware or in software if it is digital. There are many circuit topologies that give the same transfer function. However, here we will require the digital filter. Start by defining an example. Let the passband edge be 1 kHz for a second order filter. Then $\omega_c = 2000\pi = 6.28 \times 10^3$ rad/s. Substitute into (7.37) and we obtain (Fig. 7.11)

$$H(s) = \frac{3.95 \times 10^7}{s^2 + 8.886 \times 10^3 s + 3.95 \times 10^7} \tag{7.38}$$

The dc gain is unity and it has a slope of -40 dB/decade at high frequencies beyond the passband.

Now let us convert this system to an IIR digital filter by using first the Bilinear transform and then the impulse invariance method. Firstly, the Bilinear transform. Let us sample at $\times 10$ the passband edge of 1 kHz which makes a sampling frequency of $f_s = 10$ kHz. The sampling interval is therefore $T_s = 0.1$ ms. Substitute $s = \frac{2}{T_s}\left(\frac{z-1}{z+1}\right)$ into (7.38) and we get

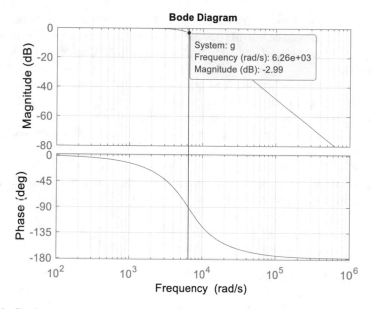

Fig. 7.11 Continuous time 2nd order butterworth filter with passband edge 1 kHz

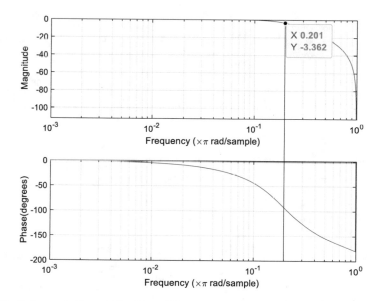

Fig. 7.12 Butterworth filter obtained from bilinear transform

$$H(z) = \frac{0.064z^2 + 0.1279z + 0.064}{z^2 - 1.1683z + 0.4241} \tag{7.39}$$

The frequency response is shown in Fig. 7.12 and has the desired characteristics, being -3.36 dB (should be -3 dB) down at a normalised frequency of $2\pi/10 = 0.2\pi$ rad/sample. The gain goes down asymptotically at -40 dB/decade as with any other two pole filter.

The impulse invariant method is a little trickier, but as an all-pole lowpass filter, is ideally suited for this method. Start with $H(s) = \frac{3.95\times 10^7}{s^2+8.886\times 10^3 s+3.95\times 10^7}$ and factorise the transfer function into two complex roots or poles.

Consider the more general second order case when the complex poles are $p_1, p_2 = -\sigma \pm j\omega$ found from $(s + \sigma - j\omega)(s + \sigma + j\omega) = s^2 + 2\sigma s + \omega^2 + \sigma^2 = s^2 + 8.886 \times 10^3 s + 3.95 \times 10^7$. Then solving gives $\sigma = 4.44 \times 10^3$, $\omega = 4.44 \times 10^3$.

Then we write the general problem as the partial fraction expansion of two complex-conjugate modes of the system

$$\frac{b}{s^2 + 2\sigma s + \omega^2 + \sigma^2} = \frac{a_1}{s + \sigma + j\omega} + \frac{a_2}{s + \sigma - j\omega}$$

By comparing coefficients of:

$$b = a_1(s + \sigma - j\omega) + a_2(s + \sigma + j\omega)$$

we get

$$a_1 + a_2 = 0$$

and

$$b = (a_1 + a_2)\sigma + (a_2 - a_1)j\omega$$
$$= (a_2 - a_1)j\omega$$

Then find $a_1 = jb/2\omega$, $a_2 = -jb/2\omega$.
Now use (7.33)

$$H(z) = T_s \sum_{i=1}^{2} \frac{za_i}{z - e^{p_i T_s}}$$

$$= T_s \left[\frac{jzb/2\omega}{z - e^{(-\sigma-j\omega)T_s}} - \frac{jzb/2\omega}{z - e^{(-\sigma+j\omega)T_s}} \right]$$

Taking a common factor

$$H(z) = T_s jb/2\omega \left[\frac{z}{z - e^{(-\sigma-j\omega)T_s}} - \frac{z}{z - e^{(-\sigma+j\omega)T_s}} \right]$$

$$= T_s jb/2\omega \left[\frac{z(z - e^{(-\sigma+j\omega)T_s}) - z(z - e^{(-\sigma-j\omega)T_s})}{(z - e^{(-\sigma-j\omega)T_s})(z - e^{(-\sigma+j\omega)T_s})} \right]$$

$$= T_s jb/2\omega \left[\frac{ze^{(-\sigma-j\omega)T_s} - ze^{(-\sigma+j\omega)T_s}}{\left(z - e^{(-\sigma-j\omega)T_s}\right)\left(z - e^{(-\sigma+j\omega)T_s}\right)} \right]$$

$$T_s jb/2\omega \left[\frac{ze^{-\sigma T_s}\left(e^{-j\omega T_s} - e^{j\omega T_s}\right)}{\left(z - e^{(-\sigma-j\omega)T_s}\right)\left(z - e^{(-\sigma+j\omega)T_s}\right)} \right]$$

$$T_s jb/2\omega \left[\frac{-2j\sin(\omega T_s)ze^{-\sigma T_s}}{z^2 - ze^{(-\sigma+j\omega)T_s} - ze^{(-\sigma-j\omega)T_s} + e^{-2\sigma T_s}} \right]$$

$$= T_s jb/2\omega \left[\frac{-2j\sin(\omega T_s)ze^{-\sigma T_s}}{z^2 - ze^{-\sigma T_s}\left(e^{j\omega T_s} + e^{-j\omega T_s}\right) + e^{-2\sigma T_s}} \right]$$

Finally, we arrive at

$$H(z) = \left[\frac{T_s b}{\omega} \right] \frac{ze^{-\sigma T_s}\sin(\omega T_s)}{z^2 - e^{-\sigma T_s}2z\cos(\omega' T_s) + e^{-2\sigma T_s}} \qquad (7.40)$$

We use the two Euler identities $e^{j\omega T_s} + e^{-j\omega T_s} = 2\cos(\omega T_s)$ and $e^{j\omega T_s} - e^{-j\omega T_s} = 2j\sin(\omega T_s)$ to obtain (7.40).

Substituting the real and imaginary values of the poles together with the sampling interval into (7.40) gives the impulse invariant Butterworth filter as

$$H(z) = \frac{0.2451\,z}{z^2 - 1.1585z + 0.4115}$$

Check the dc gain from $G(1) = 0.9688$ which is close to unity. The frequency response is shown in Fig. 7.13 and looks like the Bilinear frequency response. The passband edge frequency of $0.2\,\pi$ rad/sample gives nearly precisely the -3 dB drop as expected.

If more attenuation is needed it is tempting to use two Butterworth filters in cascade. This will of course still act as an excellent filter but can no longer count as a 4th order Butterworth. To make a 4th order we must start with the 4th order prototype. The 4th order one does not factorise into two 2nd order Butterworths. What this means is that that passband edge would be -6 dB with two 2nd order Butterworths in cascade verses -3 dB for a 4th order designed from the Butterworth polynomial. This is because the poles of a Butterworth are spaced equally depending on the order and having multiple poles (like two identical filters in cascade) is no longer Butterworth as the spacing will be the same for a higher order. Both 4th order filters will have a roll-off of -80 dB/decade however.

Fig. 7.14a shows the frequency response of an analogue 4th order filter made up of 2 s order Butterworths with 1 kHz passband edge. Note the -6 dB passband edge.

Fig. 7.14b Shows a true 4th order Butterworth digital filter found from the Butterworth polynomial prototype with 1 kHz passband edge. Note the passband edge is now -3 dB down and not -6 dB as with two identical Butterworth filters in cascade.

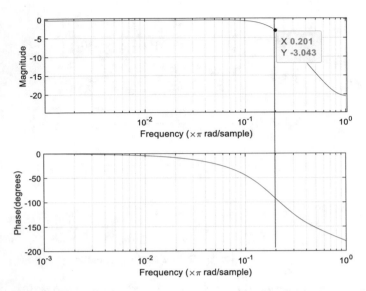

Fig. 7.13 Butterworth filter frequency response for impulse invariant method

7.6   Using the Z-Transform Tables Directly

We can use the tables instead of the other methods.

Consider a 2nd order Laplace transfer function with complex poles $s = -\sigma \pm j\omega$.

$$H(s) = \frac{b}{s^2 + 2\sigma s + \omega^2 + \sigma^2} \tag{7.41}$$

Which we write as

$$H(s) = \frac{b}{(s + \sigma)^2 + \omega^2} \tag{7.42}$$

From tables (Table 5.2) we know that from s to z we can write

$$\frac{\omega}{(s + a)^2 + \omega^2} \rightarrow \frac{e^{-aT}z\sin(\omega T)}{z^2 - 2e^{-aT}z\cos(\omega T) + e^{-2aT}} \tag{7.43}$$

But (7.42) does not have the correct numerator. So, we convert it to

$$H(s) = \left(\frac{b}{\omega}\right)\frac{\omega}{(s + \sigma)^2 + \omega^2} \tag{7.44}$$

Now we can use (7.43) and get

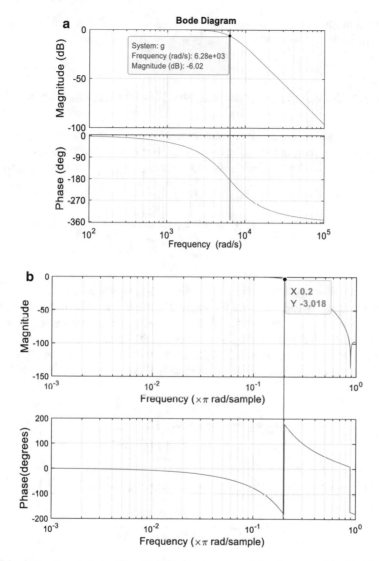

Fig. 7.14 a 4th order analogue filter frequency response made up from two identical 2nd order Butterworths with 1 kHz passband edge, **b** 4th order digital Butterworth filter frequency response with 1 kHz passband edge

$$H(z) = \left(\frac{b}{\omega}\right) \frac{e^{-\sigma T_s} z \sin(\omega T_s)}{z^2 - 2e^{-\sigma T_s} z \cos(\omega T_s) + e^{-2\sigma T_s}} \qquad (7.45)$$

This is the same result we obtained for the impulse invariant method (7.40) except for the omission of the sampling interval T_s on the numerator. Standard tables of Z-Transforms usually omit the sampling interval or make it unity. For example, the Z-transform of a Euler integrator in tables is given as $\frac{z}{z-1}$ when it should be $\frac{T_s z}{z-1}$.

Since most analysis omits the sampling interval it makes little difference until we design real filters.

7.7   Pole-Zero Mapping or Matched Z-Transform Method

Closely related to the impulse invariance method where each pole was mapped using $z_i = e^{p_i T_s}$, we do the same to the zeros as well. For m zeros and n poles (m $\leq$ n for causality) in the s-domain with transfer function $H_c(s)$ we get in the z-domain

$$H(z) = K \frac{\prod_{i=1}^{m} \left(s - e^{z_i T_s}\right)}{\prod_{i=1}^{n} \left(s - e^{p_i T_s}\right)} \tag{7.46}$$

where K is a gain to make sure the overall discrete-time dc gain H(1) matches that of the analogue gain $H_c(0)$. To make the method work, we must add a term to account for the number of zeros at infinity. We need to bring the frequency response down at half-sampling to avoid aliasing. A zero at infinity of an all-pole system is just a pole. The number of zeros at infinity of a system with poles and zeros is the number of poles—number of zeros. We can then modify (7.46) to give

$$H(z) = K(z + 1)^{n-m} \frac{\prod_{i=1}^{m} \left(s - e^{z_i T_s}\right)}{\prod_{i=1}^{n} \left(s - e^{p_i T_s}\right)} \tag{7.47}$$

Example: Consider a stable system with sampling interval $T_s = 2.1$ ms.

$$G(s) = \frac{(s + 10)10^4}{(s + 100)(s + 1000)} \tag{7.48}$$

It has a dc gain of $G(0) = 1$ or 0 dB. Two poles are at $p_1 = -100$, $p_2 = -1000$ and one zero is at $z_1 = -10$. It has 1 zero at infinity (two poles—one zero). Therefore, our discrete-time transfer function will be

$$H(z) = K(z + 1) \frac{\left(z - e^{z_1 T_s}\right)}{\left(z - e^{p_1 T_s}\right)\left(z - e^{p_2 T_s}\right)}$$

$$= K \frac{(z + 1)(z - 0.9792)}{(z - 0.8106)(z - 0.1225)}$$

We find that $H(1) = 0.25$ K which must be unity to match the continuous-time case. Therefore $K = 1/0.25 = 4$. Our discrete-time transfer function is therefore

$$H(z) = \frac{4(z+1)(z-0.9792)}{(z-0.8106)(z-0.1225)} \tag{7.49}$$

As can be seen from Figs. 7.15 and 7.16 the discrete approximation gets worse near half sampling although the step-responses are quite close. Usually this method is not the top of the list of methods used as it a bit ad-hoc in nature, especially the adding of the zeros at $z = -1$.

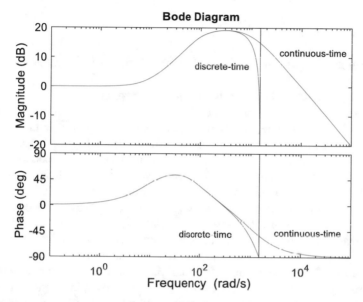

Fig. 7.15 Frequency response comparison of pole-zero mapping with continuous-time. vertical line is half sampling frequency

Fig. 7.16 Comparison of step-response for continuous and discrete-time using pole-zero mapping

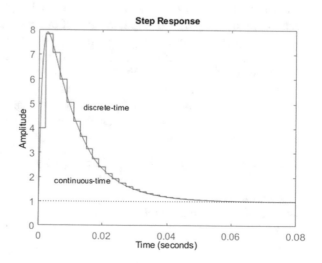

7.8   Implementing Difference Equations

Now that we have the Z-Transfer function we can implement the filter or whatever dynamic system we have designed in the z-domain. Consider a simple example of order 2. Let the transfer function be

$$G(z) = \frac{b_0 + b_1 z^{-1} + b_2 z^{-2}}{1 + a_1 z^{-2} + a_2 z^{-2}} \qquad (7.50)$$

The difference equation is given by

$$y(k) = -a_1 y(k-1) - a_2 y(k-2) + b_0 u(k) + b_1 u(k-1) + b_2 u(k-1) \quad (7.51)$$

This is often called a *recursive* digital filter or *IIR* filter.

To implement this equation in software we have a choice. We can either write code to implement it offline, in which case we are merely simulating the system, or we can write code to implement it in real-time (an embedded system). The real-time case is similar to the offline simulation except we need a real-time operating system. The real-time operating system will execute a while loop which will run forever (or as long as the embedded system is switched on). The loop must have duration equal to the sampling interval and must not change in any way. If the code for any reason cannot be executed within the time allocated the sampling interval) then there will be a real-time error and we must simplify the software in some way. Otherwise if the code runs slower than the sampling interval the processor will wait for the given duration and the execution will be completed. There are subtleties with real-time operating systems as each has its own quirks. However, in essence the method that follows will work on almost any processor. With field-programmable gate-array (FPGA) target systems the same method can be used except there is no operating system as such and the code is run in hardware gates after compilation.

Examining (7.51), we see that the current output is at time k and on the RHS of the equation we have inputs and output in terms of past values of k, specifically k−1 and k−2. The equation should always be written in this form. The a and b coefficients will of course be already calculated from theory using the methods from earlier parts of the book. Likewise, the sampling interval will be known. The difference equation does not have the sampling interval expressed explicitly within it, but it is assumed known, so that real-time is $t = kT_s$ where T_s is the sampling interval related to the sampling frequency via $f_s = 1/T_s$. For IIR filters of low order we can make the code very simple. Here it is shown in pseudo code.

Pseudo code for IIR filter *Define a1,a2,b0,b1,b2*
   % Initialise u0,u1,u2,y2,y1,y0 = 0
   Loop forever {
   % Shuffle past values of input
   u2 = u1
   u1 = u0

% Read new input sample
Read u0
% Shuffle past values of output
y2 = y1
y1 = y0
% Evaluate the latest output y(k)
y0 = −a1*y1−a2*y2 + b0*u0 + b1*u1 + b2*u2
% Output y0 as current output y(k)
out y0
}.

In the above code u0 is the current input sample u(k), u1 is u(k−1), u2 is u(k−2). Likewise, y2 = y(k−2), y1 = y(k−1), y0 = y(k) where y0 is the current output. The % symbol represents a comment statement only.

However, FIR filters are of the form

$$G(z) = b_0 + b_1 z^{-1} + b_2 z^{-2} + \ldots + b_m z^{-m} \qquad (7.52)$$

with no poles but can be of significant length. Therefore, the previous code is not suitable for lengthy filters.

The difference equation to be implemented is of the form

$$y(k) = b_0 u(k) + b_1 u(k-1) + b_2 u(k-2) + \ldots + b_m u(k-m) \qquad (7.53)$$

This is the convolution of an impulse response $\{ b_0\ b_1\ b_2\ \ldots\ b_m \}$ with the input data. It is called a *non-recursive* digital filter or *FIR* filter.

To implement the code for this kind of filter it is best to have two buffers or arrays, one for the b coefficients (often termed the weights of the filter) and one for past values of the input signal. This is like having two column vectors **W** and **X** both of length m + 1 where

$$\mathbf{W} = \begin{bmatrix} b_0\ b_1\ b_2\ \ldots\ b_m \end{bmatrix}^{\mathrm{T}} \qquad (7.54)$$

$$\mathbf{X} = \begin{bmatrix} u(k)\ u(k-1)\ u(k-2)\ \ldots\ u(k-m) \end{bmatrix}^{\mathrm{T}} \qquad (7.55)$$

and having as the output the dot product of the two vectors

$$y(k) = \mathbf{W}^{\mathrm{T}} \mathbf{X} \qquad (7.56)$$

Now the values of the **X** vector must change with time whereas **W** is constant. At each sample instant the elements of the **X** vector must shuffle to the right and the left most one u(k) which is the current value of input, must be replaced. This is like saying X(m) = X(m−1), X(m−1) = X(m−2) and so on until X(0) = u0.

Pseudo code for FIR filter
Define b0,b1,b2...bm,
   Define the array W as 0 to m elements of b coefficients in ascending order
   Initialise an array X of size m + 1 to zeros
   Loop forever {
   % Shuffle past values of input to right
   Loop: i = m to 1 step -1
   {
   X[i] = X[i-1]
   }.
   % Read most recent input and put into X[0]
   X[0] = Read current input sample u(0)
   % Multiply vectors element by element and sum to give current output
   Loop: j = 0 to m
   sum = 0.0
   {
   sum = sum + X[j]*W[j]
   }
   % current output y[k]
   y0 = sum
   out y0
   }.

There are much faster ways to implement this using a ring buffer, but the mathematical process is the same either way. Yet another way to perform this convolution is by use of the Fast Fourier Transform (FFT) since convolution in the time domain is multiplication in the frequency domain. The FFT method is a block method and gives an output of a whole vector of data at each sampling interval. When using MATLAB in later chapters there is no need to multiply vectors element by element like the above. This is because MATLAB has built in vector manipulation syntax.

7.9   The Discrete Fourier Transform (DFT)

There is a certain déjà vu when you see the name Discrete Fourier Transform (DFT) because you get the feeling you have already studied it earlier! However, we should not confuse the DFT with the discrete-time Fourier Transform (DTFT). The DTFT is a theoretical way of calculating the spectrum (magnitude and phase) properties of discrete-time signals and systems. It in no way helps us build a machine which does the same task. However, the DFT is just such an algorithm. It is designed for numerical calculation of the spectrum by dividing the spectrum into frequency "bins" or quantum chunks. To cover the entire normalised frequency-range from 0 to 2π into

N bins means that each frequency is represented as $2\pi(k/N)$ for $k = 0,1,2...N-1$. Unfortunately, we cannot go as far as N since we start at 0 (dc) and go to N−1. For example, for $N = 1024$ and a sampling frequency of 10 kHz, each frequency bin (the resolution) is (10,000/1024 or 9.76 Hz). To get more resolution means we need more samples and this in turn means more computations. If we wanted to measure frequencies to as low a resolution as 1 Hz with a sampling frequency 10 kHz we would need $N = 10,000$ points,hence increasing the observation interval. We could instead reduce the sampling frequency, but this would reduce the upper range of frequencies we can measure. With 10 kHz sampling frequency we can measure up to 5 kHz (the Nyquist frequency). The DFT is defined as

$$X(k) = \sum_{n=0}^{N-1} x(n)e^{-j2\pi kn/N}, k = 0, 1\ldots N-1 \tag{7.57}$$

where $x(n)$ and $X(k)$ are elements of vectors in time and frequency respectively of length N elements. That is for N samples of data, $\mathbf{x} = \begin{bmatrix} x(0) \ x(1) \ .. \ x(N-1) \end{bmatrix}^T$ and $\mathbf{X} = \begin{bmatrix} X(0) \ X(1) \ .. \ X(N-1) \end{bmatrix}^T$. We interpret $x(0)$ as the first sample in time and $X(0)$ represents dc in the frequency-domain. The frequency values repeat after N/2 and represent *negative frequencies* in continuous time.

The inverse DFT is found from

$$x(n) = \frac{1}{N} \sum_{k=0}^{N-1} X(k)e^{j2\pi kn/N}, n = 0, 1\ldots N-1 \tag{7.58}$$

A short-hand way of writing the complex exponential term is to define the "twiddle factor"

$$W_N = e^{-j2\pi/N} \tag{7.59}$$

Then we write the DFT as

$$X(k) = \sum_{n=0}^{N-1} x(n)W_N^{nk}, k = 0, 1\ldots N-1 \tag{7.60}$$

For example, for $N=4$ the twiddle factor is $W_4 = e^{-j2\pi/4} = e^{-j\pi/2} = \cos(\pi/2) + j\sin(-\pi/2) = -j$. Raising this to a power of zero the value is 1 or $W_4^0 = 1$. If we square the Twiddle factor we get $W_4^2 = (-j)^2 = -1$, cube it we get $W_4^3 = j$. This is just a phasor of magnitude 1 sweeping round a circle of unit radius jumping −90 degrees (clockwise) at each power of the twiddle factor. If we increase the powers of the twiddle factor, then it just repeats itself over and over. For example: $W_4^4 = 1, W_4^5 = -j, W_4^6 = -1, W_4^7 = j, W_4^8 = 1$.

For different values of N this phasor does the same thing but at different angles. For example, if $W = 8$ we have $W_8 = e^{-j2\pi/8} = e^{-j\pi/4} = \cos(-\pi/4) + j\sin(-\pi/4) = \frac{1}{\sqrt{2}} - \frac{1}{\sqrt{2}}j$ (this is a phasor at -45 degrees). Squaring it gives $W_8^2 = e^{-j4\pi/8} = e^{-j\pi/2} = j$ and the phasor has jumped another -45 degrees around a circle of unit radius. What we have is essentially the Nth square root of 1 for complex numbers. Learning this simple fact is the key to solving simple DFT problems (Fig. 7.17).

If we make N divisible by 2, then the twiddle factor has three properties. It satisfies:
Periodicity

(a)

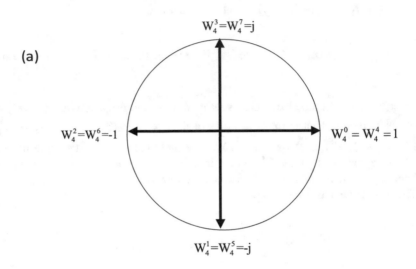

(b)

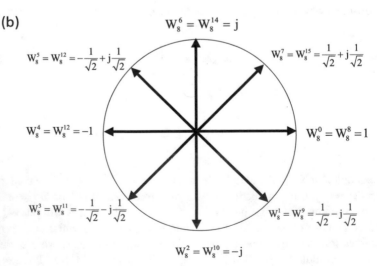

Fig. 7.17 Twiddle factor for **a** $N = 4$ and **b** $N = 8$

$$W_N^{n+N} = W_N^n \tag{7.61}$$

or

$$W_N^n = W_N^{n\pm N} \tag{7.62}$$

Decimation

$$W_{mN}^{mn} = W_N^n \tag{7.63}$$

Symmetry

$$W_N^{n+N/2} = -W_N^n \tag{7.64}$$

Usually we assign $N = 2^p$ where p is an integer, and this can exploit the twiddle factor properties to the fullest.

The DFT itself is periodic (this follows from the properties of the twiddle factor) and satisfies

$$X(k + N) = X(k) \tag{7.65}$$

and likewise

$$x(n + N) = x(n) \tag{7.66}$$

for the inverse DFT.

For (7.64), frequency values at indices greater than $k > N/2$ represent negative frequencies, and these are usually ignored since they are the same as the first $N/2$ values.

7.9.1 DFT Example 1

Find the DFT of the 4-point sequence $x(n) = \{1, 1, 1, 1\}$.
    Solution. From (7.57) or (7.59)

$$X(k) = \sum_{n=0}^{3} x(n)e^{-j2\pi kn/4}$$

$k = 0$

$$X(0) = \sum_{n=0}^{3} x(n), k = 0, 1, 2, 3$$
$$= 1 + 1 + 1 + 1$$
$$= 4$$

$k = 1$

$$X(1) = \sum_{n=0}^{3} x(n) W_4^n$$
$$= x(0) W_4^0 + x(1) W_4^1 + x(2) W_4^2 + x(3) W_4^3$$
$$= 1 + (-j) + (-1) + (j)$$
$$= 0$$

$k = 2$

$$X(2) = \sum_{n=0}^{3} x(n) W_4^{2n}$$
$$= x(0) W_4^0 + x(1) W_4^2 + x(2) W_4^4 + x(3) W_4^6$$
$$= 1 + (-1) + (1) + (-1)$$
$$= 0$$

$k = 3$

$$X(3) = \sum_{n=0}^{3} x(n) W_4^{3n}$$
$$= x(0) W_4^0 + x(1) W_4^3 + x(2) W_4^6 + x(3) W_4^9$$
$$= 1 + j + (-1) + (-j)$$
$$= 0$$

The answer is that $\mathbf{X} = \begin{bmatrix} 4 & 0 & 0 & 0 \end{bmatrix}^T$ in the frequency domain, a vector made up of the individual frequency bin components we just calculated. This tells us that at 0 Hz or dc there is magnitude 4 and the other three frequencies at $f_s/4$, $f_s/2$, $3f_s/4$ have zero power.

7.9.2 DFT Example 2

Consider a cosine wave $x(n) = \cos\left(\frac{2\pi n}{N} k_o\right)$, $n = 0, 1, 2 \ldots N - 1$. Here k_o represent the number of complete cycles of the cosine wave for N samples of data. For the frequency values $X(k)$, we can consider the maximum frequency to be last element at N-1. Let us assign k_o at one quarter the sampling frequency or $k_o = N/4$ in terms of frequency bins. Then $x(n) = \cos\left(\frac{\pi n}{2}\right)$, $n = 0, 1, 2, 3$. We have for $N = 4$, $k_o = N/4 = 1$ and the elements of the discrete time vector are

$$x(n) = \cos\left(\frac{\pi n}{2}\right)$$
$$= \{1, 0, -1, 0\}$$

Use $X(k) = \sum_{n=0}^{3} x(n) e^{-j2\pi kn/4}$ for the DFT.

$k = 0$

$$X(0) = \sum_{n=0}^{3} x(n)$$
$$= 0$$

$k = 1$

$$X(1) = \sum_{n=0}^{3} x(n) W_4^n$$
$$= x(0) W_4^0 + x(1) W_4^1 + x(2) W_4^2 + x(3) W_4^3$$
$$= 1 + 0.(-j) + (-1).(-1) + 0.(j)$$
$$= 2$$

$k = 2$

$$X(2) = \sum_{n=0}^{3} x(n) W_4^{2n}$$
$$= x(0) W_4^0 + x(1) W_4^2 + x(2) W_4^4 + x(3) W_4^6$$
$$= 1 + 0.(-1) + (-1).(1) + 0.(-1)$$
$$= 0$$

$k = 3$

$$X(3) = \sum_{n=0}^{3} x(n) W_4^{3n}$$

Fig. 7.18 Magnitude of the
4-point DFT of example 2

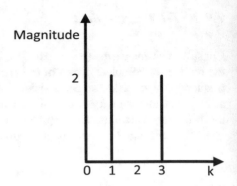

$$= x(0)W_4^0 + x(1)W_4^3 + x(2)W_4^6 + x(3)W_4^9$$
$$= 1 + 0.j + (-1).(-1) + 0.(-j)$$
$$= 2$$

Note there are no imaginary terms for this example or the previous. This indicates no phase values or zero phase-shift.

The solution is that the answer is a vector $\mathbf{X} = \begin{bmatrix} 0 & 2 & 0 & 2 \end{bmatrix}^T$. We don't need all the values in the solution since for $k = 2,3$ the values are negative frequencies and will be a repeat of what we have during $k = 0,1$. To make sense of these values we need to divide their amplitudes by $N = 4$ to give two impulses in the frequency domain of magnitude 0.5 each. If we want a single-sided spectrum we double these values to give us a single impulse of magnitude 1 at frequency bin 1. This is at a frequency ¼ of the sampling frequency (since it is the value corresponding to $k = 1$ i.e. N/4 that the impulse occurs) (Fig. 7.18).

7.9.3 DFT Example 3. Inverse DFT of an Impulse in the Frequency Domain

Consider an impulse in the frequency domain at sample bin k_o. In the frequency domain it is defined as $X(k) = \delta(k - k_o)$. Its inverse DFT is then

$$x(n) = \frac{1}{N} \sum_{k=0}^{N-1} X(k)e^{j2\pi kn/N}$$

$$= \frac{1}{N} \sum_{k=0}^{N-1} \delta(k - k_o)e^{j2\pi kn/N}$$

$$= \frac{1}{N} e^{j2\pi k_o n/N}$$

The importance of this result is that inverse of the above, namely that the direct DFT of the complex sinusoid $e^{j2\pi k_o n/N}$ is $N\delta(k - k_o)$. That is

$$\mathcal{F}\{e^{j2\pi k_o n/N}\} = N\delta(k - k_o) \tag{7.67}$$

7.9.4 DFT of a Cosine $\mathcal{F}\{\cos(\frac{2\pi n}{N}k_o)\}$, $n = 0, 1, 2 \ldots N - 1$

Where the cosine represents a data length of N samples with k_o complete samples within it.

Write using Euler's identity

$$\cos\left(\frac{2\pi n}{N}k_o\right) = \frac{1}{2}e^{j\frac{2\pi n k_o}{N}} + \frac{1}{2}e^{-j\frac{2\pi n k_o}{N}} \tag{7.68}$$

The DFT becomes

$$X(k) = \sum_{n=0}^{N-1} x(n)e^{-j2\pi kn/N}$$

$$-\frac{1}{2}\sum_{n=0}^{N-1} e^{j\frac{2\pi n(k_o-k)}{N}} + e^{-j\frac{2\pi n(k+k_o)}{N}}$$

$$= \frac{1}{2}\sum_{n=0}^{N-1} e^{-j\frac{2\pi n(k-k_o)}{N}} + e^{-j\frac{2\pi n(k+k_o)}{N}} \tag{7.69}$$

But we know from the periodicity property of the twiddle factor $W_N^n = W_N^{n\pm N}$ so we can write

$$e^{-j\frac{2\pi n(k+k_o)}{N}} = e^{-j\frac{2\pi n[(k+k_o)-N]}{N}}$$

$$= e^{-j\frac{2\pi n[k-(N-k_o)]}{N}} \tag{7.70}$$

Therefore (7.68) becomes

$$X(k) = \frac{1}{2}\sum_{n=0}^{N-1} e^{-j\frac{2\pi n(k-k_o)}{N}} + e^{-j\frac{2\pi n[k-(N-k_o)]}{N}}$$

$$= \frac{1}{2}\sum_{n=0}^{N-1} \left[e^{j\frac{2\pi n k_o}{N}} + e^{j\frac{2\pi n(N-k_o)}{N}}\right]e^{-j\frac{2\pi nk}{N}}$$

$$= \frac{1}{2}\mathcal{F}\{e^{j\frac{2\pi n k_o}{N}}\} + \frac{1}{2}\mathcal{F}\{e^{j\frac{2\pi n(N-k_o)}{N}}\} \tag{7.71}$$

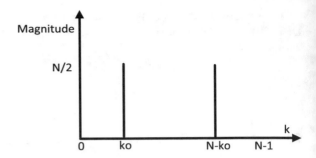

Fig. 7.19 DFT magnitude of a cosine or sine waveform

Giving the final result

$$\mathcal{F}\left\{\cos\left(\frac{2\pi n}{N}k_o\right)\right\} = \frac{N}{2}\delta(k - k_o) + \frac{N}{2}\delta(k - (N - k_o)) \qquad (7.72)$$

This is a very similar result to the continuous time apart from the scaling by N/2. See Fig. 7.19. For a single sided spectrum, the division by two would be omitted.

We can do the same for a sinewave and get

$$\mathcal{F}\left\{\sin\left(\frac{2\pi n}{N}k_o\right)\right\} = \frac{1}{2j}\mathcal{F}\left\{e^{j\frac{2\pi nk_0}{N}}\right\} - \frac{1}{2j}\mathcal{F}\left\{e^{j\frac{2\pi n(N-k_o)}{N}}\right\}$$

$$= \frac{N}{2j}\delta(k - k_o) - \frac{N}{2j}\delta(k - (N - k_o))$$

Which only effects the phase and not the magnitude. We also note that for both cases if k_o is not an integer that a problem occurs since the impulse cannot exist between samples. In practice a smearing occurs between adjacent samples which leads to inaccuracies in measurement.

7.9.5   Matrix Form of the DFT

The DFT has a matrix form. For example, for $N = 4$ we can write the relationship between frequency and time for sampled signals as

$$\mathbf{X} = \mathbf{W}_N\mathbf{x} \qquad (7.73)$$

Or

$$\begin{bmatrix} X(0) \\ X(1) \\ X(2) \\ X(3) \end{bmatrix} = \begin{bmatrix} W_N^0 & W_N^0 & W_N^0 & W_N^0 \\ W_N^0 & W_N^1 & W_N^2 & W_N^3 \\ W_N^0 & W_N^2 & W_N^4 & W_N^6 \\ W_N^0 & W_N^3 & W_N^6 & W_N^9 \end{bmatrix}\begin{bmatrix} x(0) \\ x(1) \\ x(2) \\ x(3) \end{bmatrix} \qquad (7.74)$$

If we put the numbers in

$$\mathbf{W}_4 = \begin{bmatrix} 1 & 1 & 1 & 1 \\ 1 & -j & -1 & j \\ 1 & -1 & 1 & -1 \\ 1 & j & -1 & -j \end{bmatrix} \tag{7.75}$$

This is known as a Vandermonde matrix and is a so-called unitary matrix satisfying

$$\frac{1}{N}\mathbf{W}_N\mathbf{W}_N^* = \mathbf{I} \tag{7.76}$$

where the star superscript denotes complex-conjugate transpose (Hermitian transpose) of the matrix and $\mathbf{I}$ is the identity matrix. This equation states that the columns of $\mathbf{W}_N$ are orthogonal. (unitary means orthogonal when a matrix is complex). Orthogonality implies that any column vector $\mathbf{V}_i$ when multiplied (dot multiplication or inner product) by another $\mathbf{V}_j$ gives an answer zero, or $\mathbf{V}_i^T\mathbf{V}_j = 0$ Inverting (7.72) we get

$$x = \mathbf{W}_N^{-1}\mathbf{X}$$
$$= \frac{1}{N}\mathbf{W}_N^*\mathbf{X} \tag{7.77}$$

Which is the inverse DFT. Equation (7.75) also implies that we could split the N between each of the two matrices, the first one and its conjugate transpose. Therefore, in some textbooks the DFT is defined as having $\frac{1}{\sqrt{N}}$ scaling for both forward and inverse transforms instead of what we have here.

7.10 The Fast Fourier Transform (FFT)

The reason for the name FFT isn't that the algorithm gives us anything different from the DFT as a solution, in fact it gives the same solution as the DFT. The only difference is the computational saving. This becomes of importance in modern instrumentation which use the FFT in real-time. For example, many modern oscilloscopes use the FFT for spectral analysis. There are a great many subtleties with the FFT and only the basic principle is covered here. There are two methods in common use to speed up the DFT. One is called *Decimation in time* and the other *Decimation in Frequency*. Since both give the same computational saving, we only consider the first of these. The first thing about the FFT is that we require the number of samples to be an integer power of 2. That is 2,4,8,16, 32,64 and so on. If we have 170 points of data, we then have a choice of using 128 points of data and discarding the rest by truncation or by using 256 and filling the missing data with zeros. The latter of these is the usual approach and does not waste important data. Integer powers of 2 are of course divisible by 2 and this is the key to the saving in computation. We use a divide and conquer approach which in modern times was first derived by two mathematicians Cooley and Tukey in 1965 [2]. Like all great ideas it appears that other had also had the same idea but not realised the importance of it. Carl Friedrich Gauss in 1805 had already had the idea and at least two others, Lancos and Danielson had similar ideas in the 1940s. The ordinary DFT has complexity $O(N^2)$ whereas the FFT has complexity $O(NLog_2N)$, a considerable saving in computer power. For example, for $N = 2048$ data samples we would need 4,194,304 operations for the DFT versus 22,528 for the FFT.

Consider an N point DFT

$$X(k) = \sum_{n=0}^{N-1} x(n)e^{-j2\pi kn/N}, k = 0, 1 \ldots N - 1 \tag{7.78}$$

Consider the even and odd terms separately.

$$X(k) = \sum_{\substack{n=0 \\ n\,even}}^{N-2} x(n)e^{-j2\pi kn/N} + \sum_{\substack{n=0 \\ n\,odd}}^{N-1} x(n)e^{-j2\pi kn/N}$$

$$= \sum_{n=0}^{N/2-1} x(2n)e^{-j2\pi kn/(N/2)} + \sum_{n=0}^{N/2-1} x(2n+1)e^{-j2\pi kn/(N/2)} \tag{7.79}$$

$$X(k) = \sum_{n=0}^{N/2-1} x_E(n)W_{N/2}^{kn} + W_N^k \sum_{n=0}^{N/2-1} x_O(n)W_{N/2}^{kn} \tag{7.80}$$

Note that in the above we have split the N point DFT into 2 N/2 point DFTs. One acts on even indices and the second on odd, hence the subscripts E and O. It is important to see that the twiddle factors are the same in both DFTs and the only difference is the multiplication by W_N^k before the second. This is called the Danielson-Lancos Lemma after the two scientists who made contributions to the field in 1948 [3]. Also, for a DFT of length N we need N^2 calculations whereas for two of length N/2 we need $2\left(\frac{N}{2}\right)^2 = \frac{N^2}{2}$. This is a saving already, but we can use the same idea and split the N/2 into N/4 and so on until we reach 2-point DFTs.

For example, if $N = 8$ we would perform two 4 points DFTs and then 4 two-point DFTs. The two-point DFT is of special notoriety because of its simplicity.

$$X(k) = \sum_{n=0}^{1} x(n)e^{-j2\pi kn/2}, k = 0, 1 \tag{7.81}$$

For $k = 0$

$$X(0) = x(0) + W_2^0 x(1)$$

For k = 1

$$X(0) = x(0) - W_2^0 x(1)$$

This is illustrated in Fig. 7.20 as a signal flow graph. From the look of the shape it has the appearance of a butterfly. In the diagram the circled nodes indicate summation and the arrows are multiplication. Often if there is multiplication by 1 there is no number shown. Here the numbers are shown for clarity.

This is only addition and subtraction and computationally simple to implement. Omitted from Fig. 7.20 is twiddle factor W_2^0 since $W_2^0 = 1$.

A 4-point FFT has a flow graph as shown in Fig. 7.21. Included this time are the W_2^0 and the W_4^0 twiddle factors which are both unity. This is to show the flow of data for different stages.

The 4 inputs do not appear in order but are arranged due to a special method called "bit-reversal". To find the order for the inputs we write a sequence of binary numbers from 0 to 4. We require 2 bits for a 4-point FFT and the normal binary order in ascending order is 00,01,10,11. With bit reversal this becomes 00,10,01,11 or 0,2,1,3 in decimal. Apply the same technique to any order of FFT. In Fig. 7.21 there are two stages, stage one 2-point DFTs on the left and stage two is one 4-point DFT on the right.

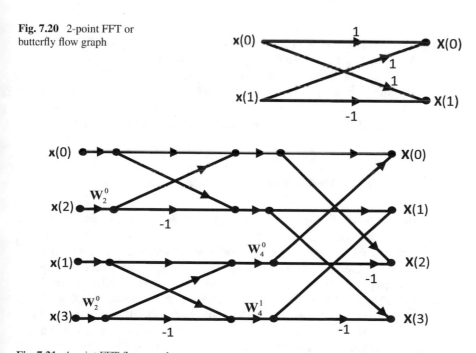

Fig. 7.20 2-point FFT or butterfly flow graph

Fig. 7.21 4-point FFT flow-graph

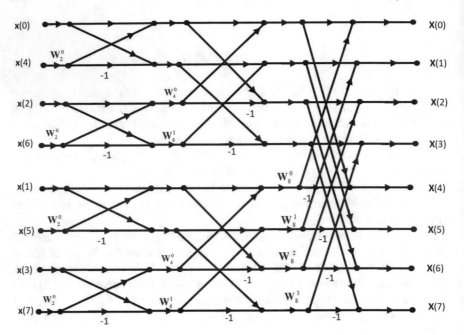

Fig. 7.22 An 8-point FFT showing three stages

For an 8-point FFT we will need 3 stages. Stage one will be four 2-point DFTs. Stage two will be two 4-point FFTs and stage three will be one 8-point DFT.

In Fig. 7.22, the input order is once again found by bit reversal. We need three bits for an 8 point FFT so with ordinary binary the numbers are 000,001,010,011,100,101,110,111 and this becomes 000,100,0010,110,001,101,011,111. In decimal the inputs are 0,4,2,6,1,5,3,7.

7.10.1   Spectral Leakage and Windowing

The Fourier transform theoretically assumes an infinite amount of data from time zero to infinity, but the FFT truncates this to some N samples. This obviously has some affect, and it turns out that if a periodic waveform is periodic within the N point time period then the results are best. This is because an integer number of cycles will occur within N samples. For example if 10 complete cycles of a sinewave are samples starting when the waveform has zero amplitude and finishing at zero amplitude, then we say the waveform is periodic within the time period. It is more likely in practise that this is not the case. When best the energy is distributed at a frequency bin which is an integer quantity. It could be any bin between 0 and N/2. (Usually we omit the upper N/2 + 1 to N frequency bins since they repeat the first half) If a fractional frequency is selected not on a bin then the FFT cannot place it precisely and smearing will be the result. This is known as *spectral leakage*. To illustrate this, consider a sampled cosine wave.

$$x(n) = \cos\left(\frac{2\pi n}{N}k_o\right), n = 0, 1, 2 \ldots N - 1 \qquad (7.82)$$

Remember that such a waveform has an impulse at the desired frequency. In the above equation we can select k_o as the frequency bin at which there is an impulse. Frequency here is normalised for convenience of the simulation and confined to any bin from $k_o = 0, 1 \ldots N/2$. Anything beyond this value will give aliasing. Let $N = 256$ for a 256-point FFT and $k_o = 40$. Evaluating the FFT in MATLAB gives the result of Fig. 7.23.

The FFT gives perfect results as expected since the frequency value is on a bin. Now change the frequency so that it is not on a bin. Let $k_o = 40.5$ and repeat the process. The result is shown in Fig. 7.24.

We can see that what should be an impulse is now quite broad and leakage occurs from the bin at 40 to adjacent bins. The amplitude is also in error.

To solve this problem a window is usually convolved with the data. There already is a window for an ordinary FFT, if the data is let through and abruptly cut off it is known as a *rectangular window* or *uniform window*. The periodic waveform must start and finish at the same point of the waveform. If instead of a rectangular window, we gradually attenuate the data at each end, so it starts at zero and ends at zero then we get a similar affect to the ideal. It may not be perfect, but it does stop the problem

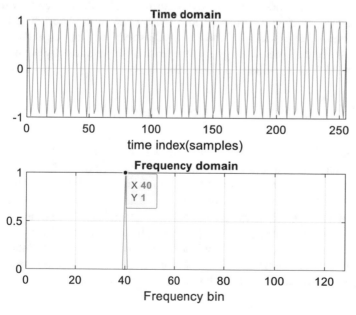

Fig. 7.23 FFT of a cosine wave at frequency bin 40 for $N = 256$ samples time record

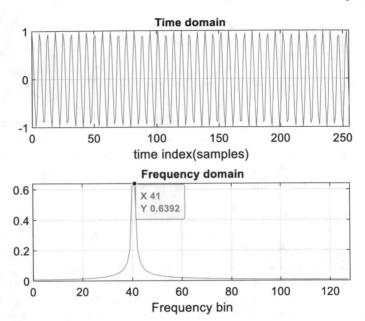

Fig. 7.24 When $k_o = 40.5$ spectral leakage occurs. $N = 256$

of spectral leakage. The same example as Fig. 7.24 was repeated but using a Hann window. The results are shown in Fig. 7.25.

Note that in the time-domain the waveform now begins and ends with amplitude zero. We can now say it is periodic within the time period. The frequency measurement is not precise, but the leakage has reduced from adjacent frequency bins. Previously there was leakage up to at least bin 60 and now by bin 42 it is very much reduced. The equation of a Hann window is

$$w(n) = \frac{1}{2}\left(1 - \cos\left(2\pi\frac{n}{N}\right)\right) \tag{7.83}$$

and has the appearance of a raised cosine waveform as shown in Fig. 7.26.

If the window is multiplied by the incoming data, then multiplication in the time domain is convolution in the frequency domain. It acts like a filtering effect as shown by the spectrum of the window in Fig. 7.26. There are many different kinds of windows in use in modern day spectrum measurement instrumentation. The chosen window will depend on the particular problem in hand. Windows affect the amplitude as well as frequency content. Some windows give better amplitude measurement and poor frequency and others vice-versa. There are a plethora of books and papers written on the subject. Perhaps one of the better references is by Harris [4].

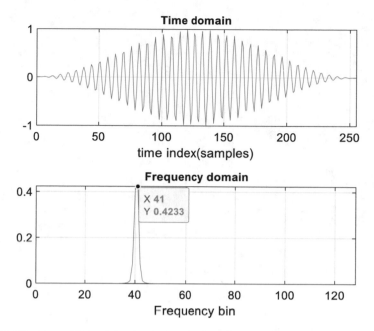

Fig. 7.25 When $k_o = 40.5$ and data is windowed with a Hann window. $N = 256$

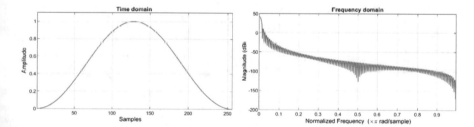

Fig. 7.26 Hann window and its spectrum

References

1. A.V. Oppenheim, R.W. Schafer, in *Digital Signal Processing*. (Prentice Hall International Ltd, 1975)
2. J.W. Cooley, J.W. Tookey, An algorithm for the machine calculation of complex fourier series. Math. Comput. **19**, 297–301 (1965)
3. G.C. Danieldson, C. Lancos, Some improvements in practical Fourier analysis and their application to X-ray scattering from liquids. J. Franklin Instit. **233**(5), 435–452 (1942)
4. F. Harris, On the use of windows for harmonic analysis with the discrete fourier transform. Proc. IEEE **66**(1), 51–83 (1978)

Chapter 8
FIR Filter Design

8.1 Definition of Linear Phase

Perfect linear phase is a property only found in digital filters. Linear phase cannot be made in the analogue world except through approximations. In continuous time linear phase is defined as a filter

$$H(j\omega) = |H(j\omega)|e^{-j\omega\tau} \tag{8.1}$$

which has amplitude (gain) as a function frequency $|H(j\omega)|$ and phase as a function of frequency $\phi(\omega) = -\omega\tau$.

By differentiating the phase term w.r.t frequency we get the group delay

$$\frac{d\phi(\omega)}{d\omega} = -\tau \tag{8.2}$$

If the system had unity gain at all frequencies and just the delay, then it would be a pure time-delay and all frequencies from a signal would pass with no change in amplitude. Their phases would be changed in such a way that the signal would appear identical at the output, it would only be delayed in time by τ seconds. Another example is that of an amplitude modulated waveform. With a linear phase filter the envelope of the waveform will be delayed but not the carrier [1]. Linear phase filters are essential for preserving edges in digital waveforms.

In discrete time systems we have a linear-phase system defined as

$$H(e^{-j\theta}) = |H(e^{-j\theta})|e^{-jk\theta/2} \tag{8.3}$$

where $|H(e^{-j\theta})|$ is the magnitude expressed as a function of normalised frequency $0 \le \theta \le 2\pi$ or $-\pi \le \theta \le \pi$. This is a periodic function which repeats after half sampling π rads/sample. The phase is expressed as

T. J. Moir, *Rudiments of Signal Processing and Systems*,
https://doi.org/10.1007/978-3-030-76947-5_8

$$\phi(\theta) = -k\theta/2 \tag{8.4}$$

Here k is an integer whose value depends on the filter. The phase is a straight line with negative gradient for positive k and its slope is the group delay.

Differentiating we find the group delay is

$$\frac{d\phi(\theta)}{d\theta} = -k/2 \tag{8.5}$$

Generalized linear phase is not true linear phase, but close to it and sometimes filters can have this as a good match. It is defined as a system

$$H(e^{-j\theta}) = \left|H(e^{-j\theta})\right|e^{-jka\theta+jc} \tag{8.6}$$

where *a and c* are constants. This can happen when the phase switches by say $\pi/2$ radians. This often happens with sinc() type functions which appear frequently in digital filters.

An important property is that if a discrete impulse response is real and symmetric (in the time domain) then it will be complex and conjugate symmetric in the frequency domain. By conjugate symmetric we mean that the phase is the complex conjugate (it flips in sign) after half sampling. Although we only use frequencies up to half sampling, we need all the information up to 2π if we are to work backwards from the frequency domain to the time domain. The simplest way to do this is by frequency sampling. Although this is not the best method, it is essential for understanding more rigorous approaches to FIR filter design.

8.2   Frequency Sampling Method of FIR Filter Design

FIR filters are important because they have only poles at the origin and therefore never unstable. With finite precision arithmetic IIR filters can sometimes have problems with rounding errors and make the filter unstable. This is not the case with FIR filters in the sense that if there are problems the filter will always be stable. The frequency sampling method is just a selection of N complex points of a desired frequency response. To get the impulse response we use the inverse DFT. Choosing the samples however must be done with care since to achieve linear phase we must design based on one of 4 basic properties. It turns out that any FIR filter with conjugate symmetric frequency response will have linear phase. We can choose from any of the properties shown in Fig. 8.1 [2–4]. These are the real filter coefficients (or impulse response) and not frequency domain samples. In what follows we use the notation that the index k is for discrete time-domain and n is for frequency-domain bins.

Suppose we have an FIR filter defined by

$$H(z^{-1}) = b_o + b_1z^{-1} + b_2z^{-2} + \ldots + b_Nz^{-(N-1)}$$

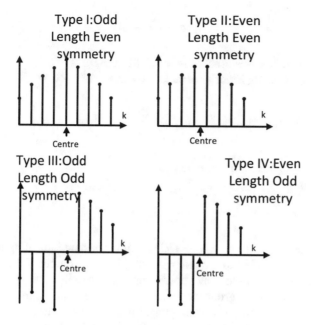

Fig. 8.1 For a filter with N coefficients $H(z^{-1}) = b_o + b_1 z^{-1} + b_2 z^{-2} + \ldots + b_{N-1} z^{-(N-1)}$. Four different kinds of FIR filter coefficient symmetry for b_k, $k = 0, 1 \ldots N - 1$ give linear phase

where N is the total number of coefficients and $N - 1$ is the order of the filter. The idea here is to know all 4 possibilities and develop design templates for each one. If we know this information, we can choose the correct filter parameters to fit the problem.

Consider the 4 cases in Fig. 8.1.

Type I Odd length even symmetry.

For N odd samples the frequency response is found from

$$H(e^{-j\theta}) = \sum_{n=0}^{N-1} b_n e^{-j\theta n} \tag{8.7}$$

$$H(e^{-j\theta}) = b_0 e^{-j\theta x 0} + b_1 e^{-j\theta} + b_2 e^{-j2\theta} + \ldots + b_{N-1} e^{-j\theta(N-1)} \tag{8.8}$$

But half the coefficients are equal to the other half, $(b_0 = b_{N-1}, b_1 = b_{N-2}$ etc.) therefore $b_{N-1-n} = b_n$ and

$$H(e^{-j\theta}) = e^{-jM\theta}[b_0(e^{jM\theta} + e^{-jM\theta}) + b_1(e^{j(M-1)\theta} + e^{-j(M-1)\theta}) + \ldots + b_M] \tag{8.9}$$

where

$$M = \frac{N-1}{2} \tag{8.10}$$

is an even number. Note also that $N - 1 = 2M$ in (8.9).

Then from (8.9) using Euler's formula $2\cos(n\theta) = e^{jn\theta} - e^{-jn\theta}$

$$H\left(e^{-j\theta}\right) = e^{-j\theta M}\left[b_M + 2\sum_{k=0}^{M-1} b_k \cos[(M-k)\theta] \right] \tag{8.11}$$

where $b_k = b_{N-1-k}$ for even symmetry.

The phase is

$$\phi(\theta) = -\theta M \tag{8.12}$$

and the group delay $- M = - (N - 1)/2$ which is constant indicating linear phase. This means the signal will be delayed by M samples at the output of the FIR filter. To make such a filter we have to select coefficients that are conjugate symmetric in the frequency domain and the group delay must be $-M$. The phase must be a straight line with a slope $- M\pi$ up to half sampling. From 0 to 2π we have each normalised frequency bin discretised as $\theta = \frac{2\pi n}{N}, n = 0, 1, 2\ldots$. So, the phase becomes from (8.12) by substitution

$$\phi(\theta) = -\theta M$$
$$= -\frac{2\pi n}{N}M$$
$$= -\left(\frac{2\pi n}{N}\right)\left(\frac{N-1}{2}\right)$$
$$= -\frac{\pi n(N-1)}{N}$$

Giving

$$\phi(\theta) = -\frac{\pi n(N-1)}{N} \tag{8.13}$$

We must therefore select phase for each frequency bin as $\phi(n) = -n\pi\frac{(N-1)}{N}, n = 0, 1, 2\ldots N/2 - 1$, and the conjugate for the other half of the samples.

If we select an ideal passband of unity amplitude, then if a signal $f(k)$ is the input to the filter within the ideal passband, at the output we get approximately $f(k - M)$, a delayed value of the input. Of course, signals outside the passband will be attenuated.

Type II Even length even symmetry.

With a similar approach to the above we can show that this case becomes

$$H(e^{-j\theta}) = e^{-j\theta M} 2 \sum_{k=0}^{N/2-1} b_k \cos[(M-k)\theta] \qquad (8.14)$$

where $b_k = b_{N-1-k}$ for even symmetry.

It has the same group delay as Type 1, however, M is not an integer here and so the problem is a little harder to generalise for some problems.

Type III Odd length odd symmetry.

This case becomes

$$H(e^{-j\theta}) = e^{-j(\theta M - \pi/2)} 2 \sum_{k=0}^{M-1} b_k \sin[(M \quad k)\theta] \qquad (8.15)$$

where $b_k = -b_{N-1-k}$ for odd symmetry.

Differentiating the phase gives the same group delay as for the other two cases. However, there is an extra 90° phase shift to be taken account of for this case.

Type IV. Even length odd symmetry.

This case is determined by

$$H(e^{-j\theta}) = e^{-j(\theta M - \pi/2)} 2 \sum_{k=0}^{N/2-1} b_k \sin[(M-k)\theta] \qquad (8.16)$$

where $b_k = -b_{N-1-k}$ for odd symmetry.

The group delay stays the same, but the phase needs an extra 90° as Type III. The extra 90° phase shift make these last two filters suitable for differentiation or as a Hilbert transformer. (A means of shifting a signal by + 90° without in theory affecting its amplitude). Typical plots for each case are shown in Fig. 8.2. Magnitude only is shown as the phase is linear for each case. A later example will show the phase response.

Some comments on filter types:

It would be good to think that we could use any of the 4 filter templates above to design any filter we choose. Unfortunately, we can only use certain types for particular cases. From Fig. 8.2 we can claim:

A Type I filter is good for all filter types, lowpass, highpass, bandpass, bandstop.

For the Type II FIR filter, its frequency response is always zero at half sampling making it unsuitable for a highpass filter. It can be used as a bandpass filter.

For a Type III filter it has a dc gain of zero and gain of zero at half sampling. Hence it cannot be used as a lowpass filter or highpass filter. It can be used as a bandpass filter.

The Type IV filter also cannot be used as a lowpass filter since it has no dc gain, but is ok as a highpass or bandpass filter.

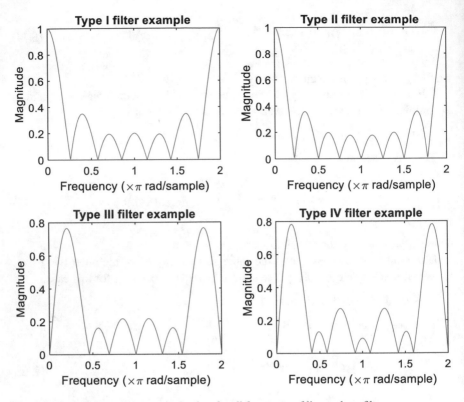

Fig. 8.2 Typical frequency magnitude plots for all four types of linear phase filter

All the filter types give rise to a delay in the output signal making them unsuitable for applications where time-delays are not allowed. A good example is in control systems where feedback is used. Any delay gives rise to a decrease in phase-margin and instability of the closed-loop system.

8.2.1 Example: Design a Lowpass Filter with a Passband Frequency of One Quarter Sampling Frequency

Solution.

Opting for filter Type II though Type I could also be used. We have even length and even symmetry. To begin with we select N = 16 as the number of samples and M = (N − 1)/2 = 7.5. One quarter sampling is therefore N/4 = 4 samples passband for an ideal filter, then the attenuation drops to zero. Since the frequency has a magnitude which is finite in length, assign a vector **h** of length N. Assign the magnitudes as follows:

$$\begin{bmatrix} 1\ 1\ 1\ 1\ 1\ 0\ 0\ 0\ 0\ 0\ 0\ 0\ 1\ 1\ 1\ 1 \end{bmatrix}$$

Note the first term is dc and for unity passband gain must also be unity. The symmetry in the frequency domain revolves around the remaining samples as the dc term has no phase. The dc term can be set to zero and a notch at dc will appear, but this is not desirable for this example. *Do not mix symmetry in the time domain with the frequency domain, this is easy to do. In the frequency domain the zeroth term or dc value stands on its own whereas in the time domain, the zero coefficient of the filter is in symmetry with the N − 1 coefficient.*

The phase is found from (8.13).

$$\phi(\theta) = -\frac{\pi n(N - 1)}{N} = -0.937n\pi, n = 0, 1, 2, 3, 4, 5, 6, 7$$

The remaining phases are the conjugate of the first 8 though we exclude the dc term as it has no natural pair to match. Beyond the first 5 phase terms (including dc) we "don't care" about their values since the amplitudes will be zero anyway. After dc only 4 more points are used and 4 from the other end for symmetry. Hence, we make a complex conjugate symmetric frequency vector:

$$\mathbf{h} = \begin{bmatrix} \mathbf{h}_1 & \mathbf{h}_2 \end{bmatrix}$$

where

$$\mathbf{h}_1 = \begin{bmatrix} 1 & -0.9808 - 0.1951i & 0.9239 + 0.3827i & -0.8315 - 0.5556i & 0.7071 + 0.7071i & 0 & 0 & 0 \end{bmatrix}$$

$$\mathbf{h}_2 = \begin{bmatrix} 0 & 0 & 0 & 0 & 0.7071 - 0.7071i & -0.8315 + 0.5556i & 0.9239 - 0.3827i & -0.9808 + 0.1951i \end{bmatrix}$$

To get the filter FIR coefficients we need only then take an inverse DFT or FFT. This gives us 16 real coefficients which are also even symmetric:

B (k) = {0.0398, − 0.0576, − 0.0206, 0.0805, − 0.0097, − 0.1269, 0.1015, 0.4929, 0.4949, 0.1015, − 0.1269, − 0.0097, 0.0805, − 0.0206, − 0.0576, 0.0398}

What we have done is construct each element of the frequency vector with complex numbers $H(n) = |H(n)|e^{j\phi(n)}$, $n = 0, 1, 2 \ldots N - 1$. Then we used the inverse DFT:

$$b_k = \frac{1}{N} \sum_{n=0}^{N-1} H(n)e^{j\frac{2\pi kn}{N}}, k = 0, 1, 2 \ldots N - 1$$

This will give the ideal filter, *but only at the selected sampling points*. They should be both real and symmetric. To find the general properties of the filter we need to take the frequency response of the FIR filter as shown in Fig. 8.3.

The filter should have a passband at quarter sampling frequency which is 0.5π radians/sample. It can be seen that the phase is linear as predicted but there is much ripple on the passband and stopband. Also, the transition band where the filter drops

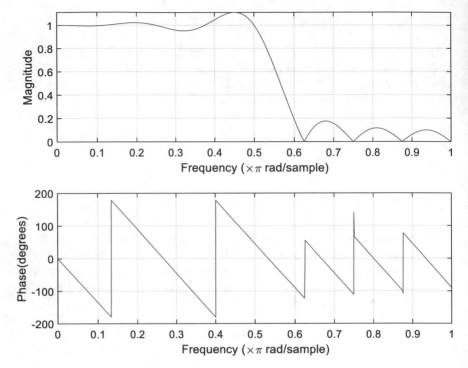

Fig. 8.3 Frequency response shown up to half sampling. FIR filter for $N = 16$ showing linear phase

down in gain is not very sharp at all. The ripple is caused by Gibbs phenomena since when taking the inverse DFT or FFT to get the filter weights the frequency samples are abruptly truncated and not smoothed with any windowing technique. Increasing the number of samples has the effect shown in Fig. 8.4.

Increasing N narrows the transition band and makes it steeper, but the Gibbs ripple still appears. The peak value of the ripple is unchanged as N increases. To overcome this problem, we multiply by a window before taking the inverse DFT to get the filter coefficients. To understand why this improves things we must look at a few windows.

Frequency response effect of windowing.

A rectangular window is really no window at all. This is the sharp transition that truncating a set of data has.

In Fig. 8.5 the window has been normalised so that 0 dB is shown as the peak value. This lets the data through and ideally it would need to be like a brick wall filter. However, it has numerous sidelobes, the first of which is around − 14 dB down. If we can find a window with greater sidelobe attenuation we can reduce the Gibbs effect.

Looking at the frequency response of the Hanning in Fig. 8.6, the window shows all the sidelobes are less than − 40 dB and significantly lower in attenuation than the

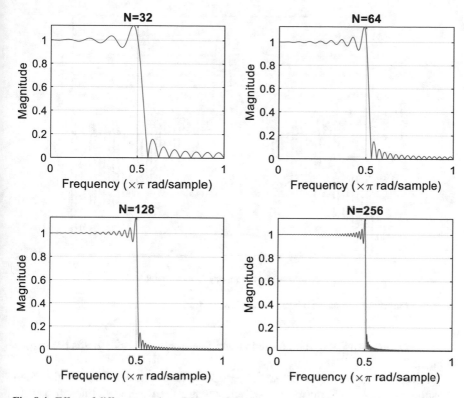

Fig. 8.4 Effect of different number of filter weights

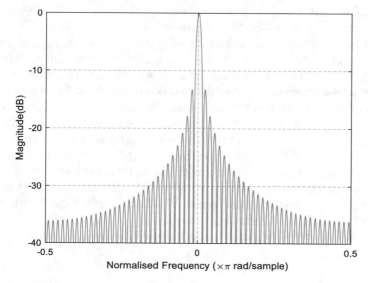

Fig. 8.5 Frequency response of a rectangular window

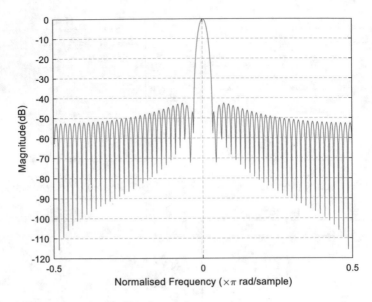

Fig. 8.6 Frequency response of a Hanning window

rectangular one. Applying the Hamming window to our example yields the results in Fig. 8.7.

We see a dramatic improvement. The ripple is hardly discernible and the steep transition when N = 256 makes it an excellent match for a near ideal filter. Too many weights will lead to difficulty in software implementation, so N = 256 is an excellent number of weights to stop at, though a little large for some real-time applications. More generally the main lobe of any window *must fit inside the passband* of our filter and the *transition bandwidth* (that's the sharp cut off) of the filter is governed by the main lobe bandwidth. The stopband attenuation is governed by the area of the sidelobes. The amount of attenuation can be seen in Fig. 8.8. This is dB magnitude which shows that attenuation is more than 50 dB in the stopband. There is a multitude of literature on FFT windows available in the literature [3].

We can simulate the FIR filter working by passing a signal plus noise through it. The filter is normalised so that for a given sampling frequency the cut off lies at one quarter of the sampling frequency. Therefore, arbitrarily choosing a sample rate of 12 kHz, let a sinusoid of amplitude 1 be at 10 Hz. Then add a noise term (another sine wave) of amplitude 1 at 400 Hz. Now 400 Hz is 0.66π radians/sample and if we look at Fig. 8.9 it should attenuate by over 60 dB at this frequency. We will use the filter when N = 256 and a Hamming window.

The MATLAB code that generated the lowpass filter is shown below.

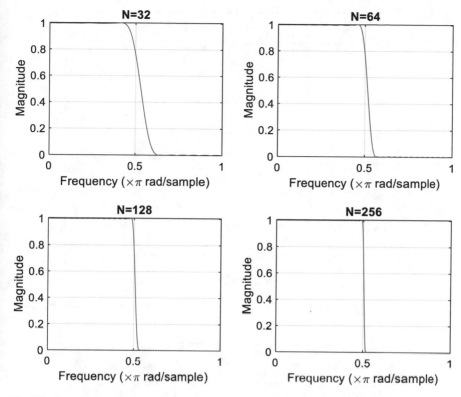

Fig. 8.7 Use of a Hamming window for various filter lengths

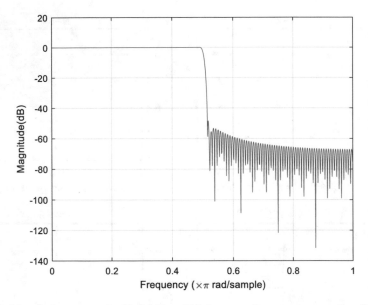

Fig. 8.8 Magnitude response in dB for N = 256. Lowpass filter, quarter sampling frequency passband

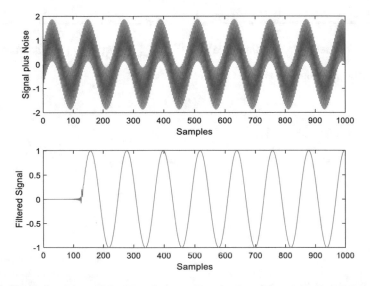

Fig. 8.9 Signal plus noise and the filtered output. Note the time-delay of the filtered signal

MATLAB Code 8.1

```
%Filter Type II
%Frequency Sampling design
%Low pass filter cutoff Ncut samples from N total
clear
close all
N=256;
M=(N-1)/2;
Ncut=N/4;%Cutoff for low pass filter up to Ncut samples
pva=zeros(N-1,1);
ha=zeros(N-1,1);
w=zeros(N-1,1);
%Phase starting from from bin 1. Bin 0 has no phase since it is dc
k=1:N/2;
pva(k)=exp(-2*j*pi*k*M/N);
%Do negative frequencies - flip side has positive phase.
pva(N-k)=exp(2*j*pi*k*M/N);
%Sample in the frequency domain to make the filter magnitude ha
 for i=1:1:Ncut
 ha(i)=1;
 ha(N-i)=1;
 end
% now add the phase to make a complex array which is conjugate symmetric
w=ha.*pva;
%add the dc value afterwards because MATLAB doesn't do arrays from zero.
%dc value
h0=1;
ha=[h0;ha];
%add dc element to complex frequency array - desired frequency response
w=[h0;w];
% time index in integers
t=0:1:N-1;
% b is the time domain coefficient array of the filter
b=ifft(w);
% Window applied here
% Now window the b coefficients
bw=b.*hamming(N);
%
% sum the values = should add to 1 for dc gain of 1 for our filter
dc=sum(b);
D=sprintf('dc gain of filter is %d.',dc);
disp(D)
% plot FIR filter coefficients, which are real
plot(t,b);
xlabel('Index number')
ylabel('Fillter coefficients')
ax = gca;
ax.XLim = [0 N];
set(gcf,'color','w');
figure
```

```
% Plot ideal sampled frequency response amplitude - real values
plot(t,ha);
xlabel('Samples in frequency')
ylabel('Amplitude')
ax = gca;
ax.XLim = [0 N];
set(gcf,'color','w');
% Frequency response
figure
a=1;
[hd,wd] = freqz(bw,a,'half',1000);
%Magnitude of frequency response
subplot(2,1,1)
plot(wd/pi,((abs(hd))),'color','blue')
xlabel('Frequency (\times\pi rad/sample)')
ylabel('Magnitude')
grid
set(gcf,'color','w');
% Phase part of filter frequency response
subplot(2,1,2)
plot(wd/pi,(angle(hd))*(180/pi),'color','blue')
ax = gca;
ax.YLim = [-200 200];
grid
xlabel('Frequency (\times\pi rad/sample)')
ylabel('Phase(degrees)')
set(gcf,'color','w');
%Apply the filter here to test it works
%Suppose the sampling frequency is 12kHz
%Signal at 10Hz and noise signal 400Hz
%Noise frequency is at 1/3rd sampling frequency or 2pi/3 radians/sample
%Desired Signal is well within passband
fs=1200;%Sampling Frequency
fl=10;%Desired Signal
fn=400;%Noise signal
t = 0:1:1000;
xs =sin(2*pi*t*(fl/fs));
xn=sin(2*pi*t*(fn/fs));
z=xs+xn;
fd=filter(bw,a,z);
figure
subplot(2,1,1)
plot(t,z)
ylabel('Signal plus Noise')
xlabel('Samples')
set(gcf,'color','w');
%
subplot(2,1,2)
plot(t,fd)
ylabel('Filtered Signal')
xlabel('Samples')
set(gcf,'color','w');
```

Observe also the filter coefficients as shown in Fig. 8.10.

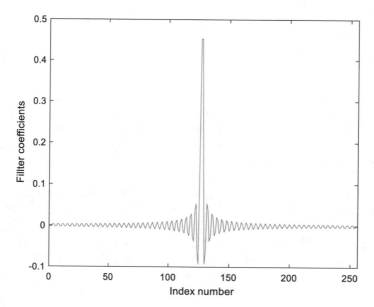

Fig. 8.10 N = 256. Coefficients of the filter for lowpass example

As the number of coefficients increase the shape of the plot looks closer to an impulse.

8.2.2 Example: Design of a Bandpass Filter

A linear phase bandpass filter can be designed using a Type I, II, II or IV FIR filter. Since we used Type II for the previous example we remain with even length even symmetry. We will design it to have a normalised passband frequency from 0.3π to 0.5π. This is one sixth of the sampling frequency to one quarter of the sampling frequency. The bandwidth will be 0.2π. We only need a few modifications to our MATLAB code, namely.

```
N=256;
%Frequency Sampling design
%Bandpass filter 0.3pi sampling to 0.5pi sampling.
%Bandwidth is 0.2pi rads/sample.
fL=2/0.3;
fH=2/0.5;
NcutL=floor(N/fL) ;%Cutoff for Flow
NcutH=floor(N/fH); %Cutoff for Fhigh
```

The floor() function in MATLAB rounds to the nearest integer value. NcutL and NcutH will be the sample numbers for the band of frequencies we wish to pass through the filter. Then we need to ensure these band of frequencies are set to unity.

```
%Sample in the frequency domain
  for i=NcutL:1:NcutH
    ha(i)=1;
    ha(N-i)=1;
  end
```

The phase is computed the same way as before for each sample.

The filter frequency response is shown in Fig. 8.11 and has linear phase as expected. This time we plot the coefficients of the filter using the *stem()* function in MATLAB (see Fig. 8.12). This gives a more aesthetic plot since these values are in fact at discrete values and not continuous.

To put some figures on this, suppose the sampling frequency is 1200 Hz and the lower and upper frequencies of the passband are 200 Hz and 300 Hz respectively. (one sixth to one quarter of the sampling frequency). We can drive the filter with white noise and look at the spectrum of the output. When dealing with random inputs we define the variance of the noise which in this case is set to unity. (it cannot have a defined amplitude since it has random amplitude values at each sample). The power spectrum is known as the *Periodogram* and defined as

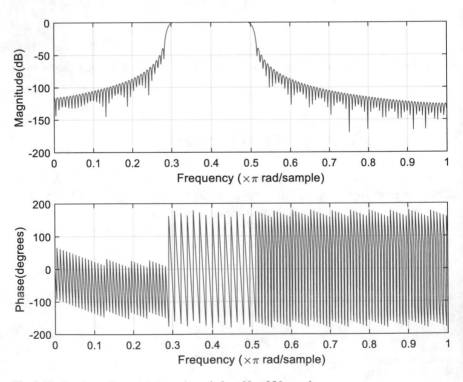

Fig. 8.11 Bandpass filter with Hamming window. N = 256 samples

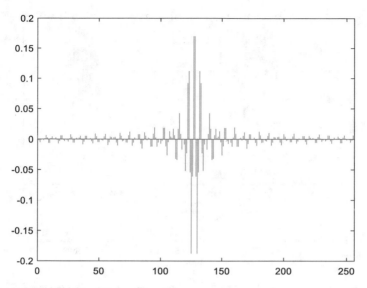

Fig. 8.12 Coefficients of the FIR bandpass filter

$$\hat{S}(n) = \frac{1}{N} \left| \sum_{k=0}^{N-1} x(k) e^{-j2\pi kn/N} \right|^2, n = 0, 1 \ldots N-1$$

$$= \frac{|X(n)|^2}{N} \tag{8.17}$$

for each frequency bin n. It is termed *Power spectrum* usually in continuous time and periodogram in discrete time. If the signal is assumed to have infinite length, then power is used otherwise we can define *energy spectral density* instead when the length of data is finite. Often power spectrum is used for discrete time, but this is a misuse of terms as periodogram is an *approximation* to continuous time power spectral density and only is the same in the limit as N → ∞.

The area under the periodogram is the total average power or energy in the signal (filtered signal in this case). Because the periodogram is discrete, its area is just the sum of the individual discrete power values at each sample. The frequency bins represent nf_s/N for n = 0,1,2...N in Hz though usually we only plot up to half sampling frequency. It is usually only applied to random signals but not always. The term *power spectrum* is used in instrumentation, embedded into many oscilloscopes as an extra function. For our example we drive the filter with white noise (variance 1) and examine the periodogram of the output. The periodogram is shown in Fig. 8.13 for the specified sampling frequency of 1200 Hz and 2560 samples.

The filter has shaped the noise so that we can see the outline of the filter. The periodogram is a bit noisy and the only way to get around this using FFTs is to average over successive frames of the FFT which smooths off the ragged edges.

Aside—verification of Parseval's theorem for the DFT.

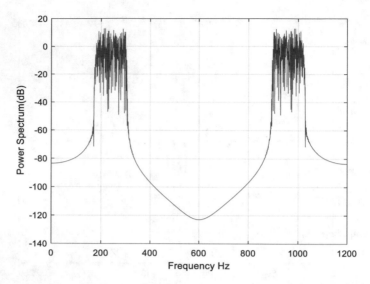

Fig. 8.13 Periodogram of the filtered white noise signal showing the filter passband

It is interesting at this stage to verify Parseval's theorem to the DFT or FFT as applied to this example. It is defined as the fact that the energy or power in the time domain and frequency domain are the same:

$$\sum_{k=0}^{N-1} |x(k)|^2 = \frac{1}{N} \sum_{n=0}^{N-1} |X(n)|^2 \qquad (8.18)$$

But we already have the squared frequency powers as obtained from the periodogram (8.17). So, we can say that

$$\sum_{k=0}^{N-1} |x(k)|^2 = \sum_{n=0}^{N-1} \hat{S}(n) \qquad (8.19)$$

The sum of squared values in the time-domain over a run of 2560 samples was found numerically as 490.5153. The sum of the periodogram values gives the same answer.

But the sample variance for any signal x (k) is $var[x(k)] = \frac{1}{N} \sum_{k=0}^{N-1} |x(k)|^2$ (ignoring any dc in the signal) and is also the average power. However, $\sum_{k=0}^{N-1} |x(k)|^2$ on the LHS of (8.19) is also a measure of the variance of the signal multiplied by the number of samples and is not a measure of power but energy. Therefore (8.19), though mathematically correct, is an equation of energies rather than power. Dividing (8.19) by N gives us the equation in power instead. The reason for this apparent mismatch is down to the basic definition of the DFT. The DFT or FFT as defined in recent years is given by $X(k) = \sum_{n=0}^{N-1} x(n)e^{-j2\pi kn/N}$, $k = 0, 1, \ldots N - 1$. If it is defined with a 1/N in front of it then the equations make more sense. For example, with a DFT when k = 0, the dc value becomes the average value when in the current definition it is N times the dc value. However, the definition as given here is the one in general use and we will continue with it. The scaling of DFTs and FFTs is by no means uniform throughout the scientific literature.

There is little real design freedom in this filter design method however and we need a more rigorous method to move further on.

8.2.3 Example. Design of an FIR Band Limited Differentiator

Despite the modern feeling of the subject area in digital signal processing, a major part of the theory was researched in the early 1960s and 1970s. The discrete-time differentiator has been researched for some time, the earliest reference appears to be in [5] and in the seminal book by Oppenheim and Schafer [6]. Differentiation is never a good thing to do with signals. The reason is that they have added noise and a differentiator amplifies high frequency. There may be the odd occasion where there is no option however and the best way is to band limit the differentiator, so it does not amplify frequencies beyond a certain designed upper limit. Perhaps the easiest way to differentiate and a method that is used by many for programming proportional plus integral plus differential (PID) controller, is to us the first difference as an approximation. That is to take the difference between two successive samples of a signal. We have $y_d = y_k - y_{k-1}$ which in Z-transforms gives us a simple FIR transfer function $1 - z^{-1}$. We can plot the frequency response of this, and it is shown in Fig. 8.14.

A continuous time differentiator has a transfer function

$$H(s) = s \tag{8.20}$$

and is physically unrealizable. However, its frequency response has a magnitude

$$|H(j\omega)| = \omega \tag{8.21}$$

and phase

$$\underline{H(j\omega)} = \pi/2 \tag{8.22}$$

The magnitude response has a positive slope of unity on a linear scale and on a log frequency vs dB magnitude scale the slope is 20 dB/decade or 6 dB/octave.

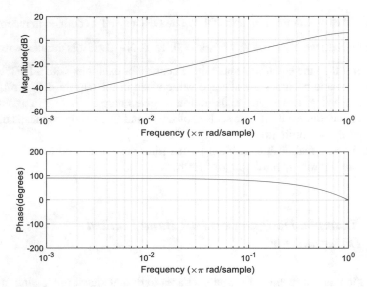

Fig. 8.14 Frequency response of first difference $1 - z^{-1}$

Figure 8.14 is quite close up to a frequency of 0.1π rad/s which is 1/20 times the sampling frequency. It has no roll-off to attenuate noise however and usually a low pass filter is used after it if it is used as a digital differentiator. We can however design our own band limited digital differentiator using an FIR filter. It must have a dc gain of zero and end with a dc gain of zero. We will design the differentiator to roll off at one quarter of the sampling frequency. We can use a Type III or Type IV filter to achieve this. Choosing Type IV means we need an even number of samples N for our frequency sampled filter. This seems a simple task, but the symmetry of the frequency response means that everything has to be just right, or we get complex filter coefficients instead of real values. The values must be placed as in Fig. 8.15, shown for 16 samples.

The amplitude of the frequency components must be set so that the gradient of the envelope of all the rising frequency components is unity. This is because we need a 20 dB/decade slope on a dB graph and that corresponds to unity slope in a linear graph. The zeroth one is dc at $n = 0$ and we set the rest as follows:

$$|H(n)| = n/(N/2), n = 0, 1, 2 \ldots N_{\text{cut}}$$

$$|H(N - n)| = |H(n)|, n = 1, 2 \ldots N_{\text{cut}}$$

where $N_{\text{cut}} = N/4$ is the inclusive last frequency bin in the first half of the samples. The other half of the samples are in conjugate symmetry i.e. the complex conjugate of the first half. But the pairing is important, bin 0 is unpaired, bin 1 is paired with bin 15, 2 with 14 and so on. If we don't require band limiting, we set $N_{\text{cut}} = N/2 - 1$.

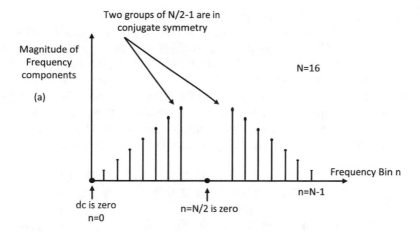

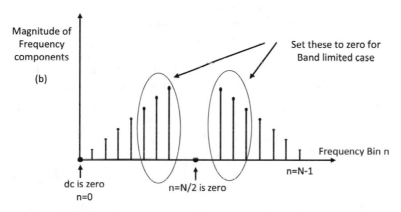

Fig. 8.15 Showing the weights of an $N = 16$ frequency sampled band limited differentiator. **a** With no band limiting. **b** With band limiting up to and including frequency bin $n = N/4$

The frequency component at N/2 has always zero magnitude and is the symmetry point in the frequency domain. For $N = 16$, it is the 8th array element which is always zero provided indexing starts at bin $k = 0$. Programming in MATLAB is a little more complicated since it can only index from 1. The phase must also allow for the extra $\exp(j\pi/2)$ in Eq. (8.16) which is needed for a Type IV filter. The phase is found from

$$\phi(n) = \frac{\pi}{2} - n\pi\frac{(N-1)}{N}, n = 0, 1, 2 \ldots N/2 - 1$$

and the phase is the negative of this for the remaining samples which gives rise to the complex conjugate of each frequency sample. Each frequency sample is therefore given by $H(n) = |H(n)|e^{j\phi(n)}, n = 0, 1, 2 \ldots N_{cut}$ and the other half are complex conjugates.

With no band limiting, the complex frequency vector becomes:

$\mathbf{H} = [0.0000 + 0.0000i, 0.0244 - 0.1226i, -0.0957 + 0.2310i, 0.2083 - 0.3118i,$
$-0.3536 + 0.3536i, 0.5197 - 0.3472i, -0.6929 + 0.2870i, 0.8582 - 0.1707i,$
$0.0000 + 0.0000i, 0.8582 + 0.1707i, -0.6929 - 0.2870i, 0.5197 + 0.3472i, -$
$0.3536 - 0.3536i, 0.2083 + 0.3118i, -0.0957 - 0.2310i, 0.0244 + 0.1226i].$

The differentiator appears to roll off itself at 0.8π radians/sample, but not very steeply. Note the slope of the differentiator is 1. Now modify the filter and insert a quarter sampling edge of a passband for the differentiator. Using a Hamming window, the following FIR filter coefficients (which are also the impulse response coefficients) were obtained:

$b(k) = \{0.0014, 0.0031, -0.0016, -0.0142, -0.0069, 0.0393, 0.0762, 0.0406,$
$-0.0406, -0.0762, -0.0393, 0.0069, 0.0142, 0.0016, -0.0031, -0.0014\}.$

They have *odd symmetry* as expected.

Using this filter, we obtain the frequency response plot in Fig. 8.16 (Fig. 8.17).

The roll off is apparent but not very sharp. Now increasing the order of the filter to $N = 128$ gives a much sharper cut off as shown in Fig. 8.18.

Finally, in Fig. 8.19 the dB magnitude is plotted vs log frequency.

We can see that the slope is 20 dB/decade and the roll off of the filter is very steep and of the order of a 50 dB drop within one frequency division of the graph.

Just to show that it really does work as a differentiator, for $N = 16$ we differentiate a triangular waveform, and the result is a square wave with some Gibbs ringing on it (see Fig. 8.20).

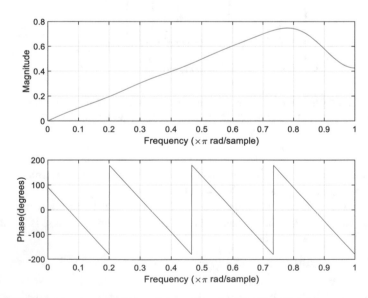

Fig. 8.16 Frequency response of differentiator with no band limiting, $N = 16$

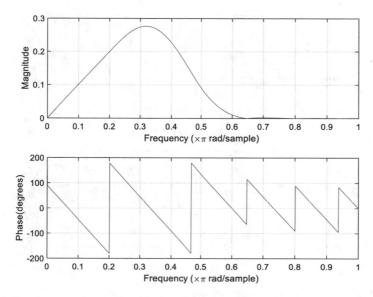

Fig. 8.17 Frequency response of band limited differentiator. N = 16

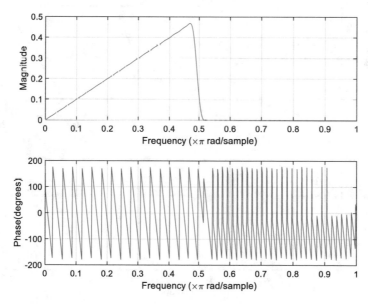

Fig. 8.18 N = 128 band limited differentiator

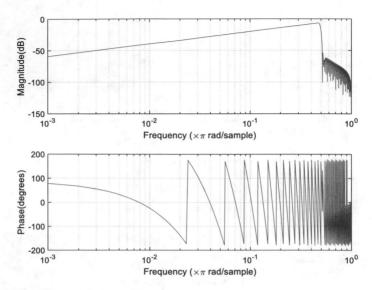

Fig. 8.19 dB plot of band limited differentiator

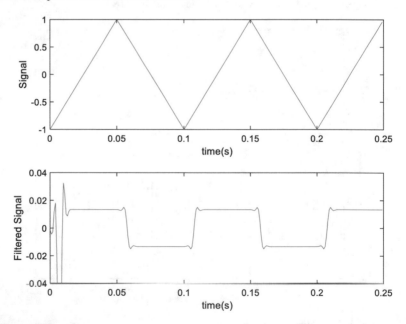

Fig. 8.20 Differentiation of a triangular waveform with an FIR filter for N = 16

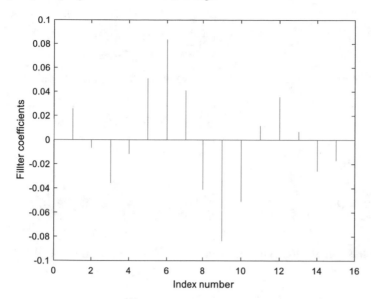

Fig. 8.21 Impulse response (filter coefficients) of FIR differentiator for N = 16. Coefficients have odd symmetry

The impulse response of the filter has 16 coefficients as shown in Fig. 8.21. They have odd symmetry as predicted. Although this works as a differentiator, it would be no use in a PID controller due to the phase delay it introduces.

8.2.4 Z-Transform of IIR Frequency Sampling Filters

The Z-transfer function of an FIR filter with N coefficients (or weights) is given by

$$H(z) = \sum_{k=0}^{N-1} b_k z^{-k} \tag{8.23}$$

From the previous section we selected frequency samples on the discrete spectrum with values $H(n), n = 0,1,2\ldots N-1$ which give rise to the symmetric filter coefficients via the inverse DFT.

$$b_k = \frac{1}{N} \sum_{n=0}^{N-1} H(n) e^{j\frac{2\pi kn}{N}}, k = 0, 1, 2 \ldots N-1 \tag{8.24}$$

What if we now substitute (8.24) into (8.23)? We get [7]

$$H(z) = \sum_{k=0}^{N-1} \left[\frac{1}{N} \sum_{n=0}^{N-1} H(n) e^{j\frac{2\pi kn}{N}} \right] z^{-k} \tag{8.25}$$

Interchange the order of summation (like we do for double integrals):

$$H(z) = \frac{1}{N} \sum_{n=0}^{N-1} H(n) \sum_{k=0}^{N-1} e^{j\frac{2\pi kn}{N}} z^{-k} \tag{8.26}$$

But

$$\sum_{k=0}^{N-1} e^{j\frac{2\pi kn}{N}} z^{-k} = 1 + e^{\frac{j2\pi n}{N}} z^{-1} + e^{\frac{j4\pi n}{N}} z^{-2} + \ldots + e^{\frac{j2\pi(N-1)}{N}} z^{-(N-1)} \tag{8.27}$$

Which is a geometric sequence summed to $N - 1$ terms. It has a closed form

$$\sum_{k=0}^{N-1} e^{j\frac{2\pi kn}{N}} z^{-k} = \frac{1 - z^{-N}}{1 - e^{\frac{j2\pi n}{N}} z^{-1}} \tag{8.28}$$

Therefore (8.26) becomes

$$H(z) = \frac{1}{N} \sum_{n=0}^{N-1} H(n) \frac{1 - z^{-N}}{1 - e^{\frac{j2\pi n}{N}} z^{-1}}$$

Tidying up, this is our desired result

$$H(z) = \frac{1 - z^{-N}}{N} \sum_{n=0}^{N-1} \frac{H(n)}{1 - e^{\frac{j2\pi n}{N}} z^{-1}} \tag{8.29}$$

The transfer function (a polynomial) interpolates the frequency response through the chosen complex values within H (k). We have

$$H(n) = H(e^{j\theta})\big|_{\theta = \frac{2\pi n}{N}} \tag{8.30}$$

If we have symmetry of the frequency samples

$$H(n) = H^*(N - n) \tag{8.31}$$

where * is complex conjugate, then we will obtain a linear phase filter. Otherwise we will obtain a filter with nonlinear phase.

What is interesting about (8.29) is that it is two filters in cascade. The first of these is a so-called *comb filter* of the form

$$H_c(z) = \frac{1 - z^{-N}}{N} \tag{8.32}$$

The comb filter is a filter that has zeros equally spaced apart exactly on the unit circle. In fact, the solution for the zeros of (8.32) is

$$1 - z^{-N} = 0 \tag{8.33}$$

This is an equation which if solved for z gives us the N complex roots of z. They all have magnitude 1 but their phase is $\varphi(i) = \frac{2\pi i}{N}$, $i = 0, 1, 2, \ldots N - 1$. They are each apart in phase by $\frac{2\pi}{N}$ radians. This gives a notch in the frequency response at every zero which if a great many zeros are used looks a bit like the elements of a comb, hence the name.

The second filter has the form

$$H_r(z) = \sum_{n=0}^{N-1} \frac{H(n)}{1 - e^{\frac{j2\pi n}{N}} z^{-1}} \tag{8.34}$$

It has a pole at every value of the summation index n found for the value of z from

$$1 - e^{\frac{j2\pi n}{N}} z^{-1} = 0 \tag{8.35}$$

Each pole is at

$$z = e^{\frac{j2\pi n}{N}}, n = 0, 1, 2 \ldots N - 1 \tag{8.36}$$

These poles lie on the unit circle spaced apart by the same angles as the zeros, so pole-zero cancellation must occur. A filter with poles on the unit circle spaced in this way is known as a *resonator*. This particular resonator is quite pure but can lead to instability due to the poles being on the unit circle. With finite precision arithmetic they can move slightly outside the stability zone. They are essentially narrowband filters (bandpass) that are valid at one frequency only. The overall filter has the form

$$H(z) = H_c(z)H_r(z) \tag{8.37}$$

Consider a simple example. Suppose $N = 4$ and we use a symmetric frequency response to obtain linear phase. Following the type II filter structure (even number N and even symmetric filter coefficients) means the phase must be conjugate symmetric and from (8.13) be

$$\phi(n) = -n\pi \frac{(N-1)}{N}, n = 0, 1, 2 \ldots N/2 - 1$$

For $N = 4$, this gives us

$$\phi(0) = 0$$
$$\phi(1) = -3\pi/4$$
$$\phi(2) = x$$
$$\phi(3) = 3\pi/4$$

where x is "don't care" for the phase of the central frequency sample because the magnitude must be zero according to the Type II template of a linear phase filter. For a lowpass filter choose:

$$|H(0)| = |H(1)| = 1, |H(2)| = 0, |H(3)| = 1$$

The complex frequency values are now

$$H(0) = 1, H(1) = e^{-\frac{3\pi j}{4}}, H(2) = 0, H(3) = e^{\frac{3\pi j}{4}}$$

Substitute into

$$H(z) = \frac{1 - z^{-N}}{N} \sum_{n=0}^{N-1} \frac{H(n)}{1 - e^{\frac{j2\pi n}{N}} z^{-1}}$$

and obtain

$$H(z) = \frac{1 - z^{-4}}{4} \sum_{n=0}^{N-1} \frac{H(n)}{1 - e^{\frac{j2\pi n}{4}} z^{-1}}$$

$$= \frac{1 - z^{-4}}{4} \left[\frac{1}{1 - z^{-1}} + \frac{e^{-\frac{j3\pi}{4}}}{1 - jz^{-1}} + \frac{e^{\frac{j3\pi}{4}}}{1 + jz^{-1}} \right] \quad (8.38)$$

Before continuing, note that we can factor

$$1 - z^{-4} = (1 - z^{-2})(1 + z^{-2})$$
$$= (1 - z^{-1})(1 + z^{-1})(1 + z^{-2})$$

And

$$(1 + z^{-2}) = (1 - jz^{-1})(1 + jz^{-1})$$

Let a $= 3\pi/4$ and add all three terms in (3.38) by forming a common denominator and using Euler's formulae for cos and sin:

$$H(z) = \frac{1 - z^{-4}}{4} \left[\frac{1}{1 - z^{-1}} + \frac{e^{-ja}}{1 - jz^{-1}} + \frac{e^{ja}}{1 + jz^{-1}} \right]$$

$$= \frac{1-z^{-4}}{4}\left[\frac{1+2\cos(a)+2z^{-1}[\sin(a)-\cos(a)]+z^{-2}[1-2\sin(a)]}{(1-z^{-1})(1-jz^{-1})(1+jz^{-1})}\right]$$

Cancelling the denominator and numerator common factors gives

$$H(z) = \frac{1+z^{-1}}{4}[1+2\cos(a)+2z^{-1}[\sin(a)-\cos(a)]+z^{-2}[1-2\sin(a)]]$$

(8.39)

Since $\cos(a) = -1/\sqrt{2}$, $\sin(a) = 1/\sqrt{2}$, we get

$$H(z) = \frac{1+z^{-1}}{4}\left[\left(1-\frac{2}{\sqrt{2}}\right)+\frac{4}{\sqrt{2}}z^{-1}+\left(1-\frac{2}{\sqrt{2}}\right)z^{-3}\right]$$

Expanding and after simplification we get

$$H(z) = 0.25\left[\left(1-\sqrt{2}\right)+z^{-1}\left(1+\sqrt{2}\right)+z^{-2}\left(1+\sqrt{2}\right)+z^{-3}\left(1-\sqrt{2}\right)\right]$$

(8.40)

Note the symmetry of the coefficients in the final result.
We can verify by taking the inverse DFT.
The frequency vector is

$$\mathbf{H} = \left[1 \; \frac{1}{\sqrt{2}} - j\frac{1}{\sqrt{2}} \; 0 \; \frac{1}{\sqrt{2}} + j\frac{1}{\sqrt{2}}\right]^{\mathrm{T}}$$

From Eq. (7.74) we have the DFT matrix

$$\mathbf{W}_4 = \begin{bmatrix} 1 & 1 & 1 & 1 \\ 1 & -j & -1 & j \\ 1 & -1 & 1 & -1 \\ 1 & j & -1 & -j \end{bmatrix}$$

The coefficients in vector form become from (7.76)

$$\mathbf{b} = \frac{1}{4}\mathbf{W}^*\mathbf{H}$$
$$= \begin{bmatrix} -0.1036 & 0.6036 & 0.6036 & -0.1036 \end{bmatrix}^{\mathrm{T}}$$

where * is the conjugate matrix. These correspond to the coefficients of the FIR filter in (8.40).
For N even, we can express any order linear phase filter as [7]

$$H(z) = \frac{1 - z^{-N}}{N}\left[\frac{H(0)}{1 - z^{-1}} + 2 \sum_{n=1}^{(N/2)-1} \frac{(-1)^n|H(n)|\cos\left(\frac{\pi n}{N}\right)\left(1 - z^{-1}\right)}{1 - 2z^{-1}\cos(2\pi n/N) + z^{-2}}\right] \quad (8.41)$$

where we have assumed that $H(N/2) = 0$. This is just combing pairs of complex poles together to give us the sum of second order IIR filters plus the dc filter which is an integrator.

Substituting our values will give the same answer as (8.40) when numerator and denominator terms are cancelled.

There appears initially to be no real advantage in this approach until we look closer at the structure of the filter. If instead of using an FIR filter, we cascade the comb and resonator filters, then because they are IIR, they have fewer computations than FIR filters. The problem is, that as it stands the structure will be close to unstable due to pole zero cancellations. As a way around this we can move the poles and zeros slightly within the unit circle and modify (8.41) thus

$$H(z) = \frac{1 - r^N z^{-N}}{N}\left[\frac{H(0)}{1 - rz^{-1}} + 2 \sum_{n=1}^{(N/2)-1} \frac{(-1)^n|H(n)|\cos\left(\frac{\pi n}{N}\right)\left(1 - rz^{-1}\right)}{1 - 2rz^{-1}\cos(2\pi n/N) + r^2 z^{-2}}\right]$$
$$(8.42)$$

where we have substituted z^{-1} with rz^{-1} and $|r| < 1$. For our example with $N = 4$, the transfer function becomes

$$H(z) = \frac{1 - r^4 z^{-4}}{4}\left[\frac{1}{1 - rz^{-1}} - \frac{\sqrt{2}\left(1 - rz^{-1}\right)}{\left(1 + r^2 z^{-2}\right)}\right] \quad (8.43)$$

The two approaches give very close frequency responses when $r = 0.95$ and this improves as r gets closer to 1. The phase is more affected than the amplitude. The phase is only marginally affected and worst near half sampling where it doesn't matter as much since the attenuation will be greatest. Equation (8.42) is only valid for even N, another formula holds for N odd (Fig. 8.22).

8.3 Interpolation Method of FIR Design

Another method that can be used (though not the modern preferred method) is as follows.

Consider for simplicity a Type I FIR filter. Type I filters are easier do deal with in general because they have a delay which is an integral number of integer samples, whereas Type II or filters with even numbers of weights have fraction delays.

The coefficients satisfy the frequency response of (8.11):

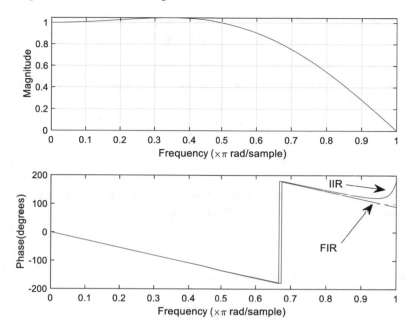

Fig. 8.22 Comparison between IIR and FIR frequency sampling methods for r = 0.95

$$H\left(e^{-j\theta}\right) = e^{-j\theta M}\left[b_M + 2\sum_{k=0}^{M-1} b_k \cos[(M-k)\theta]\right]$$

where N is an odd number of samples and M = (N − 1)/2 is even. Consider only the magnitude samples and ignore the delay term, we have

$$H(k) = b_M + 2\sum_{k=0}^{M-1} b_k \cos[(M-k)\theta]$$

This can be written in matrix form, but it is not as elegant as we would expect. Instead, substitute p = M–k into the above equation. When k = 0, p = M and when k = M − 1, p = 1. Since k = M − p, we get

$$H(k) = b_M + 2\sum_{p=1}^{M} b_{M-p}\cos(p\theta) \tag{8.44}$$

But (8.44) excludes the Mth b term from the summation. Including it we arrive at M + 1 frequency amplitude values satisfying

$$H(k) = 2 \sum_{p=0}^{M} b_{M-p} \cos(p\theta) \tag{8.45}$$

where the b_M coefficient is half of its true value, so we must double its value later. Define a frequency vector

$$\boldsymbol{\theta} = \begin{bmatrix} \theta_0 \ \theta_1 \ \theta_1 \ \dots \ \theta_M \end{bmatrix} \tag{8.46}$$

Which splits the frequency samples into $M + 1$ values from 0 to π. These values *do not have to be uniformly spaced apart* as with the other cases which use the DFT. The designer can choose which frequency values to include. Define the desired amplitude values as a column vector $\mathbf{H}$ of dimension $M + 1$.

$$\mathbf{H} = \begin{bmatrix} H_0 \ H_1 \ H_2 \ \dots \ H_M \end{bmatrix}^{\mathrm{T}} \tag{8.47}$$

Let the vector of impulse response coefficients be

$$\mathbf{b} = \begin{bmatrix} b_M/2 \ b_{M-1} \ b_{M-2} \ \dots \ b_0 \end{bmatrix}^{\mathrm{T}} \tag{8.48}$$

Then (8.45) has a matrix form

$$2 \begin{bmatrix} 1 & \cos(\theta_0) & \cos(2\theta_0) & \dots & \cos(M\theta_0) \\ 1 & \cos(\theta_1) & \cos(2\theta_1) & \dots & \cos(M\theta_1) \\ 1 & \cos(\theta_2) & \cos(2\theta_2) & \dots & \cos(M\theta_2) \\ . & & & & \\ . & & & . & \\ . & & & & \\ 1 & \cos(\theta_M) & \cos(2\theta_M) & \dots & \cos(M\theta_M) \end{bmatrix} \begin{bmatrix} (b_M)/2 \\ b_{M-1} \\ b_{M-2} \\ . \\ . \\ . \\ b_0 \end{bmatrix} = \begin{bmatrix} H_0 \\ H_1 \\ H_2 \\ . \\ . \\ . \\ H_M \end{bmatrix} \tag{8.49}$$

Write this as

$$2\mathbf{Rb} = \mathbf{H} \tag{8.50}$$

where $\mathbf{R}$ is the $M + 1$ square matrix above and $\mathbf{b}$ is the unknown coefficient vector. Provided we select the $\mathbf{H}$ vector we can solve (8.50) provided $\mathbf{R}$ is non-singular. Using matrix inversion

$$\mathbf{b} = \frac{1}{2}\mathbf{R}^{-1}\mathbf{H} \tag{8.51}$$

Consider a simple example where $N = 9$.

Then $M = 4$ and we have $M + 1$ unknown b coefficients which form the matrix equation

$$2 \begin{bmatrix} 1 & \cos(\theta_0) & \cos(2\theta_0) & \cos(3\theta_0) & \cos(4\theta_0) \\ 1 & \cos(\theta_1) & \cos(2\theta_1) & \cos(3\theta_1) & \cos(4\theta_1) \\ 1 & \cos(\theta_2) & \cos(2\theta_2) & \cos(3\theta_2) & \cos(4\theta_2) \\ 1 & \cos(\theta_3) & \cos(2\theta_3) & \cos(3\theta_3) & \cos(4\theta_3) \\ 1 & \cos(\theta_4) & \cos(4\theta_4) & \cos(3\theta_4) & \cos(4\theta_4) \end{bmatrix} \begin{bmatrix} b_4/2 \\ b_3 \\ b_2 \\ b_1 \\ b_0 \end{bmatrix} = \begin{bmatrix} H_0 \\ H_1 \\ H_2 \\ H_3 \\ H_4 \end{bmatrix} \quad (8.52)$$

To make a very simple bandpass filter define the desired amplitudes as the vector

$$\mathbf{H} = \begin{bmatrix} 0 & 1 & 1 & 1 & 0 \end{bmatrix}^T$$

And frequency vector

$$\theta = \begin{bmatrix} 0 & 0.25 & 0.5 & 0.75 & 1 \end{bmatrix}\pi$$

This means we have selected the passband with unit amplitude at frequencies $0.25\pi, 0.5\pi, 0.75\pi$ radians/sample. Here we have the frequencies uniformly spaced apart, but they need not have been chosen as such.

Solving (8.52)

$$\mathbf{b} = \begin{bmatrix} 0.375 & 0 & -0.25 & 0 & -0.125 \end{bmatrix}^T$$

We must remember that the coefficients are in reverse order and the first value must be doubled. The symmetric full impulse response of the FIR filter must be constructed from these values and becomes (Fig. 8.23).

$\{-0.125, 0, -0.25, 0, 0.75, 0, -0.25, 0, -0.125\}$.

The MATLAB code is shown below.

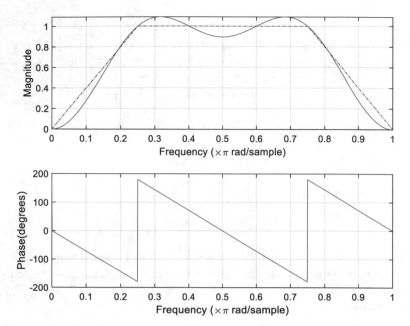

Fig. 8.23 Shows desired frequency response and actual frequency response for a simple bandpass filter with N = 9

MATLAB Code 8.2
```
%Interpolation method
N = 9;
M = (N-1)/2;
%Frequency samples up to pi
theta = [0 .25 0.6 0.75 1]'*pi;
%Magnitude of frequency response samples. M+1 values up to half sampling
H= [0 1 1 1 0]';
%create harmonic integer vector
f=[0:1:M];%M=1 integer row vector 0,1,2...M
%Form the matrix - use outer product of two vectors
% and take cosine of the matrix
R = 2*cos(theta*f);
%Invert Matrix and multiply by frequency sampled vector H
b = inv(R)*H;
% Double first value to give b(M)
% Values are in reverse order so b(1) is b(M)/2 in equations
b(1)=2*b(1);
% Form the symmetric impulse response vector
%In the equations it is [b0 b1 b2 b3 b4 b3 b2 b1 b0]
h=[b(5) b(4) b(3) b(2) b(1) b(2) b(3) b(4) b(5)];
[hd,wd] = freqz(h,1,'half',1000);
```

```
%Magnitude of frequency response
subplot(2,1,1)
%Plot ideal one and actual one
plot(theta/pi,H,'-.r',wd/pi,((abs(hd))),'-b')
xlabel('Frequency (\times\pi rad/sample)')
ylabel('Magnitude')
grid
set(gcf,'color','w');
% Phase part of filter frequency response
subplot(2,1,2)
plot(wd/pi,(angle(hd))*(180/pi),'color','blue')
ax = gca;
ax.YLim = [-200 200];
grid
xlabel('Frequency (\times\pi rad/sample)')
ylabel('Phase(degrees)')
set(gcf,'color','w');
```

Increasing the number of points leads to ripple at the passband edges and this can be reduced by windowing. An advantage of this method is that the sampling points in the frequency domain need not be uniformly spaced apart like the other methods that use the DFT.

8.4 Outline of Optimal Design of FIR Filters

Much of the early work on these methods was achieved in the early 1970s, at least a decade or more before computing power became powerful enough for real-time implementation. The problem with the other methods is that there is no real design methodology as there is with the design of analogue filters, where precise passband and stopband ripples can be specified. This changed with the work of Parks and McLellan [8]. In fact who could have guessed that an obscure paper by a French mathematician by the name of Remez as far back as 1934 would also be make contributions in engineering filter design [9]. The Remez or Remez-exchange algorithm enables us to find the best solutions to a set of equations. It doesn't solve the problem using the usual least-squares methods. Filters can be designed using ordinary least-squares too, but it was found that the Remez-exchange algorithm gave better control over the filter design. Hence the jump herein to the better method. Least-squares, known since the time of Gauss, is one of the most popular branches of mathematics and has found application in numerous scientific disciplines since its discovery. It minimises the squared error of an error function to get the so-called optimal solution. We say it is optimal in the sense of least-squares. Instead of minimising the squared error, how about minimising the maximum value of the error instead? This is known as minimax optimization and the Remez-exchange algorithm is the ideal way to solve such problems.

The algorithm uses a set of weight functions in each frequency, namely

$$W(\theta) = \begin{cases} \frac{1}{K} & B_p \\ 1 & B_s \end{cases} \tag{8.53}$$

where B_p and B_s represent the passband and stopband range of frequencies respectively. From (8.53), by adjustment of K in each within each frequency, we can adjust the relative amounts of ripple in each. The algorithm minimises the maximum of the error vector

$$\max \mathbf{W}(\theta)\|\mathbf{D}(\theta) - \mathbf{H}(\theta)\| \tag{8.54}$$

Here bold letters indicate a vector quantity. The vector $\mathbf{D}(\theta)$ is a vector of the *desired frequency response values at each frequency*. For example, for a lowpass filter, for each frequency we have

$$D(\theta) = \begin{cases} 1 & B_p \\ 0 & B_s \end{cases}$$

which makes up the desired frequency vector $\mathbf{D}(\theta)$ and $W(\theta)$ at each frequency makes up the weighting vector $\mathbf{W}(\theta)$. The vector $\mathbf{H}(\theta)$ is the frequency values based on the filter parameters $b_k, k = 0, 1, 2 \ldots M$ which we are required to find. Because the filter values are not in the same order (see previous section) we usually re-define the coefficients with another set of coefficients, say $d_k, k = 0, 1, 2 \ldots M$ and re-order afterwards. Then (8.45) becomes

$$H(k) = \sum_{p=0}^{M} d_p \cos(p\theta) \tag{8.55}$$

and we find the d coefficients instead, working back to b later. Do not confuse the D terms which are part of the desired frequency vector with the lower-case d parameter or impulse response coefficients. The H terms in (8.55) are used in the definition of frequency response, but do not appear in the set of equations we solve because we require the d coefficient terms which make up H.

We end up with a formidable set of equations to solve:

$$
\begin{bmatrix}
1 & \cos(\theta_0) & \cos(2\theta_0) & . & . & . & \cos(M\theta_0) & \frac{1}{w(\theta_0)} \\
1 & \cos(\theta_1) & \cos(2\theta_1) & . & . & . & \cos(M\theta_1) & -\frac{1}{w(\theta_1)} \\
. & & & & & & & \frac{1}{w(\theta_2)} \\
. & & & & & & & -\frac{1}{w(\theta_3)} \\
. & & & & & & & . \\
. & & & & & & & . \\
. & & & & & & & . \\
1 & \cos(\theta_{M+1}) & \cos(2\theta_{M+1}) & \cos(3\theta_{M+1}) & . & . & \cos(M\theta_{M+1}) & \frac{(-1)^{M+1}}{w(\theta_{M+1})}
\end{bmatrix}
\begin{bmatrix}
d_0 \\ d_1 \\ d_2 \\ . \\ . \\ . \\ d_M \\ \rho
\end{bmatrix}
$$

$$
=
\begin{bmatrix}
D(\theta_0) \\
D(\theta_1) \\
D(\theta_2) \\
D(\theta_3) \\
. \\
. \\
. \\
D(\theta_{M+1})
\end{bmatrix}
\tag{8.56}
$$

There are $M + 2$ terms to solve for but it is usual to compute ρ analytically beforehand. The procedure requires an initial guess of $M + 2$ points and exchanges the $M + 2$ points of alternation. (this is the key findings in the Remez solution). Hence the name *Remez exchange algorithm*. The solution gives a Chebyshev weighted solution and has quadratic convergence. The algorithm is also known as the Parks-McClellan algorithm since these authors applied the Remez exchange algorithm to filter design.

The earliest software for this method was produced in 1973 [10] and written in FORTRAN. Algorithms are available to solve this problem are obtainable on MATLAB.

Example: Use MATLAB to design a lowpass filter with the following specifications:

Sampling frequency 10 kHz.
Passband frequency 2 kHz.
Stopband frequency 22.5 kHz.
Passband ripple 0.001 or less.
Stopband ripple 0.001 or less.
Passband gain 1.

MATLAB Code 8.3

```
%Parks McClellan Algorithm
%Sampling Frequency in Hz
fs=10000;
%Passband frequency
fp=2000;
```

```
%Stopband Frequency
fstop=2500;
%Passband ripple
pr=0.001;
%Stopband ripple
sr=0.001;
%Passband amplitude
pa=1;
%Stopband amplitude
sa=0;
[n,fo,ao,w] = firpmord([fp fstop],[pa sa],[pr sr],fs);
%Filter order is n
n
b = firpm(n,fo,ao,w);
%fo are normalised frequency band edges
%Find the frequency response of the filter
[hd,fh] = freqz(b,1,1024,fs);
%Magnitude of frequency response
plot(fh,(20*log10(abs(hd))),'-b')
xlabel('Frequency (Hz)')
ylabel('Magnitude (dB)')
set(gcf,'color','w');
grid
```

The algorithm tells us that it needs an FIR filter of order 65. The frequency response is shown in Fig. 8.24.

Here we have obtained a filter which meets the design specifications and uses a relatively low number of weights for such a large amount of stopband attenuation.

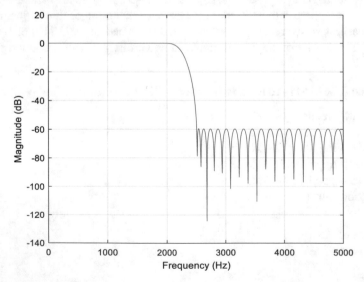

Fig. 8.24 Frequency response of Lowpass filter designed using MATLAB and the Parks McClellan algorithm with N = 65

References

1. S.K. Mitra, *Digital Signal Processing: A Computer Based Approach* (McGraw-Hill, New York, USA, 2001)
2. A. Amarder, *Analog and Digital signal processing,* 2nd edn. (Brooks/Cole Publishing company, USA, 1999)
3. F. Harris, On the use of windows for harmonic analysis with the discrete fourier transform. Proc. IEEE **66**(1), 51–83 (1978)
4. P.S.R. Diniz, E.A.B. Da Silva, S.L. Netto, *Digital Signal Processing, system analysis and design*, 2nd edn. (Cambridge University Press, Cambridge, UK, 2019)
5. L. Rabiner, K. Steiglitz, The design of wide-band recursive and nonrecursive digital differentiators. IEEE Trans. Audio Electroacoust. **18**(2), 204–209 (1970)
6. A.V. Oppenheim, R.W. Schafer, *Digital Signal Processing* (Prentice Hall International Ltd, 1975)
7. L. Rabiner, R. Schafer, Recursive and nonrecursive realizations of digital filters designed by frequency sampling techniques. IEEE Trans. Audio Electroacoust. **19**(3), 200–207 (1971). https://doi.org/10.1109/TAU.1971.1162185
8. T. Parks, J. McClellan, Chebyshev approximation for nonrecursive digital filters with linear phase. IEEE Trans. Cir. Theo. **19**(2), 189–194 (1972). https://doi.org/10.1109/TCT.1972.1083419
9. E.Y. Remez, Sur la détermination des polynômes d'approximation de degré donnée. Comm. Soc. Math. Kharkov. **10** (1934)
10. J. McClellan, T. Parks, L. Rabiner, A computer program for designing optimum FIR linear phase digital filters. IEEE Trans. Audio Electroacoust. **21**(6), 506–526 (1973)

Chapter 9
State-Space Method of System Modelling

9.1 Motivation

The previous chapters describe physical processes by means of the Laplace Transform for continuous time systems or the Z Transform for discrete time. In the case of the Laplace Transform, its origins come from a transformation of ordinary linear differential equations into transfer functions which relate output to input. The Z Transform is a similar method related to uniformly sampled systems. Usually the Z Transform is derived by conversion from the Laplace transform since most physical processes are continuous time. However, occasionally a system may be discrete in nature and a difference equation describes it rather than a differential equation. In such a case the Z Transform of the difference equation can be obtained and a mathematical Z Transfer function obtained directly. Similarly, if a system is continuous time and a sampled model is obtained, then it can be approximated by a difference equation for a given sample rate. Overall, unless a system is naturally discrete in nature, then the discrete model will always approximate its continuous time origins. In the sciences some mathematical models are described by difference equations or even nonlinear difference equations. For example, mathematical models of infectious diseases which spread on a daily rate. For a great many applications these two approaches are quite enough to solve many real-world problems. However, there are other ways which this book has not yet explored which can have advantages. The first of these is the state-variable method of system modelling.

9.2 Realizations

In terms of application, the state-variable or state-space approach to system modelling was driven by the problems in the aerospace industry in the late 1950s to early 1960s. In particular, the race to the moon and technology in the form of Polaris missiles[1]. Much of the work involved advanced inertial navigation systems and was

© The Author(s), under exclusive license to Springer Nature Switzerland AG 2022
T. J. Moir, *Rudiments of Signal Processing and Systems*,
https://doi.org/10.1007/978-3-030-76947-5_9

applied to optimal filtering and control. The state variable method is often referred to as state-space because in pure mathematics we are dealing with Euclidean vector spaces. The method is surprisingly simple in concept yet leads to some powerful results which cannot easily be solved using the transfer function methods. In the field of filter design, state-variable filters can be constructed which can realise any system configuration. A similar approach used to be used for general simulation purposes in the field of *analogue computers*. Commercial analogue computers are nowadays almost obsolete however having been replaced by the digital counterpart. A resurgence using VLSI MOS technology on a chip was more recently reported however to solve problems that can be solved in parallel such as partial differential equations[2].

Consider a linear ordinary differential equation with constant coefficients described by

$$\frac{d^2y}{dt^2} + a_1\frac{dy}{dt} + a_2y = u(t) \tag{9.1}$$

Which has output y(t) and is driven by a signal u(t). We can easily take the Laplace transform of (9.1) and arrive at

$$\frac{y(s)}{u(s)} = \frac{1}{s^2 + a_1s + a_2} \tag{9.2}$$

A transfer function which we can analyse for stability and apply various inputs to see what type of output we would get. This method is in the s-domain or frequency domain. We can however stay in the time-domain and do the following instead. Define two variable which we call state-variables $x_1(t)$, $x_2(t)$. We define as many state variables as the order of the differential equation. There are an infinite number of ways to select the state variables but only few of these are of any real interest. From (9.2) we can cross multiply and get

$$(s^2 + a_1s + a_2)y(s) = u(s)$$

Then inverse Laplace transform to get the ordinary differential equation

$$\frac{d^2y(t)}{dt^2} + a_1\frac{dy(t)}{dt} + a_2y(t) = u(t) \tag{9.3}$$

We can write in shorthand notation using the dot notation for a derivative.

$$\ddot{y}(t) + a_1\dot{y}(t) + a_2y(t) = u(t) \tag{9.4}$$

Define

$$x_1(t) = y(t)$$

$$x_2(t) = \dot{y}(t) \tag{9.5}$$

Then we have

$$\dot{x}_1(t) = \dot{y}(t)$$
$$= x_2(t) \tag{9.6}$$

Using (9.4)

$$\dot{x}_2(t) = \ddot{y}(t)$$
$$= -a_1 x_2(t) - a_2 x_1(t) + u(t) \tag{9.7}$$

We can now write the above equations in vector form

$$\begin{bmatrix} \dot{x}_1(t) \\ \dot{x}_2(t) \end{bmatrix} = \begin{bmatrix} 0 & 1 \\ -a_2 & -a_1 \end{bmatrix} \begin{bmatrix} x_1(t) \\ x_2(t) \end{bmatrix} + \begin{bmatrix} 0 \\ 1 \end{bmatrix} u(t) \tag{9.8}$$

$$y(t) = \begin{bmatrix} 1 & 0 \end{bmatrix} \begin{bmatrix} x_1(t) \\ x_2(t) \end{bmatrix} \tag{9.9}$$

We define $\mathbf{x}(t) = \begin{bmatrix} x_1(t) \\ x_2(t) \end{bmatrix}$ as the state vector, which has as many states as the order of the system. We use bold letter for vectors and matrices. Define the matrix. $\mathbf{A} = \begin{bmatrix} 0 & 1 \\ -a_2 & -a_1 \end{bmatrix}$, the vector $\mathbf{B} = \begin{bmatrix} 0 \\ 1 \end{bmatrix}$ and the vector $\mathbf{C} = \begin{bmatrix} 1 & 0 \end{bmatrix}$. We then write (9.8) is a standard-state-space format.

$$\dot{x}(t) = \mathbf{A}x(t) + \mathbf{B}u(t)$$
$$y(t) = \mathbf{C}x(t) \tag{9.10}$$

This is the most common nomenclature used in the literature. We can extend the idea to any order systems. For example, the third order system

$$\dddot{y}(t) + a_1 \ddot{y}(t) + a_2 \dot{y}(t) + a_3 y(t) = u(t) \tag{9.11}$$

We can define

$$x_1(t) = y(t)$$
$$x_2(t) = \dot{y}(t)$$
$$x_3(t) = \ddot{y}(t) \tag{9.12}$$

and obtain the third order state-space description

$$\begin{bmatrix} \dot{x}_1(t) \\ \dot{x}_2(t) \\ \dot{x}_3(t) \end{bmatrix} = \begin{bmatrix} 0 & 1 & 0 \\ 0 & 0 & 1 \\ -a_3 & -a_2 & -a_1 \end{bmatrix} \begin{bmatrix} x_1(t) \\ x_2(t) \\ x_3(t) \end{bmatrix} + \begin{bmatrix} 0 \\ 0 \\ 1 \end{bmatrix} u(t)$$

$$y(t) = \begin{bmatrix} 1 & 0 & 0 \end{bmatrix} \begin{bmatrix} x_1(t) \\ x_2(t) \\ x_3(t) \end{bmatrix} \tag{9.13}$$

The states were defined arbitrarily in the above examples and although their assignments are not unique, the relationship from output to input must be unique. We can choose the states differently, for example we could have chosen:

$$x_1(t) = \ddot{y}(t)$$
$$x_2(t) = \dot{y}(t)$$
$$x_3(t) = y(t) \tag{9.14}$$

and arrived at an alternative state-space description

$$\begin{bmatrix} \dot{x}_1(t) \\ \dot{x}_2(t) \\ \dot{x}_3(t) \end{bmatrix} = \begin{bmatrix} -a_1 & -a_2 & -a_3 \\ 1 & 0 & 0 \\ 0 & 1 & 0 \end{bmatrix} \begin{bmatrix} x_1(t) \\ x_2(t) \\ x_3(t) \end{bmatrix} + \begin{bmatrix} 1 \\ 0 \\ 0 \end{bmatrix} u(t)$$

$$y(t) = \begin{bmatrix} 0 & 0 & 1 \end{bmatrix} \begin{bmatrix} x_1(t) \\ x_2(t) \\ x_3(t) \end{bmatrix} \tag{9.15}$$

These two state-space results are known as *realisations* since either is correct and from output to input the systems will behave the same. The realisations are known as phase-variable canonical form. A canonical form is like a pattern you can follow and is usually known as a standard format. We can see from (9.13) or (9.15) that a pattern emerges in the form of the **A** matrix. It is known as a *companion* matrix. The states themselves are defined differently so the state-responses will be different but the **C** and **B** vectors ensure that from input to output they behave the same. Another interesting realisation is found from the differential equation.

$$\ddot{y}(t) + a_1\ddot{y}(t) + a_2\dot{y}(t) + a_3y(t) = u(t)$$

Take Laplace transforms and write

$$s^3y(s) + s^2a_1y(s) + sa_2y(s) + a_3y(s) = u(s)$$

Now divide both sides with the highest power of s

$$y(s) + \frac{1}{s}a_1y(s) + \frac{1}{s^2}a_2y(s) + \frac{1}{s^3}a_3y(s) = \frac{1}{s^3}u(s) \tag{9.16}$$

Re-arrange

$$y(s) = \frac{1}{s^3}[u(s) - a_3 y(s)] - \frac{1}{s^2}[a_2 y(s)] - \frac{1}{s}[a_1 y(s)] \qquad (9.17)$$

Now define $x_1(s) = y(s)$ (we could choose another state of course but the result would be yet another realisation). Now consider the terms in square brackets. Moving left to right through the integrators:

For the triple integrator output, label it as

$$x_3(s) = \frac{1}{s}[u(s) - a_3 y(s)] \qquad (9.18)$$

The double integrator output is

$$x_2(s) = \frac{1}{s}\{-[a_2 y(s)] + x_3(s)\} \qquad (9.19)$$

This includes the previous integrator output as an input.
And for the single integrator, label its output

$$x_1(s) = \frac{1}{s}\{-[a_1 y(s)] + x_2(s)\} \qquad (9.20)$$

If we start at (9.20) and substitute the states from (9.19) and (9.18), we arrive back at (9.17). But (9.20) can be written in vector form once transformed into the time-domain.

From (9.20)

$$\dot{x}_1(t) = -a_1 y(t) + x_2(t)$$
$$= -a_1 x_1(t) + x_2(t) \qquad (9.21)$$

From (9.19)

$$\dot{x}_2(t) = -a_2 x_1(t) + x_3(t) \qquad (9.22)$$

And from (9.18)

$$\dot{x}_3(t) = -a_3 x_1(t) + u(t) \qquad (9.23)$$

Combining

$$\begin{bmatrix} \dot{x}_1(t) \\ \dot{x}_2(t) \\ \dot{x}_3(t) \end{bmatrix} = \begin{bmatrix} -a_1 & 1 & 0 \\ -a_2 & 0 & 1 \\ -a_3 & 0 & 0 \end{bmatrix} \begin{bmatrix} x_1(t) \\ x_2(t) \\ x_3(t) \end{bmatrix} + \begin{bmatrix} 0 \\ 0 \\ 1 \end{bmatrix} u(t)$$

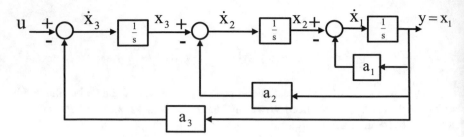

Fig. 9.1 Block diagram of state-space description (9.24)

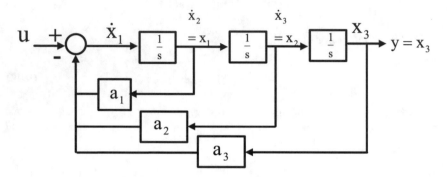

Fig. 9.2 Block diagram of state-space description (9.15)

$$y(t) = \begin{bmatrix} 1 & 0 & 0 \end{bmatrix} \begin{bmatrix} x_1(t) \\ x_2(t) \\ x_3(t) \end{bmatrix} \tag{9.24}$$

Drawing a block diagram, we see in Fig. 9.1 that each integrator is labelled with a state.

For Eq. (9.15) we get a different block diagram. (Fig. 9.2)

By re-labelling the output of the integrators, we can arrive at a different state-space description. One of the most popular state-space descriptions is the phase-variable canonical form where the second state is the derivative of the first, second state is derivative of the third state and so on. This is convenient for electro-mechanical systems where the states can take on physical values that are the derivative of each other. For example, position, velocity and acceleration.

9.3   Solution of the State Equation

The point of this approach is that no matter how high the order of the system, the vector representation of (9.10) is always of order one. Knowing it is of order one means that we use a standard method to manipulate it and solve the equation. For example, if the system had only one state

$$\dot{x}(t) = ax(t) + bu(t)$$
$$y(t) = cx(t)$$

Re-write with an initial condition on the state

$$\dot{x}(t) - ax(t) = bu(t); \; x(0) = x_0 \tag{9.25}$$

A well-known technique to solve such equations is to introduce an integrating factor to make the left-hand side of (9.25) a differential of the form.

$$e^{-at}(\dot{x}(t) - ax(t)) = \frac{d}{dt}\left(e^{-at}x(t)\right) \tag{9.26}$$

Multiply (9.25) by e^{-at}

$$\frac{d}{dt}\left(e^{-at}x(t)\right) = e^{-at}bu(t) \tag{9.27}$$

Now integrate both sides of (9.27) with limits from 0 to t. A change of variable is needed

$$e^{-at}x(t) - x(0) - \int_0^t e^{-a\tau}bu(\tau)d\tau \tag{9.28}$$

and obtain

$$e^{-at}x(t) - x(0) = \int_0^t e^{-a\tau}bu(\tau)d\tau \tag{9.29}$$

Then

$$x(t) = e^{at}x(0) + e^{at}\int_0^t e^{-a\tau}bu(\tau)d\tau \tag{9.30}$$

and finally

$$x(t) = e^{at}x_0 + \int_0^t e^{a(t-\tau)}bu(\tau)d\tau \tag{9.31}$$

The first part of the RHS of (9.31) is the response to the initial condition of the state. It is sometimes called the zero-input response or unforced response. The integral on the RHS of (9.31) is a convolution integral and known as the zero-state response.

For example, suppose $a = -2$, $b = 1$, $c = 2$. Let the initial condition on the state be $x_0 = 0.1$. Then (9.31) becomes

$$x(t) = 0.1e^{-2t} + \int_0^t e^{-2(t-\tau)}u(\tau)d\tau \tag{9.32}$$

and the output equation is

$$y(t) = 2x(t) \tag{9.33}$$

Suppose a unit step is applied to our system. Then $u(t) = 1$ and (9.21) is

$$x(t) = 0.1e^{-2t} + \int_0^t e^{-2(t-\tau)}d\tau$$

$$= 0.1e^{-2t} + e^{-2t}\int_0^t e^{2\tau}d\tau$$

$$= 0.1e^{-2t} + e^{-2t}\frac{1}{2}e^{2\tau}\Big|_0^t$$

$$= 0.1e^{-2t} + e^{-2t}\frac{1}{2}\left(e^{2t} - 1\right)$$

$$= 0.1e^{-2t} + \frac{1}{2}\left(1 - e^{-2t}\right)$$

The initial condition will die away to zero and the convolution part responds like a typical first order system. The state must reach a steady-state value of 0.5, but the output of the system will be twice as large since from (9.33)

$$y(t) = 2\left[0.1e^{-2t} + \frac{1}{2}\left(1 - e^{-2t}\right)\right] \tag{9.34}$$

If we treat the vector state equation case in a similar way, we get the general solution to the state as

$$\mathbf{x}(t) = e^{At}\mathbf{x}_0 + \int_0^t e^{A(t-\tau)}\mathbf{B}u(\tau)d\tau \tag{9.35}$$

with output equation

$$y(t) = \mathbf{C}\mathbf{x}(t) \tag{9.36}$$

The exponential for the integrating factor now is applied to a matrix quantity. Although it is possible to solve (9.35) for a few simple cases, it is not much practical use in general and discrete-time methods are preferred. There are a few methods of finding the matrix exponential $e^{\mathbf{A}t}$, usually term called *state-transition matrix* in (9.35) without digitising, these include:

Laplace method

$$e^{\mathbf{A}t} = \mathcal{L}^{-1}[s\mathbf{I} - \mathbf{A}]^{-1}$$

For example, if

$$\mathbf{A} = \begin{bmatrix} 0 & 1 \\ -a_2 & -a_1 \end{bmatrix}$$

$$s\mathbf{I} - \mathbf{A} = \begin{bmatrix} s & -1 \\ a_2 & s + a_1 \end{bmatrix}$$

And its inverse is

$$[s\mathbf{I} - \mathbf{A}]^{-1} = \frac{1}{s^2 + a_1 s + a_2} \begin{bmatrix} s + a_1 & 1 \\ -a_2 & s_1 \end{bmatrix}$$

To find the matrix exponential $e^{\mathbf{A}t}$, we then need to take inverse Laplace transforms of each element in the matrix.

Power-series expansion method

$$e^{\mathbf{A}t} = \mathbf{I} + \mathbf{A}t + \frac{\mathbf{A}^2 t^2}{2!} + \ldots + \frac{\mathbf{A}^n t^n}{n!} + \ldots$$

For some large enough n. This method is a little messy, since each term in the matrix is itself a power-series and we must recognise each term to convert the terms to a closed form solution. Although through the decades a number of alternative solutions have been found, perhaps the most direct way is to use computer algebra [3].

Consider a second-order system with state-space matrices

$$\begin{bmatrix} \dot{x}_1(t) \\ \dot{x}_2(t) \end{bmatrix} = \begin{bmatrix} 0 & 1 \\ -12 & -7 \end{bmatrix} \begin{bmatrix} x_1(t) \\ x_2(t) \end{bmatrix} + \begin{bmatrix} 0 \\ 1 \end{bmatrix} u(t)$$

$$y(t) = \begin{bmatrix} 12 & 0 \end{bmatrix} \begin{bmatrix} x_1(t) \\ x_2(t) \end{bmatrix}$$

With initial conditions

$$\mathbf{x}_0 = \begin{bmatrix} 0.1 \\ 0.1 \end{bmatrix}$$

Find the unit step response for the states and the output.
Solution.
First find the state-transition matrix $e^{At} = \mathcal{L}^{-1}[s\mathbf{I} - \mathbf{A}]^{-1}$

$$[s\mathbf{I} - \mathbf{A}]^{-1}$$

$$= \begin{bmatrix} s & -1 \\ 12 & s+7 \end{bmatrix}^{-1}$$

$$= \frac{1}{s^2 + 7s + 12} \begin{bmatrix} s+7 & 1 \\ -12 & s \end{bmatrix}$$

$$= \frac{1}{(s+3)(s+4)} \begin{bmatrix} s+7 & 1 \\ -12 & s \end{bmatrix}$$

And we require the inverse Laplace transform of the matrix element by element. We can use the cover-up method for partial-fraction expansion to speed the process up. There are two simple poles at s = -3 and s = -4. Using partial fractions, we get

$$\frac{1}{(s+3)(s+4)} \begin{bmatrix} s+7 & 1 \\ -12 & s \end{bmatrix} = \begin{bmatrix} \frac{4}{(s+3)} - \frac{3}{(s+4)} & \frac{1}{(s+3)} - \frac{1}{(s+4)} \\ \frac{-12}{(s+3)} + \frac{12}{(s+4)} & \frac{-3}{(s+3)} + \frac{4}{(s+4)} \end{bmatrix}$$

Now take inverse Laplace transforms from tables

$$e^{At} = \mathcal{L}^{-1} \begin{bmatrix} \frac{4}{(s+3)} - \frac{3}{(s+4)} & \frac{1}{(s+3)} - \frac{1}{(s+4)} \\ \frac{-12}{(s+3)} + \frac{12}{(s+4)} & \frac{-3}{(s+3)} + \frac{4}{(s+4)} \end{bmatrix}$$

$$= \begin{bmatrix} 4e^{-3t} - 3e^{-4t} & e^{-3t} - e^{-4t} \\ -12e^{-3t} + 12e^{-4t} & -3e^{-3t} + 4e^{-4t} \end{bmatrix}$$

The zero-input response becomes from (9.25)

$$\mathbf{x}(t) = e^{At}\mathbf{x}_0$$

$$= \begin{bmatrix} 4e^{-3t} - 3e^{-4t} & e^{-3t} - e^{-4t} \\ -12e^{-3t} + 12e^{-4t} & -3e^{-3t} + 4e^{-4t} \end{bmatrix} \begin{bmatrix} 0.1 \\ 0.1 \end{bmatrix}$$

$$= \frac{1}{10} \begin{bmatrix} 5e^{-3t} - 4e^{-4t} \\ -15e^{-3t} + 16e^{-4t} \end{bmatrix}$$

Therefore, the zero-input response due to initial conditions is

$$\begin{bmatrix} x_1(t) \\ x_2(t) \end{bmatrix} = \begin{bmatrix} 0.5e^{-3t} - 0.4e^{-4t} \\ -1.5e^{-3t} + 1.6e^{-4t} \end{bmatrix}$$

To find the zero-state response we use the convolution integral

$$\mathbf{x}(t) = \int_0^t e^{A(t-\tau)} \mathbf{B} u(\tau) d\tau$$

With an input $u = 1$ for a unit step.
Substituting we get

$$\mathbf{x}(t) = \int_0^t \left[\begin{bmatrix} 4e^{-3(t-\tau)} - 3e^{-4(t-\tau)} & e^{-3(t-\tau)} - e^{-4(t-\tau)} \\ -12e^{-3(t-\tau)} + 12e^{-4(t-\tau)} & -3e^{-3(t-\tau)} + 4e^{-4(t-\tau)} \end{bmatrix} \right] \begin{bmatrix} 0 \\ 1 \end{bmatrix} d\tau$$

Now because the **B** vector has a zero as the first element, this help us in simplification.

$$\mathbf{x}(t) = \int_0^t \begin{bmatrix} e^{-3(t-\tau)} - e^{-4(t-\tau)} \\ -3e^{-3(t-\tau)} + 4e^{-4(t-\tau)} \end{bmatrix} d\tau$$

This integration is with respect to τ, so we can treat any exponential terms with t as constants.

$$\mathbf{x}(t) = \begin{bmatrix} \frac{e^{-3t}}{3} e^{3\tau} - \frac{e^{-4t}}{4} e^{4\tau} \\ -e^{-3t} e^{3\tau} + e^{-4t} e^{4\tau} \end{bmatrix}_{\tau=0}^{\tau=t}$$

$$= \begin{bmatrix} \frac{1}{3} - \frac{1}{4} \\ 0 \end{bmatrix} - \begin{bmatrix} \frac{e^{-3t}}{3} - \frac{e^{-4t}}{4} \\ -e^{-3t} + e^{-4t} \end{bmatrix}$$

$$= \begin{bmatrix} \frac{1}{12} - \frac{e^{-3t}}{3} + \frac{e^{-4t}}{4} \\ e^{-3t} - e^{-4t} \end{bmatrix}$$

The output follows from

$$y(t) = \begin{bmatrix} 12 & 0 \end{bmatrix} \begin{bmatrix} x_1(t) \\ x_2(t) \end{bmatrix}$$

$$= 6e^{-3t} - 4.8e^{-4t} + 1 - 4e^{-3t} + 3e^{-4t}$$

In steady-state the initial-condition dies out and the final value goes to 1.

A quick check on the zero-state response $\begin{bmatrix} \frac{1}{12} - \frac{e^{-3t}}{3} + \frac{e^{-4t}}{4} \\ e^{-3t} - e^{-4t} \end{bmatrix}$ reveals that the second state is the derivative of the first. (same applies to the zero-input response). This is because of the canonical form of the state-space description where the second state is the derivative of the first.

9.4 Poles of the State-Space System

If we know the state-space description we an work back to the transfer-function. Take Laplace transforms of

$$\dot{x}(t) = Ax(t) + Bu(t)$$
$$y(t) = Cx(t)$$

$$sx(s) = Ax(s) + Bu(s)$$
$$y(s) = Cx(s) \tag{9.37}$$

Collecting terms (note that I is the identity matrix of the same order as the A matrix, a matrix of diagonal 1's).

$$(sI - A)x(s) = Bu(s)$$

Therefore

$$x(s) = (sI - A)^{-1}Bu(s) \tag{9.38}$$

and

$$y(s) = C(sI - A)^{-1}Bu(s) \tag{9.39}$$

The transfer function is therefore given by the relationship from output to input which is

$$G(s) = C(sI - A)^{-1}B \tag{9.40}$$

But the inverse of the matrix in (9.40) can be written as

$$(sI - A)^{-1} = \frac{\text{adj}(sI - A)}{|sI - A|} \tag{9.41}$$

where $\text{adj}(sI - A)$ is the adjoint matrix and $|sI - A|$ is the determinant. The determinant is a polynomial of the same degree as the number of states. Therefore, the poles of the system are defined by the solution of

$$|s\mathbf{I} - \mathbf{A}| = 0 \qquad (9.42)$$

Equation (9.42) is recognisable as the equation for eigenvalues of the matrix $\mathbf{A}$. Therefore, the poles of the system are the eigenvalues of the $\mathbf{A}$ matrix. As a simple example consider $\mathbf{A} = \begin{bmatrix} 0 & 1 \\ -a_2 & -a_1 \end{bmatrix}$. Then the eigenvalues of $\mathbf{A}$ are found from

$$|s\mathbf{I} - \mathbf{A}| = \begin{vmatrix} s & -1 \\ a_2 & s + a_1 \end{vmatrix}$$

$$= s^2 + a_1 s + a_2 \qquad (9.43)$$

And the roots of (9.43) are the poles of the system.

9.5   State-Space Descriptions with Zeros

So far, the examples have had poles only in the description. Suppose we have an example with a numerator polynomial. The numerator has one less degree polynomial than the denominator as called a *strictly proper* system.

$$\frac{y(s)}{u(s)} = \frac{b_3 + b_2 s + b_1 s^2}{s^3 + a_1 s^2 + a_2 s + a_3} \qquad (9.44)$$

Here the numerator term is one less degree than the denominator. We can deal with this case in a number of ways. First method is to multiply out and get

$$s^3 y(s) + s^2 a_1 y(s) + s a_2 y(s) + a_3 y(s) = b_3 u(s) + b_2 s u(s) + b_1 s^2 u(s) \qquad (9.45)$$

Divide by the highest power of s and collect terms, obtaining

$$y(s) = \frac{1}{s^3}\big[b_3 u(s) - a_3 y(s)\big] + \frac{1}{s^2}\big[b_2 u(s) - a_2 y(s)\big] + \frac{1}{s}\big[b_1 u(s) - a_1 y(s)\big] \qquad (9.46)$$

Define $y(s) = x_1(s)$. For the triple integrator output, label it as

$$x_3(s) = \frac{1}{s}\big[b_3 u(s) - a_3 y(s)\big] \qquad (9.47)$$

The double integrator output is

$$x_2(s) = \frac{1}{s}\big\{\big[b_2 u(s) - a_2 y(s)\big] + x_3(s)\big\} \qquad (9.48)$$

And for the single integrator, label its output

$$x_1(s) = \frac{1}{s}\{[b_1 u(s) - a_1 y(s)] + x_2(s)\} \tag{9.49}$$

Inverse Laplace transforming each of the states and using $y(s) = x_1(s)$ gives us

$$\begin{bmatrix} \dot{x}_1(t) \\ \dot{x}_2(t) \\ \dot{x}_3(t) \end{bmatrix} = \begin{bmatrix} -a_1 & 1 & 0 \\ -a_2 & 0 & 1 \\ -a_3 & 0 & 0 \end{bmatrix} \begin{bmatrix} x_1(t) \\ x_2(t) \\ x_3(t) \end{bmatrix} + \begin{bmatrix} b_1 \\ b_2 \\ b_3 \end{bmatrix} u(t)$$

$$y(t) = \begin{bmatrix} 1 & 0 & 0 \end{bmatrix} \begin{bmatrix} x_1(t) \\ x_2(t) \\ x_3(t) \end{bmatrix} \tag{9.50}$$

Now suppose the numerator has the same order as the denominator. This is the limit of the numerator order in a continuous-time system since the degree of the numerator must be no larger than the degree of the denominator for physical realizability.

$$\frac{y(s)}{u(s)} = \frac{s^3 + b_1 s^2 + b_2 s + b_3}{s^3 + a_1 s^2 + a_2 s + a_3} \tag{9.51}$$

Rewrite (9.51) using a power of s which is one less than the denominator

$$\frac{y(s)}{u(s)} = \frac{d_0 s^2 + d_1 s + d_2}{s^3 + a_1 s^2 + a_2 s + a_3} + d_3 \tag{9.52}$$

Now solve for the unknown d coefficients by comparing powers of s after multiplying across. We can also use long division, which is equivalent to what we are doing above.

We get

$$\frac{y(s)}{u(s)} = \frac{(b_1 - a_1)s^2 + (b_2 - a_2)s + (b_3 - a_3)}{s^3 + a_1 s^2 + a_2 s + a_3} + 1$$

Now we can use one of our previous patterns. (canonical forms).

$$\begin{bmatrix} \dot{x}_1(t) \\ \dot{x}_2(t) \\ \dot{x}_3(t) \end{bmatrix} = \begin{bmatrix} -a_1 & 1 & 0 \\ -a_2 & 0 & 1 \\ -a_3 & 0 & 0 \end{bmatrix} \begin{bmatrix} x_1(t) \\ x_2(t) \\ x_3(t) \end{bmatrix} + \begin{bmatrix} b_1 - a_1 \\ b_2 - a_2 \\ b_3 - a_3 \end{bmatrix} u(t)$$

$$y(t) = \begin{bmatrix} 1 & 0 & 0 \end{bmatrix} \begin{bmatrix} x_1(t) \\ x_2(t) \\ x_3(t) \end{bmatrix} + u(t) \tag{9.53}$$

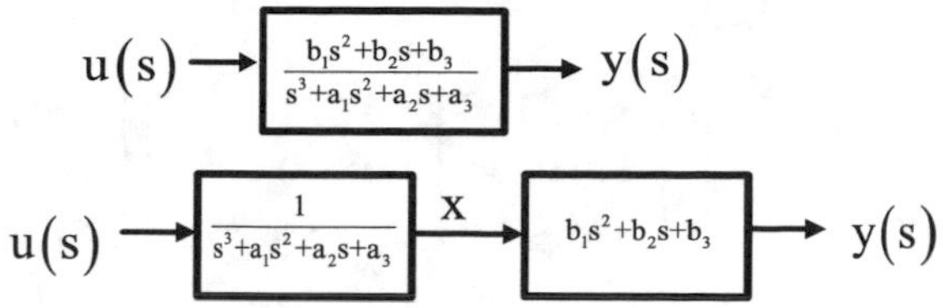

Fig. 9.3 Using a
state-defining variable for
systems with zeros

$$u(s) \longrightarrow \boxed{\frac{b_1s^2+b_2s+b_3}{s^3+a_1s^2+a_2s+a_3}} \longrightarrow y(s)$$

$$u(s) \longrightarrow \boxed{\frac{1}{s^3+a_1s^2+a_2s+a_3}} \xrightarrow{\text{X}} \boxed{b_1s^2+b_2s+b_3} \longrightarrow y(s)$$

Note the extra term in the output equation for y(t). In general, for these kinds of systems the general form is

$$\dot{x}(t) = Ax(t) + Bu(t)$$
$$y(t) = Cx(t) + Du(t) \tag{9.54}$$

If the numerator is the same degree as the denominator the system is called a *biproper* system.

Taking Laplace transforms we can find an expression for the transfer function as

$$G(s) = C(sI - A)^{-1}B + D \tag{9.55}$$

The **D** vector term is scalar here but if the system has more than one output or input it becomes a vector or a matrix. As an alternative to (9.53) we could use a phase-variable canonical form. To do this for a strictly proper system we split the numerator and denominator and define an intermediate state-defining variable which here is called x(s).

We then get two expressions from Fig. 9.3, namely

$$s^3x(s) + s^2a_1x(s) + sa_2x(s) + a_3x(s) = u(s) \tag{9.56}$$

and

$$y(s) = b_3x(s) + b_2sx(s) + b_1s^2x(s) \tag{9.57}$$

These can be put into the time-domain as follows

$$\dddot{x}(t) + a_1\ddot{x}(t) + a_2\dot{x}(t) + a_3x(t) = u(t) \tag{9.58}$$

$$y(t) = b_3x(t) + b_2\dot{x}(t) + b_1\ddot{x}(t) \tag{9.59}$$

Now define states as

$$x_1(t) = x(t)$$
$$x_2(t) = \dot{x}(t)$$
$$x_3(t) = \ddot{x}(t) \tag{9.60}$$

and arrive at

$$
\begin{bmatrix} \dot{x}_1(t) \\ \dot{x}_2(t) \\ \dot{x}_3(t) \end{bmatrix} = \begin{bmatrix} 0 & 1 & 0 \\ 0 & 0 & 1 \\ -a_3 & -a_2 & -a_1 \end{bmatrix} \begin{bmatrix} x_1(t) \\ x_2(t) \\ x_3(t) \end{bmatrix} + \begin{bmatrix} 0 \\ 0 \\ 1 \end{bmatrix} u(t)
$$

$$
y(t) = \begin{bmatrix} b_3 & b_2 & b_1 \end{bmatrix} \begin{bmatrix} x_1(t) \\ x_2(t) \\ x_3(t) \end{bmatrix} + u(t) \tag{9.61}
$$

For our *biproper* system

$$
\frac{y(s)}{u(s)} = \frac{(b_1 - a_1)s^2 + (b_2 - a_2)s + (b_3 - a_3)}{s^3 + a_1 s^2 + a_2 s + a_3} + 1
$$

The state-space description becomes

$$
\begin{bmatrix} \dot{x}_1(t) \\ \dot{x}_2(t) \\ \dot{x}_3(t) \end{bmatrix} = \begin{bmatrix} 0 & 1 & 0 \\ 0 & 0 & 1 \\ -a_3 & -a_2 & -a_1 \end{bmatrix} \begin{bmatrix} x_1(t) \\ x_2(t) \\ x_3(t) \end{bmatrix} + \begin{bmatrix} 0 \\ 0 \\ 1 \end{bmatrix} u(t)
$$

$$
y(t) = \begin{bmatrix} b_3 - a_3 & b_2 - a_2 & b_1 - a_1 \end{bmatrix} \begin{bmatrix} x_1(t) \\ x_2(t) \\ x_3(t) \end{bmatrix} + u(t) \tag{9.62}
$$

Given a state-space description with zeros, to find the zeros we need the eigenvalues of the following matrix:

$$
\begin{vmatrix} sI - A & -B \\ C & D \end{vmatrix} = 0 \tag{9.63}
$$

Suppose we have a transfer function

$$
\frac{y(s)}{u(s)} = \frac{s + 3}{s^2 + 7s + 10}
$$

And a realisation

$$
\begin{bmatrix} \dot{x}_1(t) \\ \dot{x}_2(t) \end{bmatrix} = \begin{bmatrix} 0 & 1 \\ -10 & -7 \end{bmatrix} \begin{bmatrix} x_1(t) \\ x_2(t) \end{bmatrix} + \begin{bmatrix} 0 \\ 1 \end{bmatrix} u(t)
$$

$$
y(t) = \begin{bmatrix} 3 & 1 \end{bmatrix} \begin{bmatrix} x_1(t) \\ x_2(t) \end{bmatrix}
$$

The **D** matrix term is zero here. The matrix determinant (9.63) becomes

$$\begin{vmatrix} s & -1 & 0 \\ 10 & s+7 & -1 \\ 3 & 1 & 0 \end{vmatrix} = 0 \qquad (9.64)$$

To find the determinant it is easiest to work down a row or column with majority zeros in it. This means that any minors (matrices within a matrix) multiply zero. Working down the 3rd column we need only calculate the second term which is

$$\begin{vmatrix} s & -1 \\ 3 & 1 \end{vmatrix} = 0 \qquad (9.65)$$

This gives us

$$s + 3 = 0 \qquad (9.66)$$

The zero is therefore at s = -3.

9.6 Controllability and Observability

Consider the transfer function

$$G(s) = \frac{s^2 + 3s + 2}{s^3 + 6s^2 + 11s + 6} \qquad (9.67)$$

Convert it to state-space. First let's use the phase-variable canonical form first

$$\begin{bmatrix} \dot{x}_1(t) \\ \dot{x}_2(t) \\ \dot{x}_3(t) \end{bmatrix} = \begin{bmatrix} 0 & 1 & 0 \\ 0 & 0 & 1 \\ -6 & -11 & -6 \end{bmatrix} \begin{bmatrix} x_1(t) \\ x_2(t) \\ x_3(t) \end{bmatrix} + \begin{bmatrix} 0 \\ 0 \\ 1 \end{bmatrix} u(t)$$

$$y(t) = \begin{bmatrix} 2 & 3 & 1 \end{bmatrix} \begin{bmatrix} x_1(t) \\ x_2(t) \\ x_3(t) \end{bmatrix} \qquad (9.68)$$

The controllability of a state-space description is the number of states that can be driven from the input. There is a simple rank test to determine if a system is controllable. For a system of order n, it defines a matrix $\mathcal{C}$ where

$$\mathcal{C} = \begin{bmatrix} \mathbf{B}, \mathbf{AB}, \mathbf{A}^2\mathbf{B} \ldots \mathbf{A}^{n-1}\mathbf{B} \end{bmatrix} \qquad (9.69)$$

For a system to be completely controllable the above matrix must be full row rank. This means that the number of independent rows must be n. Forming the controllability matrix, we find it is

$$\mathcal{C} = \begin{bmatrix} 0 & 0 & 1 \\ 0 & 1 & -6 \\ 1 & -6 & 25 \end{bmatrix} \tag{9.70}$$

For a matrix that is square, full rank implies that the determinant is non-zero. Equation (9.68) has rank 3 so it is a completely controllable system.

Now define a different state-space description

$$\begin{bmatrix} \dot{x}_1(t) \\ \dot{x}_2(t) \\ \dot{x}_3(t) \end{bmatrix} = \begin{bmatrix} -6 & 1 & 0 \\ -11 & 0 & 1 \\ -6 & 0 & 0 \end{bmatrix} \begin{bmatrix} x_1(t) \\ x_2(t) \\ x_3(t) \end{bmatrix} + \begin{bmatrix} 1 \\ 3 \\ 2 \end{bmatrix} u(t)$$

$$y(t) = \begin{bmatrix} 1 & 0 & 0 \end{bmatrix} \begin{bmatrix} x_1(t) \\ x_2(t) \\ x_3(t) \end{bmatrix} \tag{9.71}$$

and form the controllability matrix from (9.69). we get

$$\mathcal{C} = \begin{bmatrix} 1 & -3 & 9 \\ 3 & -9 & 27 \\ 2 & -6 & 18 \end{bmatrix} \tag{9.72}$$

We can see from (9.72) that if we multiply the second row by 3 we get the first row, so it cannot be full rank. Also, the third row is the first row multiplied by 2. The matrix is therefore rank 1. Two states are uncontrollable. The reason the first realisation in (9.68) does not show the uncontrollable states is because it has a special canonical form which is always controllable. It is known as *controllable canonical form*. But the reason for the problem in the first place is because the transfer function can be factorised thus

$$G(s) = \frac{(s+1)(s+2)}{(s+1)(s+2)(s+3)} \tag{9.73}$$

The numerator and denominator polynomials have two common factors. Polynomials with no common factors are known as *relatively prime* polynomials. We require our transfer function polynomials to be relatively prime. This is only a first order system pretending to be third order.

An observable system is a system from which the states can be reconstructed from measurements of the output.

The observability matrix is defined as $\boldsymbol{O}$, where

$$\boldsymbol{O} = \begin{bmatrix} \mathbf{C} \\ \mathbf{CA} \\ \mathbf{CA}^2 \\ . \\ . \\ . \\ \mathbf{CA}^{n-1} \end{bmatrix}$$

and must have row rank n for a system to be fully observable. For example, for the system described by (9.68) the observability matrix is

$$\boldsymbol{O} = \begin{bmatrix} 2 & 3 & 1 \\ -6 & -9 & -3 \\ 18 & 27 & 9 \end{bmatrix}$$

and has rank 1 indicating that two of the states are unobservable. Now apply the same method to the state-space description of (9.71) and get

$$\boldsymbol{O} = \begin{bmatrix} 1 & 0 & 0 \\ -6 & 1 & 0 \\ 25 & -6 & 1 \end{bmatrix}$$

Which has full rank 3. Therefore, this system is fully observable. The state space description of (9.71) is known as *observable canonical form*. Systems described in this way are always observable but not necessarily controllable. Observability is essential to create a filter to estimate or reconstruct the states.

9.7 Discrete-Time State-Space

Much of the discrete-time case follows similar patterns to continuous-time. Instead of splitting an nth order differential equation into n 1st order differential equations we split an nth order difference equation into n first order difference equations and write in vector form.

Consider the 3rd order transfer function

$$\frac{y(z)}{u(z)} = \frac{1}{z^3 + a_1 z^2 + a_2 z + a_3} \tag{9.74}$$

which has difference equation

$$y(k+3) + a_1 y(k+2) + a_2 y(k+1) + a_3 y(k) = u(k) \tag{9.75}$$

Define the states as

$$x_1(k) = y(k)$$
$$x_2(k) = y(k + 1)$$
$$x_3(k) = y(k + 2)$$

Then we get the following state-space description in phase-variable canonical form:

$$\begin{bmatrix} x_1(k+1) \\ x_2(k+1) \\ x_3(k+1) \end{bmatrix} = \begin{bmatrix} 0 & 1 & 0 \\ 0 & 0 & 1 \\ -a_3 & -a_2 & -a_1 \end{bmatrix} \begin{bmatrix} x_1(k) \\ x_2(k) \\ x_3(k) \end{bmatrix} + \begin{bmatrix} 0 \\ 0 \\ 1 \end{bmatrix} u(k)$$

$$y(k) = \begin{bmatrix} 1 & 0 & 0 \end{bmatrix} \begin{bmatrix} x_1(k) \\ x_2(k) \\ x_3(k) \end{bmatrix} \tag{9.76}$$

This is a vector difference equation of order 1 and we can write it as

$$x(k + 1) = Fx(k) + Gu(k)$$
$$y(k) = Hx(k) + Du(k) \tag{9.77}$$

where $D = 0$. We use different symbols for the other matrices to distinguish them from the continuous time case.

Similarly, when there are zeros, we use the same technique as for continuous time.

$$\frac{y(z)}{u(z)} = \frac{b_1 z^2 + b_2 z + b_3}{z^3 + a_1 z^2 + a_2 z + a_3} \tag{9.78}$$

Which results in a difference equation of the form

$$y(k + 3) + a_1 y(k + 2) + a_2 y(k + 1) + a_3 y(k) = b_1 u(k + 2) + b_2 u(k + 1) + b_3 u(k)$$

We split the numerator from the denominator and define an intermediate variable giving two difference equations:

$$x(k + 3) + a_1 x(k + 2) + a_2 x(k + 1) + a_3 x(k) = u(k)$$
$$y(k) = b_1 x(k + 2) + b_2 x(k + 1) + b_3 x(k) \tag{9.79}$$

Then define states according to

$$x_1(k) = x(k)$$
$$x_2(k) = x(k + 1)$$
$$x_3(k) = x(k + 2)$$

And we get the state-space realisation

$$
\begin{bmatrix} x_1(k+1) \\ x_2(k+1) \\ x_3(k+1) \end{bmatrix} = \begin{bmatrix} 0 & 1 & 0 \\ 0 & 0 & 1 \\ -a_3 & -a_2 & -a_1 \end{bmatrix} \begin{bmatrix} x_1(k) \\ x_2(k) \\ x_3(k) \end{bmatrix} + \begin{bmatrix} 0 \\ 0 \\ 1 \end{bmatrix} u(k)
$$

$$
y(k) = \begin{bmatrix} b_3 & b_2 & b_1 \end{bmatrix} \begin{bmatrix} x_1(k) \\ x_2(k) \\ x_3(k) \end{bmatrix} \tag{9.80}
$$

Alternatively, we can start again with the difference equation

$$
y(k+3) + a_1 y(k+2) + a_2 y(k+1) + a_3 y(k) = b_1 u(k+2) + b_2 u(k+1) + b_3 u(k)
$$

And take Z-Transforms

$$
z^3 y(z) + z^2 a_1 y(z) + z a_2 y(z) + a_3 y(z) = z^2 b_1 u(z) + z b_2 u(z) + b_3 u(z) \tag{9.81}
$$

Then write $y(z)$ on the LHS of (9.81) and collect negative powers of z thus

$$
y(z) = \big[b_1 u(z) - a_1 y(z)\big]z^{-1} + \big[b_2 u(z) - a_2 y(z)\big]z^{-2} + \big[b_3 u(z) - a_3 y(z)\big]z^{-3} \tag{9.82}
$$

Define $y(z) = x_1(z)$ and the following states in the Z-domain

$$
\begin{aligned}
x_3(z) &= z^{-1}\big[b_3 u(z) - a_3 y(z)\big] \\
x_2(z) &= z^{-1}\big\{\big[b_2 u(z) - a_2 y(z)\big] + x_3(z)\big\} \\
x_1(z) &= z^{-1}\big\{\big[b_1 u(z) - a_1 y(z)\big] + x_2(z)\big\}
\end{aligned} \tag{9.83, a,b,c}
$$

Note, as a check, by back substituting we have

$$
y(z) = z^{-1}\big\{\big[b_1 u(z) - a_1 y(z)\big] + z^{-1}\big\{\big[b_2 u(z) - a_2 y(z)\big] + z^{-1}\big[b_3 u(z) - a_3 y(z)\big]\big\}\big\} \tag{9.84}
$$

Which is (9.82) written in a nested form.

Multiply (9.83, a,b,c) by z and then take inverse Z-transforms and obtain three state-variable equations

$$
\begin{aligned}
x_3(k+1) &= \big[b_3 u(k) - a_3 y(k)\big] \\
x_2(k+1) &= \big\{\big[b_2 u(k) - a_2 y(k)\big] + x_3(k)\big\} \\
x_1(k+1) &= \big\{\big[b_1 u(k) - a_1 y(k)\big] + x_2(k)\big\}
\end{aligned}
$$

Also substitute in the above

$$y(k) = x_1(k)$$

In vector format we get the observable canonical form

$$\begin{bmatrix} x_1(k+1) \\ x_2(k+1) \\ x_3(k+1) \end{bmatrix} = \begin{bmatrix} -a_1 \ 1 \ 0 \\ -a_2 \ 0 \ 1 \\ -a_3 \ 0 \ 0 \end{bmatrix} \begin{bmatrix} x_1(k) \\ x_2(k) \\ x_3(k) \end{bmatrix} + \begin{bmatrix} b_1 \\ b_2 \\ b_3 \end{bmatrix} u(k)$$

$$y(k) = \begin{bmatrix} 1 \ 0 \ 0 \end{bmatrix} \begin{bmatrix} x_1(k) \\ x_2(k) \\ x_3(k) \end{bmatrix}$$

When the numerator and denominator have the same order polynomial

$$\frac{y(z)}{u(z)} = \frac{z^3 + b_1 z^2 + b_2 z + b_3}{z^3 + a_1 z^2 + a_2 z + a_3}$$

We can use long division again to obtain

$$\frac{y(z)}{u(z)} = \frac{(b_1 - a_1)z^2 + (b_2 - a_2)z + (b_3 - a_3)}{z^3 + a_1 z^2 + a_2 z + a_3} + 1$$

and obtain a discrete-time state-space equation

$$\begin{bmatrix} x_1(k+1) \\ x_2(k+1) \\ x_3(k+1) \end{bmatrix} = \begin{bmatrix} -a_1 \ 1 \ 0 \\ -a_2 \ 0 \ 1 \\ -a_3 \ 0 \ 0 \end{bmatrix} \begin{bmatrix} x_1(k) \\ x_2(k) \\ x_3(k) \end{bmatrix} + \begin{bmatrix} b_1 - a_1 \\ b_2 - a_2 \\ b_3 - a_3 \end{bmatrix} u(k)$$

$$y(k) = \begin{bmatrix} 1 \ 0 \ 0 \end{bmatrix} \begin{bmatrix} x_1(k) \\ x_2(k) \\ x_3(k) \end{bmatrix} + u(k)$$

The transfer function can be obtained from state-space by taking Z-Transforms of the state equations

$$x(k+1) = \mathbf{F}x(k) + \mathbf{G}u(k)$$
$$y(k) = \mathbf{H}x(k) + \mathbf{D}u(k)$$

$$zx(z) = \mathbf{F}x(z) + \mathbf{G}u(z)$$
$$y(z) = \mathbf{H}x(z) + \mathbf{D}u(z)$$

Collecting terms

$$(z\mathbf{I} - \mathbf{F})x(z) = \mathbf{G}u(z)$$

Or

$$\mathbf{x}(z) = (z\mathbf{I} - \mathbf{F})^{-1}\mathbf{G}u(z)$$

And then

$$\frac{y(z)}{u(z)} = \mathbf{H}(z\mathbf{I} - \mathbf{F})^{-1}\mathbf{G} + \mathbf{D}$$

$$= G(z)$$

Sometimes this is written in backward shift notation

$$G(z) = \mathbf{H}(\mathbf{I} - \mathbf{F}z^{-1})^{-1}z^{-1}\mathbf{G} + \mathbf{D} \qquad (9.85)$$

The eigenvalues of the $\mathbf{F}$ matrix are the poles of the system and are found from

$$|z\mathbf{I} - \mathbf{F}| = 0 \qquad (9.86)$$

The poles of the system must have magnitude less than unity for stability. For a input $u(k)$ we can find the state output from

$$\mathbf{x}(k) = \mathbf{F}^k\mathbf{x}(0) + \sum_{i=0}^{k-1} \mathbf{F}^{k-i-1}\mathbf{G}u(i) \qquad (9.87)$$

The first part of the RHS of (9.87) is the zero-input response to initial conditions and the second part is the discrete convolution summation which describes the zero-state response.

Controllability and observability are the same as for continuous time with a change of matrices. The controllability and observability matrices

$$\mathcal{C} = [\mathbf{G}, \mathbf{FG}, \mathbf{F}^2\mathbf{G} \ldots \mathbf{F}^{n-1}\mathbf{G}] \qquad (9.88)$$

$$\mathcal{O} = \begin{bmatrix} \mathbf{H} \\ \mathbf{HF} \\ \mathbf{HF}^2 \\ . \\ . \\ . \\ \mathbf{HF}^{n-1} \end{bmatrix} \qquad (9.89)$$

must both be full row rank.

9.8   Similarity Transformations

For continuous or discrete-time state-space systems we can transform the system into an alternative coordinate system by using a matrix transformation. For example, consider a discrete state-space representation

$$\mathbf{x}(k+1) = \mathbf{Fx}(k) + \mathbf{Gu}(k)$$
$$y(k) = \mathbf{Hx}(k) \qquad\qquad\qquad (9.90)$$

Define a new state vector $\mathbf{z}(k)$ where

$$\mathbf{x}(k) = \mathbf{Tz}(k) \qquad\qquad\qquad (9.91)$$

Here bold z, that is $\mathbf{z}$ should not be confused with the Z-Transform operator. Here $\mathbf{z}$ is a vector and $\mathbf{T}$ a non-singular arbitrary matrix. Substitute (9.91) into (9.90) and obtain

$$\mathbf{Tz}(k+1) = \mathbf{FTz}(k) + \mathbf{Gu}(k)$$
$$y(k) = \mathbf{HTz}(k) \qquad\qquad\qquad (9.92)$$

Solving for $\mathbf{z}$

$$\mathbf{z}(k+1) = \mathbf{T}^{-1}\mathbf{FTz}(k) + \mathbf{T}^{-1}\mathbf{Gu}(k)$$
$$y(k) = \mathbf{HTz}(k) \qquad\qquad\qquad (9.93)$$

Example:

Suppose $\mathbf{F} = \begin{bmatrix} 0 & 1 \\ -0.5 & 1 \end{bmatrix}$, $\mathbf{G} = \begin{bmatrix} 0 \\ 1 \end{bmatrix}$, $\mathbf{H} = \begin{bmatrix} 1 & 0 \end{bmatrix}$. Choose $\mathbf{T} = \begin{bmatrix} 1 & -1 \\ 1 & 1 \end{bmatrix}$ and

$\mathbf{T}^{-1}\mathbf{FT} = \begin{bmatrix} 0.75 & 1.25 \\ -0.25 & 0.25 \end{bmatrix}$, $\mathbf{T}^{-1}\mathbf{G} = \begin{bmatrix} 0.5 \\ 0.5 \end{bmatrix}$, $\mathbf{HT} = \begin{bmatrix} 1 & -1 \end{bmatrix}$. We now have a new state-space description

$$\begin{bmatrix} z_1(k+1) \\ z_2(k+1) \end{bmatrix} = \begin{bmatrix} 0.75 & 1.25 \\ -0.25 & 0.25 \end{bmatrix} \begin{bmatrix} z_1(k) \\ z_2(k) \end{bmatrix} + \begin{bmatrix} 0.5 \\ 0.5 \end{bmatrix} u(k) \qquad (9.94)$$

$$y(k) = \begin{bmatrix} 1 & -1 \end{bmatrix} \begin{bmatrix} z_1(k) \\ z_2(k) \end{bmatrix} \qquad\qquad\qquad (9.95)$$

Equations (9.94) and (9.95) are said to be similar to our original system. The transformation is known as a *similarity transformation*.

There are clearly an infinity of state-space realisations for any system since $\mathbf{T}$ can be any non-singular matrix. We can prove that the similar system has the same poles as the original system by taking the eigenvalues of $\mathbf{T}^{-1}\mathbf{FT}$.

$$|\lambda I - T^{-1}FT|$$
$$= |\lambda T^{-1}I - T^{-1}F||T|$$
$$= |T^{-1}||\lambda I - F||T|$$
$$= |\lambda I - F|$$

In the above we use the fact that for any matrix, its inverse, which is scalar, has the same determinant as the original matrix.

A special case of similarity transform is the transform to *diagonal form*. Consider a matrix **F** which has n distinct *real* eigenvalues. Then we construct a matrix of these eigenvalues and call it **Λ**, where

$$
\Lambda = \begin{bmatrix}
\lambda_1 & 0 & 0 & \ldots & 0 \\
0 & \lambda_2 & & & 0 \\
0 & & \lambda_3 & & 0 \\
\cdot & & & \cdot & \cdot \\
\cdot & & & & \cdot & \cdot \\
\cdot & & & & \\
0 & 0 & 0 & & \lambda_n
\end{bmatrix}
\tag{9.96}
$$

(we can write this in shorthand notation as $\Lambda = \text{diag}\{\lambda_1 \ \lambda_2 \ \lambda_3 \ \ldots \ \lambda_n\}$).

Define the *modal* matrix **T** which is made up of the *eigenvectors* of **F**. Suppose an eigenvalue λ_i has a corresponding eigenvector v_i. Then by definition

$$Fv_i = \lambda_i v_i, i = 1, 2 \ldots n \tag{9.97}$$

and we construct the modal matrix from these eigenvectors

$$T = \begin{bmatrix} v_1 & v_2 & v_3 & \ldots & v_n \end{bmatrix} \tag{9.98}$$

In the same order as the diagonal matrix (9.96). Then this transformation will diagonalize the system.

Consider a system with real distinct eigenvalues (poles)

$$F = \begin{bmatrix} 0 & 1 \\ -0.02 & 0.3 \end{bmatrix}, G = \begin{bmatrix} 0 \\ 1 \end{bmatrix}, H = \begin{bmatrix} 1 & 0 \end{bmatrix}$$

Finding the eigen values of **F**, we have $\Lambda = \begin{bmatrix} -0.1 & 0 \\ 0 & -0.2 \end{bmatrix}$. Then we find the eigenvectors and construct the modal matrix.

$$\mathbf{T} = \begin{bmatrix} -0.995 & -0.9806 \\ -0.0995 & -0.1961 \end{bmatrix}. \text{ Then we find } \mathbf{T^{-1}FT} = \begin{bmatrix} -0.1 & 0 \\ 0 & -0.2 \end{bmatrix}, \mathbf{T^{-1}G} =$$

$$\begin{bmatrix} 10.0499 \\ -10.1980 \end{bmatrix}, \mathbf{HT} = \begin{bmatrix} -0.995 & -0.9806 \end{bmatrix}. \text{ We now have a diagonal state-space system.}$$

$$\begin{bmatrix} z_1(k+1) \\ z_2(k+1) \end{bmatrix} = \begin{bmatrix} -0.1 & 0 \\ 0 & -0.2 \end{bmatrix} \begin{bmatrix} z_1(k) \\ z_2(k) \end{bmatrix} + \begin{bmatrix} 10.0499 \\ -10.1980 \end{bmatrix} u(k) \qquad (9.99)$$

$$y(k) = \begin{bmatrix} -0.995 & -0.9806 \end{bmatrix} \begin{bmatrix} z_1(k) \\ z_2(k) \end{bmatrix} \qquad (9.100)$$

A signal-flow graph is shown in Fig. 9.4. A signal-flow graph is like a block diagram but is more common in certain fields than others. The nodes act as summing junctions always and have no negative sign. The sign is taken from the arrow in the preceding line of the diagram. Note how the input signal splits into two first-order system and then recombines at the output. This is the nature of the parallel or diagonal form. The inverse z-operator symbols represent unit step time-delays.

Figure 9.5 shows the signal-flow graph for the same system in its original phase-variable canonical form.

If $\mathbf{F}$ is symmetric then this is a special case again and the modal matrix has columns which are orthogonal. Suppose $\mathbf{F} = \begin{bmatrix} 0.2 & 0.5 \\ 0.5 & 0.1 \end{bmatrix}$, then $\mathbf{\Lambda} = \begin{bmatrix} -0.3525 & 0 \\ 0 & 0.6525 \end{bmatrix}$ and $\mathbf{T} = \begin{bmatrix} 0.6710 & -0.7415 \\ -0.7415 & -0.6710 \end{bmatrix}$. We find that the $\mathbf{T}$ matrix is orthogonal because

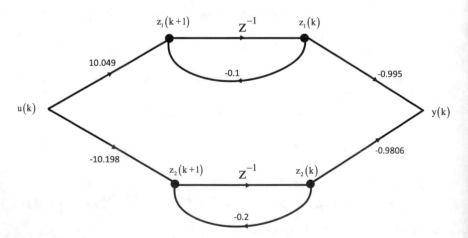

Fig. 9.4 Signal-flow graph for diagonalised system

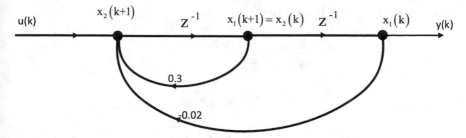

Fig. 9.5 Signal-flow graph for phase-variable canonical form

$$\mathbf{TT}^T = \begin{bmatrix} 0.6710 & -0.7415 \\ -0.7415 & -0.6710 \end{bmatrix} \begin{bmatrix} 0.6710 & -0.7415 \\ -0.7415 & -0.6710 \end{bmatrix} = \begin{bmatrix} 1 & 0 \\ 0 & 1 \end{bmatrix}, \text{ the identity matrix.}$$

An orthogonal matrix satisfies $\mathbf{TT}^T = \mathbf{I}$.

We can also use the similarity transformation to move from one canonical form to another. For example, if a system is controllable then we can transform it to controllable canonical form from another canonical form.

9.9 States as Measurable Signals

The previous section has shown a number of ways to represent states of a system, but they are all canonical forms of one variety or another. These are handy in some theoretical work, but in many physical systems we can use real measurements as the states. This is useful as a mathematical analysis method for systems. For example, consider the circuit shown in Fig. 9.6.

Before converting to state-space we can do some classical analysis on it and find the transfer function. The circuit is a lowpass filter. It is formed by an impedance divider of the RC network in parallel and the inductor in series. The RC network is found from the parallel Laplace form of the impedances giving

$$\frac{R/sc}{R + \frac{1}{sc}} = \frac{R}{1 + sCR}$$

Fig. 9.6 Second order passive analogue filter

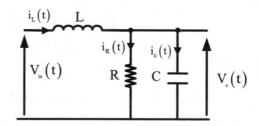

The transfer function is therefore

$$\frac{V_c}{V_{in}}(s) = \frac{\frac{R}{1+sCR}}{\frac{R}{1+sCR} + sL}$$

$$= \frac{R}{R + sL(1 + sCR)}$$

$$= \frac{R}{s^2LCR + sL + R}$$

$$= \frac{1/LC}{s^2 + s/RC + 1/LC}$$

At this stage we could create a canonical form for the transfer function. We could have for instance

$$\begin{bmatrix} \dot{x}_1(t) \\ \dot{x}_2(t) \end{bmatrix} = \begin{bmatrix} 0 & 1 \\ -1/LC & -1/RC \end{bmatrix} \begin{bmatrix} x_1(t) \\ x_2(t) \end{bmatrix} + \begin{bmatrix} 0 \\ 1/LC \end{bmatrix} V_{in}(t)$$

$$V_c(t) = \begin{bmatrix} 1 & 0 \end{bmatrix} \begin{bmatrix} x_1(t) \\ x_2(t) \end{bmatrix}$$

In this form however the states have no physical meaning. If instead we define the inductor current and capacitor voltages as state variables:

$$x_1(t) = i_L(t)$$
$$x_2(t) = V_c(t)$$

From Fig. 9.6, using Kirchhoff's voltage law

$$V_{in}(t) - L\frac{di_L(t)}{dt} - V_c(t) = 0$$

Giving

$$V_{in}(t) - L\dot{x}_1(t) - V_c(t) = 0$$

Then

$$\dot{x}_1(t) = \frac{1}{L}V_{in}(t) - \frac{1}{L}x_2(t)$$

By Kirchhoff's current law

$$i_L(t) = i_R(t) + i_C(t)$$
$$= \frac{V_c(t)}{R} + C\frac{dV_c}{dt}$$

Substituting for the state variables

$$x_1(t) = \frac{x_2(t)}{R} + C\dot{x}_2(t)$$

This results in a state-space realisation

$$\begin{bmatrix} \dot{x}_1(t) \\ \dot{x}_2(t) \end{bmatrix} = \begin{bmatrix} 0 & -1/L \\ 1/C & -1/RC \end{bmatrix} \begin{bmatrix} x_1(t) \\ x_2(t) \end{bmatrix} + \begin{bmatrix} 1/L \\ 0 \end{bmatrix} V_{in}(t)$$

$$V_c(t) = \begin{bmatrix} 0 & 1 \end{bmatrix} \begin{bmatrix} x_1(t) \\ x_2(t) \end{bmatrix}$$

Putting some values into the circuit. Suppose the input voltage is $V_{in}(t) = 1V, R = 100\Omega, L = 1mH, C = 50nF$. The state-space description becomes

$$\begin{bmatrix} \dot{x}_1(t) \\ \dot{x}_2(t) \end{bmatrix} = 10^7 \begin{bmatrix} 0 & -0.0001 \\ 2 & -0.02 \end{bmatrix} \begin{bmatrix} x_1(t) \\ x_2(t) \end{bmatrix} + \begin{bmatrix} 1000 \\ 0 \end{bmatrix} V_{in}(t)$$

$$V_c(t) = \begin{bmatrix} 0 & 1 \end{bmatrix} \begin{bmatrix} x_1(t) \\ x_2(t) \end{bmatrix}$$

Now we can simulate the system and examine the behaviour of the inductor current and capacitor voltage. Using the MATLAB code below, the following step responses we obtained as shown in Fig. 9.7 and Fig. 9.8.

MATLAB has functions that will simulate any continuous-time system. Transparent to the user is the method that it uses. No digital computer can exactly simulate a continuous-time system. In order to do so it must convert the system to discrete time using some form of integration method. The simulation must therefore have a

Fig. 9.7 Inductor current for unit step input

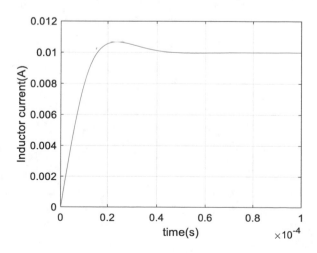

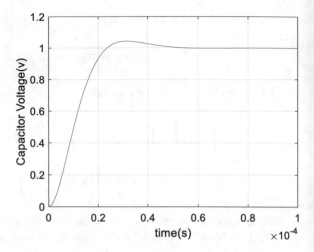

Fig. 9.8 Capacitor voltage for unit step input

sampling frequency. In this case the time array vector is chosen as t = 0:1e-6:0.1e-3; which is time zero to time 0.1 ms in steps of 1μ s. The sampling frequency is therefore 1 MHz.

MATLAB Code 9.1

```
%State-space example
R=100;
C1=50e-9;
L=1e-3;
A=[0 -1/L;1/C1 -1/(R*C1)];
B=[1/L 0]';
C=[0 1];
D=0;
%No initial conditions
x0=[0 0]';
% time vector
t=0:1e-6:0.1e-3;
%Step input
Vin=ones(size(t));
gc=ss(A,B,C,D)
[y,t,x]=lsim(gc,Vin,t,x0);
plot(t,x(:,1))
xlabel('time(s)')
ylabel('Inductor current(A)')
set(gca,'FontSize',14);
grid
set(gcf,'color','w');
figure
plot(t,x(:,2))
xlabel('time(s)')
ylabel('Capacitor Voltage(v)')
set(gca,'FontSize',14);
grid
set(gcf,'color','w');
```

9.10 Conversion from Continuous to Discrete Time State-Space

Conversion from continuous to discrete time state-space is a well-researched area for some time and is important from the point of view of computer simulation and digital control systems [4–6].

The continuous time state-space solution was found earlier from:

$$\mathbf{x}(t) = e^{At}\mathbf{x}_0 + \int_{\tau=0}^{\tau=t} e^{A(t-\tau)}\mathbf{B}u(\tau)d\tau \tag{9.101}$$

For a sampling interval T_s, substitute $t = kT_s$ into (9.101) and obtain

$$\mathbf{x}(kT_s) = e^{AkT_s}\mathbf{x}_0 + \int_{\tau=0}^{\tau=kT_s} e^{A(kT_s-\tau)}\mathbf{B}u(\tau)d\tau \tag{9.102}$$

If we increment time by one sample, (9.102) becomes

$$\mathbf{x}(k+1) = e^{A(k+1)T_s}\mathbf{x}_0 + \int_{\tau=0}^{\tau=(k+1)T_s} e^{A((k+1)T_s-\tau)}\mathbf{B}u(\tau)d\tau \tag{9.103}$$

where for clarity we drop the dependence on the sampling interval, i.e. $\mathbf{x}(kT_s) = \mathbf{x}(k)$.

Multiply (9.102) throughout by e^{AT_s}, subtract from (9.103) and rearrange. We get

$$\mathbf{x}(k+1) = e^{AT_s}\mathbf{x}(k) + \int_{\tau=kT_s}^{\tau=(k+1)T_s} e^{A(kT_s+T_s-\tau)}\mathbf{B}u(\tau)d\tau \tag{9.104}$$

Substitute $\sigma = (k+1)T_s - \tau$. Now when $\sigma = 0$, $\tau = (k+1)T_s$. When $\sigma = T_s$, $\tau = kT_s$ and $d\sigma = -d\tau$. The second term in the RHS of (9.104) now becomes

$$-\int_{T_s}^{0} e^{A(kT_s+T_s-\tau)}\mathbf{B}u(kT_s)d\sigma = \int_{0}^{T_s} e^{A\sigma}\mathbf{B}u(kT_s)d\sigma$$

$$= \int_{0}^{T_s} e^{A\sigma}\mathbf{B}d\sigma u(kT_s) \tag{9.105}$$

The formula for discrete-time state-space is therefore

$$\mathbf{x}(k+1) = e^{AT_s}\mathbf{x}(k) + \left[\int_{0}^{T_s} e^{A\sigma}\mathbf{B}d\sigma\right]u(k) \tag{9.106}$$

If we make a new matrix

$$\mathbf{F} = e^{AT_s} \tag{9.107}$$

and

$$G = \int_0^{T_s} e^{A\sigma} B d\sigma \qquad (9.108)$$

Then

$$x(k+1) = Fx(k) + Gu(k) \qquad (9.109)$$

For the output equation we define $H = C$ and get

$$y(k) = Hx(k) \qquad (9.110)$$

Equation (9.107) can be calculated from Laplace Transforms, but for larger problems this is not practical. Instead we can use a power-series expansion for F. Expanding (9.107) in a Maclaurin series gives:

$$F = e^{AT_s}$$
$$= I + AT_s + \frac{(AT_s)^2}{2!} + \frac{(AT_s)^3}{3!} + \dots \qquad (9.111)$$

Also beware that to a first order approximation only

$$F \approx I + AT_s$$

Making it a Euler approximation which can have problems with stability as discussed in a previous chapter. This approximation should therefore be used with great care if at all. Higher order terms in the Taylor expansion must therefore be used.

Write (9.111) as

$$F = I + AT_s \Psi \qquad (9.112)$$

where

$$\Psi = I + \frac{1}{2!} AT_s + \frac{1}{3!} A^2 T_s^2 + \dots$$
$$= \sum_{n=0}^{\infty} \frac{A^n T_s^n}{(n+1)!} \qquad (9.113)$$

Now from (9.106).

$\int_0^{T_s} e^{A\sigma} B d\sigma$ can also be expanded in a Maclaurin series and integrated term by term. This gives us

$$G = T_s \Psi B \tag{9.114}$$

Note that if A is invertible, then we can write

$$G = \int_0^{T_s} e^{A\sigma} B d\sigma$$

$$= A^{-1}[F - I]B \tag{9.115}$$

Yet another way is to use

$$F = T e^{A T_s} T^{-1} \tag{9.116}$$

where $e^{A T_s} = \text{diag}\{e^{\lambda_1 T_s}, e^{\lambda_2 T_s}, \ldots e^{\lambda_n T_s}\}$ is a diagonal matrix of eigenvalues of A and T is the modal matrix of eigenvectors. This is a similarity transformation as discussed in a previous section. The stability of a discrete-time state space system is found by the eigenvalues of the F matrix. For stability the eigenvalues must all have magnitude less than unity.

References

1. C. E. Hutchinson, "The Kalman Filter Applied to Aerospace and Electronic Systems," *IEEE Transactions on Aerospace and Electronic Systems,* vol. AES-20, no. 4, pp. 500–504, 1984.
2. G.E.R. Cowan, R.C. Melville, Y.P. Tsividis, A VLSI analog computer/digital computer accelerator. IEEE J. Solid-State Circuits **41**(1), 42–53 (2006)
3. J. Watkins, S. Yurkovich, *"Calculation of the State Transition Matrix for Linear Time Varying Systems,"* in *Mathematical Computation with Maple V: Ideas and Applications, T,* Lee. (Birkhauser, Boston, 1993)
4. W. Everlins, "On the evaluation of E^AT by power series " *Proc. IEEE(Lett.),* vol. 55, p. 413, March 1967.
5. T.A. Bickart, Matrix exponential: Approximation by truncated power series. Proc. IEEE **56**(5), 872–873 (1968). https://doi.org/10.1109/PROC.1968.6423
6. J. Johnson, C. Phillips, An algorithm for the computation of the integral of the state transition matrix. IEEE Trans. Autom. Control **16**(2), 204–205 (1971). https://doi.org/10.1109/TAC.1971.1099689

Chapter 10
Toeplitz Convolution Matrix Method

10.1 Preliminaries

This method only applies to discrete time systems and is useful mainly for systems described by finite impulse response (FIR) models. Having said this, the FIR approach is the commonest in the digital signal processing field in any case due to the guaranteed stability and linear phase when the impulse response is symmetric. We need to consider discrete-time linear systems as applied to lower triangular-Toeplitz (LTT) matrices. We will show that the properties are almost identical to that of polynomials.

We consider finite impulse response (FIR) transfer functions defined in negative powers of z, with polynomials of the form

$$w(z^{-1}) = w_0 + w_1 z^{-1} + \cdots + w_n z^{-n}, w_0 \neq 0 \tag{10.1}$$

The condition above that $w_0 \neq 0$ can be relaxed provided the inverse of the polynomial in terms of its power series is not required. This is a causal FIR transfer function with all $(n + 1)$ zeros of z lying within or on the unit circle of the z-plane. Likewise, define the adjoint system

$$w(z) = w_0 + w_1 z + \ldots + w_n z^n \tag{10.2}$$

defined as having all $(n + 1)$ zeros of z outside of the unit circle. The z-transform operator z^{-1} is often used interchangeably as the backwards-shift operator. Thus for a sampled signal at time sample instant k, we can say $y_{k-1} = z^{-1} y_k$. Conversely the z operator shifts forward in time one step.

Define the output y_k of a linear time-invariant system in terms of an input u_k

$$y_k = w(z^{-1}) u_k \tag{10.3}$$

or in terms of the convolution summation

© The Author(s), under exclusive license to Springer Nature Switzerland AG 2022
T. J. Moir, *Rudiments of Signal Processing and Systems*,
https://doi.org/10.1007/978-3-030-76947-5_10

$$y_k = \sum_{i=0}^{k} w_{k-i} u_i \tag{10.4}$$

We can then write for $k = 0, 1, 2\ldots$ the equations $y_0 = w_0 u_0, y_1 = w_1 u_0 + w_0 u_1$ etc. and we build a matrix notation

$$y(k) = \mathbf{W}u(k) \tag{10.5}$$

where

$$
\begin{bmatrix} y_0 \\ y_1 \\ . \\ . \\ y_m \end{bmatrix} =
\begin{bmatrix}
w_0 & 0 & . & . & 0 \\
w_1 & w_0 & 0 & . & 0 \\
w_2 & w_1 & w_0 & . & 0 \\
. & . & . & . & . \\
w_m & w_{m-1} & . & . & w_0
\end{bmatrix}
\begin{bmatrix} u_0 \\ u_1 \\ . \\ . \\ u_m \end{bmatrix}
\tag{10.6}
$$

and

$$
\mathbf{W} =
\begin{bmatrix}
w_0 & 0 & . & . & 0 \\
w_1 & w_0 & 0 & . & 0 \\
w_2 & w_1 & w_0 & . & 0 \\
. & . & . & . & . \\
w_m & w_{m-1} & . & . & w_0
\end{bmatrix}
\tag{10.7}
$$

Bold letters here denote either vectors of matrices. Sometimes for convenience we omit the time dependence on k in the vectors. Note that $\mathbf{W}$ is a lower triangular square Toeplitz matrix of dimension $m > n$ which is sometimes named the *convolution matrix*. Such a matrix is characterized by the fact that each row is the previous one shifted to the right and a new value added in the preceding space of each row. The diagonal elements are all the same and all other elements are zero. Note that using this LTT matrix method a polynomial with a pure time-delay cannot be represented if the inverse is required (because its inverse is singular). The following properties are found.

10.2   LTT Matrix Properties for Dynamic Systems

Property 1 Inverse Systems

The inverse of a LTT matrix is another matrix of the same type. For example, suppose $w(z^{-1}) = w_0 + w_1 z^{-1} = 1 + 0.6z^{-1}$. Then choosing the order of the LTT matrix as $m = 4$

$$W = \begin{bmatrix} 1 & 0 & 0 & 0 & 0 \\ 0.6 & 1 & 0 & 0 & 0 \\ 0 & 0.6 & 1 & 0 & 0 \\ 0 & 0 & 0.6 & 1 & 0 \\ 0 & 0 & 0 & 0.6 & 1 \end{bmatrix}$$

We invert and get

$$W^{-1} = \begin{bmatrix} 1 & 0 & 0 & 0 & 0 \\ -0.6 & 1 & 0 & 0 & 0 \\ 0.36 & -0.6 & 1 & 0 & 0 \\ -0.2160 & 0.36 & -0.6 & 1 & 0 \\ 0.1296 & -0.2160 & 0.36 & -0.6 & 1 \end{bmatrix}$$

By reading the elements of the bottom row from *right to left*, or the first column *from top to bottom*, the inverse clearly represents the inverse system $1/w(z^{-1}) = 1 - 0.6z^{-1} + 0.36z^{-2} - 0.2160z^{-3} + 0.1296z^{-4} + \cdots$ truncated up to a power series of four delays. A LTT matrix has eigen-values given by its diagonal elements. The diagonal elements here are the coefficient $w_0 \neq 0$. The determinant of any square matrix is given by the product of its eigen-values $|W| = w_0^m$ which is also none-zero. Therefore, the inverse of this Toeplitz convolution matrix always exists provided $w_0 \neq 0$. Note that if $w(z)$ has zeros outside of the unit circle, its inverse is a divergent power series and W will have large numbers within it.

Property 2 The Adjoint system

The adjoint system of a transfer function $w(z^{-1})$ is denoted $w(z)$. This is often written as $w(z^{-1}) = w^*$ for brevity.

From the system LTT of (9.123) we can transpose it

$$W^T = \begin{bmatrix} w_0 & w_1 & . & . & w_m \\ 0 & w_0 & w_1 & . & w_{m-1} \\ 0 & 0 & w_0 & . & w_{m-2} \\ . & . & . & . & . \\ 0 & 0 & . & . & w_0 \end{bmatrix} \tag{10.8}$$

Using this in (10.6) instead of W we see that the output is non-causal because present values of $y(k)$ depend on future values of $u(k)$. Therefore, an easy method of obtaining the adjoint system is to merely transpose the convolution matrix. Moreover, we can easily recognise a causal or non-polynomial just by observation of whether the convolution matrix is lower or upper triangular Toeplitz. If it has elements of both then the corresponding polynomial will have both negative and positive powers of z.

Property 3 Causal and Non-causal System separation

In linear estimation problems a special notation is used for a z-transform with a two-sided impulse-response. For a Laurent polynomial (a polynomial with positive and negative powers of z) g of order m where the negative and positive coefficients are not identical

$$g = g_{-m}z^m + \cdots + g_{-3}z^3 + g_{-2}z^2 + g_{-1}z + g_0 + g_1z^{-1} + g_2z^{-2} + g_3z^{-3} + \cdots + g_m z^{-m}$$

We write $[g]_+ = g_0 + g_1z^{-1} + g_2z^{-2} + g_3z^{-3} + \cdots + g_m z^{-m}$ as the z-transform of the function g(z) over the positive-time interval $k \geq 0$ and $[g]_- = g_{-m}z^m + \cdots + g_{-3}z^3 + g_{-2}z^2 + g_{-1}z$ as the z-transform over the negative-time interval $k < 0$.

As an example of a Laurent series that has both positive and negative powers of z that can be separated easily, the matrix below represents the following Laurent series of order three where the coefficients for positive power of z are different from the negative ones, i.e. the Laurent series is *not* symmetric

$$g = g_{-3}z^3 + g_{-2}z^2 + g_{-1}z + g_0 + g_1z^{-1} + g_2z^{-2} + g_3z^{-3}$$

$$\mathbf{G} = \begin{bmatrix} g_0 & g_{-1} & g_{-2} & g_{-3} \\ g_1 & g_0 & g_{-1} & g_{-2} \\ g_2 & g_1 & g_0 & g_{-1} \\ g_3 & g_2 & g_1 & g_0 \end{bmatrix} = \mathbf{G_+} + \mathbf{G_-}$$

Now the causal polynomial is represented by the lower triangular Toeplitz matrix

$$\mathbf{G_+} = \begin{bmatrix} g_0 & 0 & 0 & 0 \\ g_1 & g_0 & 0 & 0 \\ g_2 & g_1 & g_0 & 0 \\ g_3 & g_2 & g_1 & g_0 \end{bmatrix}$$

and the non-causal one by

$$\mathbf{G_-} = \begin{bmatrix} 0 & g_{-1} & g_{-2} & g_{-3} \\ 0 & 0 & g_{-1} & g_{-2} \\ 0 & 0 & 0 & g_{-1} \\ 0 & 0 & 0 & 0 \end{bmatrix}$$

We make sure that $\mathbf{G_-}$ has zero as the leading diagonal since the non-causal terms of a Laurent series has no term in z^0. That is $g_+ = g_0 + g_1z^{-1} + g_2z^{-2} + g_3z^{-3}$ and $g_- = g_{-3}z^3 + g_{-2}z^2 + g_{-1}z$.

Property 4 Cascaded Systems

Systems in cascade are represented by the convolution of their individual impulse-responses. This follows from standard system theory where convolution in the time-domain is multiplication in the frequency or z-domain. This can be achieved for example by multiplying the two polynomials associated with each impulse response. In the Toeplitz format however, we need only multiply their individual LTT matrices in any order. For example, for a causal system represented by its convolution matrix W_1 cascaded with a causal system W_2, the result is $W_3 = W_1 W_2 = W_2 W_1$.

Note that *the product of two lower(upper) triangular matrices is a lower(upper) triangular matrix*.

Clearly for equivalent LTT matrices the order of the cascaded matrix has to be large enough to fit the product of the other two, otherwise the product polynomial contained therein is truncated. Note that when we multiply two polynomials of order n and m then the results is a polynomial of order (n + m). This can be achieved by padding the individual matrices with zeros to give them a higher order.

If a causal and non-causal system are cascade (say $W_3 = W_1 W_2^T$) then the result will have causal and non-causal terms and a matrix which is neither lower or upper triangular Toeplitz. It is important to note that noncausal and causal products *do not commute*. That is $W_1 W_2^T \neq W_2^T W_1$. The causal term can then be separated using the method above.

Property 5 ARMA or pole-zero models

In many filtering and control problems an autoregressive moving average (ARMA) model is used to model the plant or noise characteristic. An ARMA model is just another way of saying a pole-zero model (except it is driven usually by noise and not a deterministic signal). To modify the approach here is not too difficult. We convert our ARMA into an all-zero (FIR) model by long division and truncate the answer to a length dependant on the amount of accuracy needed. The division is done using LTT matrices. Consider a pole-zero model

$$y(k) = \frac{D(z^{-1})}{A(z^{-1})} u(k) \tag{10.9}$$

Define the two polynomials as a numerator polynomial

$$D(z^{-1}) = d_0 + d_1 z^{-1} + \cdots + d_{nd} z^{-nd} \tag{10.10}$$

and a denominator polynomial

$$A(z^{-1}) = 1 + a_1 z^{-1} + \cdots + a_{na} z^{-na} \tag{10.11}$$

where we assume nd $\leq$ na and the zeros of both $A(z^{-1})$ and $B(z^{-1})$ are assumed to lie inside the unit circle on the z-plane. Although we could consider the case where

$D(z^{-1})$ has zeros outside the unit circle (has non-minimum phase zeros), we omit the case at this time as this requires further factorization of the $D(z^{-1})$ polynomial. We see that $a(0) = 1$, since any other non-unity term can be divided out and absorbed by the numerator polynomial. So, we have

$$y_k + a_1 y_{k-1} + \cdots + a_{na} y_{k-na} = d_0 u_k + d_1 u_{k-1} + \cdots + d_{nd} u_{k-nd} \qquad (10.12)$$

Now we write (9.128) in a vector form

$$\mathbf{A}y(k) = \mathbf{D}u(k) \qquad (10.13)$$

with

$$\mathbf{A} = \begin{bmatrix} 1 & 0 & 0 & 0 & 0 & 0 & 0 \\ a_1 & 1 & 0 & 0 & 0 & 0 & 0 \\ a_2 & a_1 & 1 & 0 & 0 & 0 & 0 \\ . & a_2 & a_1 & 1 & 0 & 0 & 0 \\ . & . & a_2 & a_1 & 1 & 0 & 0 \\ a_{na-1} & . & . & a_2 & a_1 & 1 & 0 \\ a_{na} & a_{na-1} & . & . & a_2 & a_1 & 1 \end{bmatrix} \qquad (10.14)$$

$$\mathbf{D} = \begin{bmatrix} d_0 & 0 & 0 & 0 & 0 & 0 & 0 \\ d_1 & d_0 & 0 & 0 & 0 & 0 & 0 \\ d_2 & d_1 & d_o & 0 & 0 & 0 & 0 \\ . & d_2 & d_1 & d_0 & 0 & 0 & 0 \\ . & . & d_2 & d_1 & d_o & 0 & 0 \\ d_{nd-1} & . & . & d_2 & d_1 & d_o & 0 \\ d_{nd} & d_{nd-1} & . & . & d_2 & d_1 & d_o \end{bmatrix} \qquad (10.15)$$

If we re-define the order of $\mathbf{A}$, as a LTT matrix of order $m > na$ by setting all higher-order coefficients to zero and $\mathbf{D}$ likewise with dimension m then we can write

$$y(k) = \mathbf{A}^{-1}\mathbf{D}u(k) = \mathbf{D}\mathbf{A}^{-1}u(k) = \mathbf{W}u(k) \qquad (10.16)$$

Property 6 Polynomial with pure time-delay

Consider a polynomial with a pure integer time-delay ℓ

$$\begin{aligned} y_k &= z^{-\ell} w(z^{-1}) u_k \\ &= z^{-\ell} (w_0 + w_1 z^{-1} + \cdots + w_n z^{-n}) u_k \end{aligned} \qquad (10.17)$$

We represent this in LTT form as follows

$$y(k) = \mathbf{W}_D \mathbf{W} u(k) \qquad (10.18)$$

Suppose m $= 4$, and the delay $\ell = 2$. Then $\mathbf{W}$ is

$$
\mathbf{W} = \begin{bmatrix}
w_0 & 0 & 0 & 0 & 0 \\
w_1 & w_0 & 0 & 0 & 0 \\
w_2 & w_1 & w_0 & 0 & 0 \\
w_3 & w_2 & w_1 & 0 & 0 \\
w_4 & w_3 & w_2 & w_1 & w_0
\end{bmatrix}
$$

We define the *delay* matrix as

$$
\mathbf{W_D} = \begin{bmatrix}
0 & 0 & 0 & 0 & 0 \\
0 & 0 & 0 & 0 & 0 \\
1 & 0 & 0 & 0 & 0 \\
0 & 1 & 0 & 0 & 0 \\
0 & 0 & 1 & 0 & 0
\end{bmatrix}
$$

If we multiply from the left or right, we obtain

$$
\mathbf{W_D W} = \mathbf{W W_D} = \begin{bmatrix}
0 & 0 & 0 & 0 & 0 \\
0 & 0 & 0 & 0 & 0 \\
w_0 & 0 & 0 & 0 & 0 \\
w_1 & w_0 & 0 & 0 & 0 \\
w_2 & w_1 & w_0 & 0 & 0
\end{bmatrix}
$$

Which has shifted the columns of $\mathbf{W}$ to the left by two columns and left two zero columns at the far right. We can do this for any order delay but clearly, we must choose $m >> \ell$ or too much information is lost. Also note that $\mathbf{W_D}$ is singular and cannot be inverted. Of some minor theoretical interest is that $\mathbf{W_D W_D^T}$ is the identity matrix with the first two uppermost elements of the diagonal equal to zero representing the delay of 2 steps. Thus $\mathbf{W_D}$ is a special matrix but not a pure unitary matrix. We continue with the following properties of systems represented with LTT matrices (Table 10.1).

The spectral factorization problem will be covered in another chapter related to optimal filters. It is important to stress that the above relationship for cascaded polynomials is for causal polynomials only or for noncausal. We cannot multiply a causal times a noncausal polynomial and get commutativity with their corresponding Toeplitz forms because this will be a lower times an upper Toeplitz matrix. The same applies for division of polynomials.

Example of using the Toeplitz method

This problem is closely related to the deconvolution problem except it is deterministic (there are no noise terms in this analysis). Suppose a signal passes through a communication channel of some form. The channel will cause amplitude and phase distortion in the signal. If we could find the inverse of the channel, then we can use

Table 10.1 Summary of system theory mathematical properties of the LTT matrix

Property	Transfer function approach	LT Toeplitz matrix approach
Minimum-phase system	$w(z^{-1}) = w$	$\mathbf{W}$
Inverse	$1/w(z^{-1})$	$\mathbf{W}^{-1}$
Adjoint system	$w(z) = w^*$	$\mathbf{W}^T$
Cascaded causal polynomials	$w_1(z^{-1})w_2(z^{-1})w_3(z^{-1})$ $= w_1(z^{-1})w_3(z^{-1})w_2(z^{-1}) = w_2(z^{-1})w_3(z^{-1})w_1(z^{-1})$ etc	$\mathbf{W}_1\mathbf{W}_2\mathbf{W}_3$ $= \mathbf{W}_1\mathbf{W}_3\mathbf{W}_2 = \mathbf{W}_2\mathbf{W}_3\mathbf{W}_1$ etc
Division of causal polynomials	$w_1(z^{-1})/w_2(z^{-1})$	$\mathbf{W}_2^{-1}\mathbf{W}_1 = \mathbf{W}_1\mathbf{W}_2^{-1}$
Spectral Factorization	$\Delta(z^{-1})\Delta(z)$	**Cholesky factorization:** LL^T or LD_eL^T
Causal selection from Laurent series	$\left[w_1(z^{-1})w_2(z^{-1})w_3(z)\right]_+ = \alpha_0+\alpha_1z^{-1}+\alpha_2z^{-2}+\cdots$	**Lower triangular part of** $\mathbf{W}_1\mathbf{W}_2\mathbf{W}_3^T$
ARMA models	$\frac{D(z^{-1})}{A(z^{-1})}$	$\mathbf{A}^{-1}\mathbf{D} = \mathbf{D}\mathbf{A}^{-1} = \mathbf{W}$
FIR system with explicit Time-delay	$z^{-\ell}w(z^{-1})$	$\mathbf{W_D}\mathbf{W}$

that as a filter to convolve with the received signal and get near perfect quality. If we assume the channel transfer function is given by an FIR transfer function b, where

$$b(z^{-1}) = b_0 + b_1 z^{-1} + b_2 z^{-2} + \cdots + b_n z^{-n} \tag{10.19}$$

Then our filter is simply

$$H(z^{-1}) = \frac{1}{b(z^{-1})} \tag{10.20}$$

Unfortunately, this naïve approach would seldom work because the zeros of the polynomial will have some roots that lie outside of the unit circle. This makes the polynomial nonminimum phase. It makes little difference to the channel transfer function in as much as it is stable, but its inverse will be unstable because zeros are converted into poles via (10.20). The solution to the problem is given in reference [1]. First, we need some mathematical preliminaries for nonminimum phase polynomials.

If a polynomial defined as

$$P(z^{-1}) = p_0 + p_1 z^{-1} + p_2 z^{-2} + \cdots + p_n z^{-n} \tag{10.21}$$

of degree n with real coefficients has all its roots within the unit circle in the z plane, then it is termed *strict sense minimum phase* and no zeros are assumed to be on the unit circle. For simplicity $P(z^{-1})$ is often written as P, omitting the complex argument z^{-1}. The conjugate polynomial

$$P^*(z^{-1}) = p_0 + p_1 z + p_2 z^2 + \cdots + p_n z^n \tag{10.22}$$

is *strict sense non-minumum* phase having all of its roots outside the unit circle on the z plane.

Now define a special polynomial that has its coefficients in reverse order. The reciprocal polynomial is defined as

$$\begin{aligned} \tilde{P}(z^{-1}) &= p_0 z^{-n} + p_1 z^{-(n-1)} + p_2 z^{-(n-2)} + \cdots + p_n \\ &= z^{-n} P^*(z) \end{aligned} \tag{10.23}$$

which has all its roots outside the unit circle provided $P(z^{-1})$ is strict sense minimum phase. (and vice-versa if $P(z^{-1})$ is nonminimum phase). The zeros of $\tilde{P}$ are the zeros of P reflected in the unit circle. Similarly, $\tilde{P}^*(z^{-1}) = z^n P(z^{-1})$ has all its roots within the unit circle.

For polynomials which are strict sense non-minimum phase, we can factorise

$$P(z^{-1}) = P_1(z^{-1}) P_2(z^{-1}) \tag{10.24}$$

where P_1 is strict sense minimum phase and P_2 is strict sense non-minimum phase.

The solution to this problem is to first factorise the b polynomial into minimum and nonminimum phase polynomial thus:

$$b(z^{-1}) = b_1(z^{-1})b_2(z^{-1}) \tag{10.25}$$

We require an inverse transfer–function

$$H(z^{-1}) = \frac{1}{b_1(z^{-1})b_2(z^{-1})} \tag{10.26}$$

To proceed further we note that $b_2 b_2^* = \tilde{b}_2 \tilde{b}_2^*$ and substitute $b_2 = \frac{\tilde{b}_2 \tilde{b}_2^*}{b_2^*}$ into (10.26) giving

$$H(z^{-1}) = \frac{b_2^*}{b_1 \tilde{b}_2 \tilde{b}_2^*} \tag{10.27}$$

Now the reciprocal polynomial $\tilde{b}_2$ is stable having all of its roots within the unit circle, but $\tilde{b}_2^*$ is not. Therefore, expand in positive powers of z by long division

$$\frac{b_2^*}{\tilde{b}_2^*} = d_0 + d_1 z + d_2 z^2 + \cdots \tag{10.28}$$

This is noncausal, so to make it causal we must delay the resulting impulse response to make the positive powers of z negative. This is where the Toeplitz method can be used. We can make LTT Toeplitz matrices out of b_2^* and $\tilde{b}_2^*$. To do long division we only then need to invert $\tilde{b}_2^*$ and multiply it by b_2^* using these LTT matrices. Suppose the LTT matrices for polynomials b_2^* and $\tilde{b}_2^*$ are given a notation in bold as $\mathbf{B}_2^*$ and $\tilde{\mathbf{B}}_2^*$ respectively, and likewise for the other relevant polynomials. All the LTT matrices will be of the same order. The inverse filter is then found from (10.27) but given in LTT matrix form. The denominator is found from

$$\mathbf{H_d} = \mathbf{B}_1 \tilde{\mathbf{B}}_2 \tag{10.29}$$

and the numerator is found from

$$\mathbf{H_n} = \left[\left[\tilde{\mathbf{B}}_2^* \right]^{-1} \mathbf{B}_2^* \right]^T \tag{10.30}$$

This will give an IIR filter. If an FIR is needed then one extra step will do the long division:

$$\mathbf{H} = \mathbf{H_d}^{-1} \mathbf{H_n} \tag{10.31}$$

The transpose in (10.30) is needed because the two Toeplitz matrices are in fact upper triangular Toeplitz since they are representing polynomials in positive powers of z (10.28). Taking the transpose here is like delaying them by the highest power into negative powers of z and a zeroth power of z.

The MATLAB code to calculate the IIR filter is shown below. Consider the same example as given in the reference [1] where $b(z^{-1}) = b_1(z^{-1})b_2(z^{-1})$. With $b_1(z^{-1}) = 1 - 0.6z^{-1}$ and $b_2(z^{-1}) = 1 - 2z^{-1}$.

MATLAB Code 10.1

```
%Find an inverse transfer function of a nonminimum phase polynomial
%Using Lower triangular Toeplitz matrices
clear
close all
m=7 %order of Toeplitz matrices >> system order
%Here m is the delay required (m-1) is numerator filter order
%b1 is minimum phase,b2 is nonminimum phase
b1=[1 -0.6];b2=[1 -2];
b=conv(b1,b2);
bd=[1 0 0];
sys=tf(b,bd,1)
%B2 is the nonminimum phase polynomial
%Create B2 Toeplitz matrix from its polynomial
tr=zeros(1,m);
tr(1)=b2(1);
nul=zeros(1,(m-length(b2)));
tc=horzcat(b2,nul);
B2=toeplitz(tc,tr);
%B2t is B2~ the reciprocal B2 polynomial
%reverse the b2 polynomial coefficients to give b2~ reciprocal polynomial
b2t=fliplr(b2);
%Create B2~ Toeplitz matrix from its polynomial
tr=zeros(1,m);
tr(1)=b2t(1);
nul=zeros(1,(m-length(b2t)));
tc=horzcat(b2t,nul);
B2t=toeplitz(tc,tr);
%B1 is the minimum phase polynomial
%Create B1 Toeplitz matrix from its polynomial
tr=zeros(1,m);
tr(1)=b1(1);
nul=zeros(1,(m-length(b1)));
tc=horzcat(b1,nul);
B1=toeplitz(tc,tr);
```

```
%
%B2* is the transpose or adjoint of B2
B2s=B2';
%B2~* is the transpose or adjoint of B2~
B2ts=B2t';
X=(inv(B2ts)*B2s)';
%Transpose above makes the X system causal, converting from anticausal
%Polynomial extraction is xn numerator polynomial inverse filter
xn=fliplr(X(m,:));
%
XD=B1*B2t;%Denominator is b1*b2~
%denominator polynomial

xd=fliplr(XD(m,:));
%xn needs flipping to take account of the delay
xn1=fliplr(xn);
%Disply Z-Transfer function of inverse filter
sysf=tf(xn1,xd,1)
%
%Combine filter with the original polynomial - numerator only is needed
%should give unity gain and a delay
num=conv(b,xn1);
den=xd;
%Should have unity passband
[hd,wd] = freqz(num,den,'half',1000);
%Magnitude of frequency response
subplot(2,1,1)
plot(wd/pi,(20*log10(abs(hd))),'color','blue')
xlabel('Frequency (\times\pi rad/sample)')
ylabel('Magnitude (dB)')
ax = gca;
ax.YLim = [-5 2];
set(gcf,'color','w');

grid
set(gcf,'color','w');
% Phase part of filter frequency response
subplot(2,1,2)
plot(wd/pi,(angle(hd))*(180/pi),'color','blue')
%ax = findall(gcf, 'Type', 'axes');
%set(ax, 'XScale', 'log');
ax = gca;
ax.YLim = [-200 200];
grid
xlabel('Frequency (\times\pi rad/sample)')
ylabel('Phase(degrees)')
set(gcf,'color','w');
```

Then by our earlier polynomial definitions, $b_2^* = 1 - 2z, \tilde{b}_2 = z^{-1} - 2, \tilde{b}_2^* = z - 2$. Choosing a Toeplitz matrix order of m $= 7$ gives us a solution for the filter as

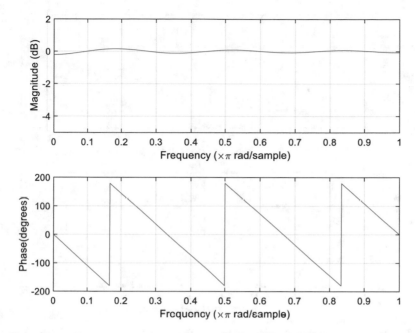

Fig. 10.1 Filter and polynomial combined has a unity gain passband with linear phase for m = 6

$$H\left(z^{-1}\right) = \frac{-\left[0.0234 + 0.0469z^{-1} + 0.0938z^{-2} + 0.1875z^{-3} + 0.3750z^{-4} + 0.7500z^{-5} - 0.5000z^{-6}\right]}{2 - 2.2z^{-1} + 0.6z^{-2}}$$

This filter does not have linear phase, but when it is convolved with the original polynomial, the combination becomes an approximation to a time-delay. Therefore, the frequency response of the filter and polynomial is shown for m = 6 in Fig. 10.1.

Note the small amount of ripple deviating from 0 dB. Increasing m to m = 16 gives the result shown in Fig. 10.2.

Although this method needs matrix inversion, there are a number of papers which use fast methods to invert LTT matrices due to their special nature. For example, Bini's algorithm [2–4] and its variants.

10.3 Inverse of a LTT Matrix Using Two FFTs

Since we are dealing with FFTs, assume that n + 1 is an integer power of 2. We have a LTT matrix defined as

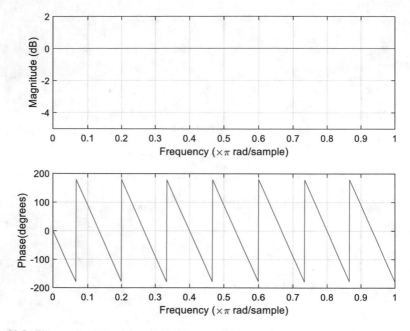

Fig. 10.2 Filter and polynomial combined for m = 16

$$\mathbf{T}_{n+1} = \begin{bmatrix} t_0 & 0 & 0 & 0\ 0\ 0 \\ t_1 & t_0 & 0 & 0\ 0\ 0 \\ t_2 & t_1 & t_0 & 0\ 0\ 0 \\ . & t_2 & t_1 & t_0\ 0\ 0 \\ . & . & . & .\ t_0\ 0 \\ t_n & t_{(n-1)} & t_{(n-2)} & .\ .\ t_0 \end{bmatrix}$$

Then the inverse of $\mathbf{T}_{n+1}$ can be approximated by using two FFT's. The algorithm is given below. The input to the algorithm is the first column vector $\mathbf{t} = (t_0, t_1, \ldots, t_n)^T$ of $\mathbf{T}_{n+1}$. It returns $\mathbf{b}$ the first column vector of the inverse of $\mathbf{T}_{n+1}$ and is illustrated in Fig. 10.3.

Step 1:

Choose $0 < \varepsilon < 1$, a small constant experimentally found to give best results when $\varepsilon = 10^{\frac{-5}{(n+1)}}$.

Compute the vector

$$\tilde{\mathbf{t}} = \mathbf{t}\varepsilon^k, \ \ k = 0, 1, \ldots, n,$$

Step 2:

Define a vector $\mathbf{d} = (d_0, d_1, \ldots, d_n)^T$

Compute

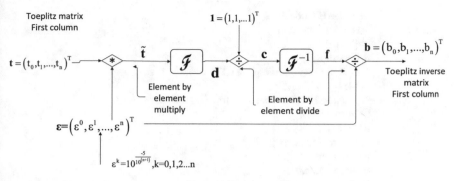

Fig. 10.3 Bini's algorithm to invert a lower triangular Toeplitz matrix

$$d = \mathcal{F}[\tilde{t}]$$

Step 3:
Compute $\mathbf{c} = \left(\frac{1}{d_0}, \frac{1}{d_1}, \ldots, \frac{1}{d_n}\right)^{\mathrm{T}}$ a pointwise division of the two vectors $\mathbf{c} = 1./\mathbf{d}$ with $\mathbf{1} = (1, 1, \ldots, 1)^{\mathrm{T}}$
Step 4:

$$\mathbf{f} = \mathcal{F}^{-1}[\mathbf{c}]$$

Step 5:
Compute $\mathbf{b} = \left(\frac{f_0}{\varepsilon^0}, \frac{f_2}{\varepsilon^1}, \ldots, \frac{f_n}{\varepsilon^n}\right)^{\mathrm{T}}$, a pointwise division of two vectors $\mathbf{b} = \mathbf{f}./\mathbf{e}$
In the above, $\mathcal{F}$ and $\mathcal{F}^{-1}$ represent the Fourier and inverse Fourier transforms as used in this textbook. The vector $\mathbf{b} = (b_0, b_1, \ldots, b_n)^{\mathrm{T}}$ is then the first column of the inverse LTT matrix.
The MATLAB code is shown below.

MATLAB Code 10.2

```
%Bini's Algorithm
a=[1 -1 0.5];
n=length(a);
%m is dimension of square LTT Toeplitz matrix >=n
m=4;
%Create aToeplitz matrix A from its polynomial a
tr=zeros(1,m);
tr(1)=a(1);
nul=zeros(1,(m-n));
tc=horzcat(a,nul);
A=toeplitz(tc,tr);
%Choose epsilon, a small constant
eps=10^(-5/(m))
%Create t~ vector named tt
tt=zeros(1,m);
for i=1:m
    tt(i)=tc(i)*eps^(i-1);
end
%find d vector from FFT
d=fft(tt);
%Compute c vector
one=ones(1,m);
c=one./d;
%Compute f vector
f=ifft(c);
%Compute b polynomial which is the inverse first column
b=zeros(1,m);
for i=1:m
    b(i)=f(i)/eps^(i-1);
end
%Reconstruct the inverse Toeplitz matrix
%Create aToeplitz matrix A from its polynomial a
n=length(b);
tr=zeros(1,m);
tr(1)=b(1);
nul=zeros(1,(m-n));
tc=horzcat(b,nul);
%Inverse LTT Matrix is below
Ai=toeplitz(tc,tr);
```

For the MATLAB example: $\mathbf{A} = \begin{bmatrix} 1 & 0 & 0 & 0 \\ -1 & 1 & 0 & 0 \\ 0.5 & -1 & 1 & 0 \\ 0 & 0.5 & -1 & 1 \end{bmatrix}$ and it returns the first column

vector of the inverse $\mathbf{b} = (1, 1, 0.5, 0)^T$ and the inverse Matrix itself is $\mathbf{A}^{-1} = \begin{bmatrix} 1 & 0 & 0 & 0 \\ 1 & 1 & 0 & 0 \\ 0.5 & 1 & 1 & 0 \\ 0 & 0.5 & 1 & 1 \end{bmatrix}$.

References

1. D. Bini, Parallel solution of certain Toeplitz linear systems. SIAM J. Comput. **13**, 244–255 (1984)
2. B. Skander, D. Marwa, A note on computing the inverse of a triangular Toeplitz matrix. Appl. Math. Comput. **236**, 512–523 (2014)
3. G.E.P. Box, J.M. Jenkins, G.C. Reinsel, G.M. Ljung, *Time Series Analysis, Forcasting and Control*, 5th edn. (Wiley, NJ, 2016)
4. A. Takemura, Exponential decay rate of partial autocorrelation coefficients of ARMA and short-memory processes. Statist. Probab. Lett. **110**, 207–210 (2015)

Chapter 11
FIR Wiener Filters and Random Signals

11.1 Motivation

Up to this chapter we have considered deterministic signals only. These are signals that we know the future of the signal based on its present measurement. For example, a sinewave: if we know its value at a certain point in time we can tell what it will do in the future as it is periodic. There are a great many signals however of which this is not the case. For example, white noise. Take any sample of white noise and we cannot predict what the next sample will be based on the past and present measurement. There are other signals which although we cannot tell exactly what the value will be, we can say what probability there is that it will lie within certain bounds. We need a statistical framework in order to deal with such signals.

11.2 Mathematical Preliminaries

In what follows we will usually consider only the discrete-time case, since it is proving by far the most useful in today's world and although an understanding of continuous time is mandatory, the design of equipment that is analogue in nature is becoming far less common. For a discrete-time random signal x(k), it must belong to some defined probability function. The most common of these is the *Gaussian* probability density function (PDF), though we are far from restrained to this PDF. For example, a simple analysis of a sampled selection of a human speech signal will reveal it has a *Laplacian* PDF. Most of the theory will work on approximating signals to have Gaussian distributions, and we can in special circumstances capitalise on non-Gaussian distributions and make them work in our favour for some specialised algorithms which separate a mixture of random signals. The Gaussian PDF (often called the *Normal distribution*) is given as

© The Author(s), under exclusive license to Springer Nature Switzerland AG 2022
T. J. Moir, *Rudiments of Signal Processing and Systems*,
https://doi.org/10.1007/978-3-030-76947-5_11

$$p(x) = \frac{1}{\sigma\sqrt{2\pi}}e^{-\frac{1}{2}(\frac{x-\mu}{\sigma})^2} \tag{11.1}$$

This looks like a classic bell curve centred on the mean or average value μ. We are used to taking the average value of say N samples given by

$$\mu = \frac{1}{N}\sum_{k=1}^{N}x(k) \tag{11.2}$$

In the statistical sense we usually write the sample mean:

$$E[x(k)] = \mu \tag{11.3}$$

where $E[.]$ represents *expected value* or first moment of the distribution. It should be commented that for small values of N (10.2) will give slightly biased estimates and it is usual to divide by N–1 instead. However, for large N we usually stick with (10.2) as it stands since in signal processing, we have no shortage of data in terms of numbers of samples. For most real signals the mean is just a dc term which is readily removed either by subtraction or ac-coupling in the electronic circuit. Much of the work assumes a zero mean signal and zero mean noise terms. Distinguishing between a signal and noise can be difficult because both are random signals. Therefore, much work goes into a related problem of separating them rather than classical filtering. For example, if two audio human speech signals are mixed together then they may well have very similar statistical properties and frequency content. This makes a filter in the tradition sense impossible since the spectrum of both signals overlap.

The second statistical measurement that is used by engineers is the *variance* of a signal. We cannot measure the amplitude of a signal since it varies from sample to sample. Instead we define the variance. In terms of statistical measurements, *if the mean is zero*, it becomes

$$E[x^2(k)] = \sigma^2$$
$$= \frac{1}{N}\sum_{k=1}^{N}x^2(k) \tag{11.4}$$

This is the sample variance, and (10.2) was the sample mean. Theoretically they are found from the PDF as moment generating functions

$$E[x^n(k)] = \sum_{k=-\infty}^{\infty}x^n(k)p(x(k)) \tag{11.5}$$

For value of n greater than 2 the expectations are zero provided the PDF is Gaussian. This may not be the case with many other distributions. For example, when $n = 3$

a measure of skewness of the distribution is found and for n = 4 the so-called *Kurtosis* which is a measure of the tails and flatness of the top of the distribution. Kurtosis is not really a moment, but a moment divided by information from another moment and this is usually called a *cumulant*. Most of the basic classical filter theory is based around Gaussian distributions, though a whole new world of signal processing can be uncovered when non-Gaussian distributions are allowed for, provided the signal is of the right matching type.

Of use to engineers is the variance, since although it cannot tell us the amplitude of a random signal it can tell us the "span" or range of the signal. The square-root of variance is standard-deviation σ, and for a Gaussian distribution we can say that it is very unlikely that we find measurements in size beyond 3σ. Up to 3σ, 99.7% of all values are present. This does not mean that we will not get values of the signal which are larger (or smaller for negative values -3σ), only the vast majority of samples will lie there. Therefore, looking at a display of white noise on an oscilloscope and engineer can make a very rough empirical guess at the standard deviation. In fact, 99% of all values lie within 2.58σ. Of course, with sampling instrumentation, we can find the exact measure of variance via (11.4). The variance in statistics is also the average power in the waveform assuming zero dc (or mean). If it does have a dc value, then it has dc power as well as ac power.

11.2.1 Autocovariance or Autocorrelation

The *sample autocovariance* of a zero-mean signal is another useful measure given for some delay integer ℓ between the signal and itself from

$$R_{xx}(\ell) = E[x(k)x(k + \ell)]$$

$$= \frac{1}{N} \sum_{k=1}^{N-|\ell|} x(k)x(k + |\ell|), \ell = \ldots -2, -1, 0, 1, 2\ldots \tag{11.6}$$

The sample autocovariance function is valid for positive and negative values of delay and is symmetric i.e. $R_{xx}(\ell) = R_{xx}(-\ell)$. When the delay $\ell = 0$, the autocovariance with zero lag is the variance of the signal. It is a good measure of how white a signal is. White noise is a random signal that is only related with itself and no other signal. It is said to be *uncorrelated* with any other signal. It has only a relationship with itself at the same point of time. If it is shifted in time, the next sample bears no resemblance with the previous. Covariance and *convolution* are related in as much that if a signal is convolved with its negative-time self we get the autocorrelation.

For a zero-mean white noise signal $\xi(k)$

$$E[\xi(k)\xi(k+\ell)], \ell=...-2,-1,0,1,2...$$
$$=E[\xi(k)\xi(k-\ell)] \tag{11.7}$$
$$= \sigma_\xi^2 \delta(\ell)$$

The impulse in (10.7) tells us that if a drawing is made of $R_{xx}(\ell)$ verses positive and negative ℓ, then it only exists at zero lag $\ell=0$ and nowhere else. Note that autocovariance is often used interchangeably with *autocorrelation* in engineering literature. In statistical texts the autocorrelation is usually normalised by the variance to give the Pearson correlation coefficient. Cross-covariance or cross-correlation is defined in a similar way to auto-covariance except with two signals. It tells us the relationship between two random signals.

$$R_{xy}(\ell) = E[x(k)y(k+\ell)], \ell=...-2,-1,0,1,2...$$
$$= \frac{1}{N} \sum_{k=1}^{N-|\ell|} x(k)y(k+|\ell|) \tag{11.8}$$

Signals that do not appear like an impulse are said to be correlated signals. Two signals are uncorrelated if

$$E[x(k)y(k)] = 0 \tag{11.9}$$

This means they are not in any way related in a statistical sense.

11.2.2   Autoregressive (AR) Time-Series Model

In statistics, one of the most basic random signals is the *autoregressive* signal [1] which has a difference equation

$$y(k) = ay(k-1) + \xi(k) \tag{11.10}$$

where $|a| < 1$ and $\xi(k)$ is white noise with zero mean and variance σ_ξ^2. Equation (10.10) is termed an AR(1) time-series or process. In engineering we refer to it as a first order all-pole filter or system, since on taking Z-Transforms we get a transfer function driven by white noise

$$y(k) = \frac{1}{1-az^{-1}}\xi(k) \tag{11.11}$$

In statistics the AR terminology stands for autoregressive since it involves past values (*regressors*) of the output (auto here meaning with respect to itself as in autocovariance). Taking expectations of the squared value of (11.10) gives us

$$E[y^2(k)] = E[ay(k-1) + \xi(k)]^2 \tag{11.12}$$

Expanding and taking expectations gives

$$R_{yy}(0) = a^2 R_{yy}(0) + \sigma_\xi^2 \tag{11.13}$$

Note that in the above we take it that $E[y^2(k)] = E[y^2(k-1)]$ and $E[y(k)\xi(k)] = 0$. This is because our time-series is said to be *wide-sense* stationary, and this means the statistical properties do not vary with time (the variance at time k and k-1 must be the same after statistical steady-state). Also, the white noise is uncorrelated with any other signal. Therefore from (11.13)

$$R_{yy}(0) = \frac{\sigma_\xi^2}{1-a^2} \tag{11.14}$$

This is the variance or average power of the signal that comes out of an AR(1) filter driven by zero-mean white noise variance σ_ξ^2. We can calculate the autocovariance at different lag values and show it to be

$$R_{yy}(\ell) = \frac{\sigma_\xi^2}{1-a^2} a^{|\ell|} \tag{11.15}$$

Another interesting result is found if we consider the AR(1) process as white noise passing through a digital filter as in (10.11). define the transfer function

$$W(z) = \frac{1}{1-az^{-1}} \tag{11.16}$$

We can find the *Z-transform spectral density* of the output as

$$\Phi_{yy}(z) = \sum_{k=-\infty}^{\infty} R_{yy}(k)z^{-k} \tag{11.17}$$
$$= |W(z)|^2 \sigma_\xi^2$$

In the above though the terms spectral density is often used for discrete systems, it should be noted that the correct term is *Periodogram*. The Periodogram approximates continuous time spectral density. Hence the above can also be referred to as Z-transform periodogram.

The above also follows by substituting $z = e^{j\theta}$ into (10.17), which is the equivalent DTFT result

$$\Phi_{yy}(\theta) = \sum_{k=-\infty}^{\infty} R_{yy}(k)e^{-jk\theta} \tag{11.18}$$

Equation (11.17) can also be written as

$$\Phi_{yy}(z) = W(z)W(z^{-1})\sigma_\xi^2 \tag{11.19}$$

Equation (11.19) is a causal and anticausal system in cascade. This comes about because autocorrelation is convolution of a signal with its time-reversed self. i.e. if $\mathcal{Z}\{x(k)\} = X(z)$ then $\mathcal{Z}\{x(-k)\} = X(z^{-1})$ and $\mathcal{Z}\{x(k) * x(-k)\} = X(z)X(z^{-1}) = |W(z)|^2$.

This is a convenient way of expressing the power-spectral density. When systems are cascaded, the power-spectra just get multiplied. The input power in (10.19) is just white noise and its power spectrum is the variance which is flat across the whole spectrum.

Now the *Wiener-Khinchen* theorem states that the Fourier transform of autocorrelation is power-spectral density. Also, the reverse is true, namely the inverse Fourier transform of spectral density is autocovariance. The same applies to Z-transforms. If we take the inverse Z-transform of (11.18) we get autocovariance. Perhaps the easiest way to do this is to use Cauchy's residue theorem.

$$R_{yy}(k) = \frac{1}{2\pi j} \oint_{|z|=1} W(z)W(z^{-1})\sigma_\xi^2 z^{k-1} dz \tag{11.20}$$

The solution to this integral is the sum of the residues at poles that lie *within* the unit circle on the z plane. Write (11.20) as

$$R_{yy}(k) = \frac{\sigma_\xi^2}{2\pi j} \oint_{|z|=1} \left(\frac{z}{z-a}\right)\left(\frac{1}{1-az}\right) z^{k-1} dz, |a| < 1$$

$$= \frac{\sigma_\xi^2}{2\pi j} \oint_{|z|=1} \frac{z^k}{(z-a)(1-az)} dz \tag{11.21}$$

When $k \geq 0$ there is only one simple pole at $z = a$ within the contour. The other is outside the unit circle. The residue is only taken at poles within the unit circle and therefore

$$R_{yy}(k) = \frac{a^k \sigma_\xi^2}{1-a^2}, k \geq 0 \tag{11.22}$$

When $k < 0$ we have multiple poles at $z = 0$ of order k. For k $= -1$ we have a pole at $z = 0$ and another at $z = a$ that are within the unit circle. The sum of the residues is

$$R_{yy}(-1) = \frac{\sigma_\xi^2}{2\pi j} \oint_{|z|=1} \left(\frac{1}{z-a}\right)\left(\frac{1}{1-az}\right)\frac{dz}{z}$$

$$= \left[-a^{-1} + \frac{1}{(1-a)a}\right]\sigma_\xi^2$$

$$= \frac{a\sigma_\xi^2}{1-a^2}$$

For $k = -2$ we get

$$R_{yy}(-2) = \frac{\sigma_\xi^2}{2\pi j} \oint_{|z|=1} \left(\frac{1}{z-a}\right)\left(\frac{1}{1-az}\right)\frac{dz}{z^2}$$

$$= \left[\frac{1}{(1-a^2)a^2} - \frac{(1+a^2)}{a^2}\right]\sigma_\xi^2$$

$$= \frac{a^2\sigma_\xi^2}{1-a^2}$$

Continuing in the same way we get

$$R_{yy}(k) = \frac{a^{|k|}\sigma_\xi^2}{1-a^2}, k = ...-3, -2, 1, 0, 1, 2, 3... \tag{11.23}$$

This is illustrated in Fig. 11.1.
Equation (11.20) is used a lot in optimal filtering theory when $k = 0$.

$$R_{yy}(0) = \frac{\sigma_\xi^2}{2\pi j} \oint_{|z|=1} W(z)W(z^{-1})\frac{dz}{z} \tag{11.24}$$

It is an expression for the variance or average power obtained from spectral density.
Substitute $z = e^{j\theta}$ and $dz = je^{j\theta}d\theta$ where $-\pi \le \theta \le \pi$. (11.24) can be written as

Fig. 11.1 Autocorrelation
function for AR(1) example

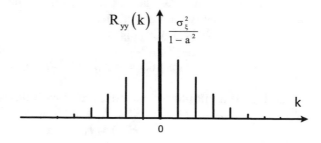

$$R_{yy}(0) = \frac{\sigma_\xi^2}{2\pi} \int\limits_{-\pi}^{\pi} W(e^{j\theta}) W(e^{-j\theta}) d\theta \tag{11.25}$$

This is just Parseval's theorem:

$$R_{yy}(0) = \frac{\sigma_\xi^2}{2\pi} \int\limits_{-\pi}^{\pi} \Phi_{yy}(e^{j\theta}) d\theta$$

$$= \frac{1}{N} \sum_{k=1}^{N} x^2(k) \tag{11.26}$$

A statement that the power in the time and frequency domain is the same. We could also write more generally:

$$R_{yy}(k) = \frac{\sigma_\xi^2}{2\pi} \int\limits_{-\pi}^{\pi} \Phi_{yy}(e^{j\theta}) e^{jk\theta} d\theta \tag{11.27}$$

the inverse DTFT of spectral density is autocorrelation.

It is also of some minor interest from (11.22) above that

$$R_{yy}(k) = a R_{yy}(k-1)$$

In fact, the autocovariance coefficients satisfy the characteristic difference equation of the system. They decay to zero for some initial value (Fig. 11.1). This is true in general [1] for all AR systems and the more general case for other models is shown in [2].

11.2.3   Moving Average Time-Series Model

For systems with only zeros, they are termed FIR systems in electrical engineering, *but moving average* systems in statistics.

For example, a MA(1) system is

$$y(k) = b_0 \xi(k) + b_1 \xi(k-1) \tag{11.28}$$

This is just an FIR filter of order one driven by white-noise. The filter is given by

$$H(z) = b_0 + b_1 z^{-1} \tag{11.29}$$

This type of process plays an important role in engineering systems due to the fact it has no poles and is always guaranteed stable. (of course, it does have poles but they are all at the origin which is pretty safe).

We can find the variance of y(k) from (11.28)

$$
\begin{aligned}
E\left[y^2(k)\right] &= E(b_0\xi(k) + b_1\xi(k-1))^2 \\
&= \sigma_\xi^2(b_0^2 + b_1^2)
\end{aligned}
\tag{11.30}
$$

The above follows since the white noise is not correlated except at the same sample instant of time. So $E\left[\xi^2(k)\right] = E\left[\xi^2(k-1)\right] = \sigma_\xi^2$, but $E[\xi(k)\xi(k-1)] = 0$. The autocovariance of (11.28) is found from

$$
R_{yy}(\ell) = E[y(k)y(k + |\ell|)], \ell = \ldots -2, -1, 0, 1, 2 \ldots
$$

For zero delay this gives us

$$
R_{yy}(0) = E\left[(b_0\xi(k) + b_1\xi(k-1))^2\right]
$$

Which we already have as (11.30) above.
For a delay of $\ell = \pm 1$.

$$
\begin{aligned}
R_{yy}(1) &= E[(b_0\xi(k) + b_1\xi(k-1))(b_0\xi(k+1) + b_1\xi(k))] \\
&= b_0 b_1 \sigma_\xi^2 \\
&= R_{yy}(-1)
\end{aligned}
\tag{11.31}
$$

For delays beyond this the answer is always zero. A MA process has a finite number of autocovariance terms unlike the AR process which has an infinite number. This is in analogy to the impulse responses which are finite in length for an FIR filter and infinite for an IIR filter.

11.2.4 Autoregressive Moving Average Time-Series Model

The autoregressive moving average model (ARMA) as it is known in time-series analysis is also known as a pole-zero transfer function in electrical engineering. It is a simple extension of the AR and MA time-series. An ARMA(1) system driven by white noise would be of the form of the difference equation

$$
y(k) = ay(k-1) + b_0\xi(k) + b_1\xi(k-1)
\tag{11.32}
$$

More generally it can be written as

$$y(k) = \frac{b(z^{-1})}{a(z^{-1})} \xi(k) \tag{11.33}$$

This kind of model is often used by engineers, though for many practical applications the FIR (MA) approach is preferred due to stability concerns. Any system with poles when used in embedded applications is susceptible to finite word limitations, though this is less of a problem than it was in the past. Poles can drift by small amounts and wander outside the unit circle. For this reason, the all-zero or FIR methods are preferred. Added to this is the fact that many FIR filters can be designed to be linear phase. However, the theory for this case is well developed.

11.2.5   Colouring by Filtering White Noise

When white noise is passed through a transfer function its spectrum is changed from being flat to having a defined shape defined by the frequency response of the transfer function. Passing white noise through any LTI filter will result in *coloured noise*. This is the term used in engineering though in statistics the term *correlated noise* is used instead. The variance (average power) of a signal $\xi(k)$ *after* it has passed through an FIR filter the transfer function $b(z^{-1})$ of order n is found from Parseval's theorem:

$$\sigma_y^2 = \frac{\sigma_\xi^2}{2\pi j} \oint_{|z|=1} b(z^{-1}) b(z) \frac{dz}{z} \tag{11.34}$$

For this particular case of an FIR filter the variance is easily calculated using Cauchy's residue theorem. Assume that the contour of integration is the unit circle. Multiply out

$$b(z^{-1}) b(z) = x_n z^n + x_{n-1} z^{n-1} + \dots + x_0 + x_1 z^{-1} + x_2 z^{-2} + \dots + x_n z^{-n} \tag{11.35}$$

This is a *Laurent series* with identical positive and negative powers of z. It is of course symmetric. If we integrate this term by term, we can easily show that the only integral term that is nonzero is the x_0 value. Its value can be calculated by multiplication of $b(z^{-1}) b(z)$ to be

$$x_0 = \sum_{i=0}^{n} b_i^2 \tag{11.36}$$

Essentially, we have from (10.35)

$$b(z^{-1}) b(z)/z = x_n z^{n-1} + x_{n-1} z^{n-2} + \dots + \frac{x_0}{z} + \frac{x_1}{z^2} \dots + \frac{x_n}{z^{n+1}} \tag{11.37}$$

The positive powers of z have no poles at all so their integral is zero when the unit circle is the path of integration. The integral of

$$\sigma_y^2 = \frac{\sigma_\xi^2}{2\pi j} \oint_{|z|=1} \frac{x_0 dz}{z}$$

$$= \sigma_\xi^2 x_0$$

If we integrate the other terms, they have multiple poles at the origin which means differentiation needs to happen with respect to z after the multiple pole is cancelled. The differential will put those terms to zero as they have no coefficient of z. For example

$$\frac{\sigma_\xi^2}{2\pi j} \oint_{|z|=1} \frac{x_1 dz}{z^2} = 0$$

and so on for the other terms.

Hence

$$\sigma_y^2 = \frac{\sigma_\xi^2}{2\pi j} \oint_{|z|=1} b(z^{-1}) b(z) \frac{dz}{z}$$

$$- \sigma_\xi^2 \sum_{i=0}^{n} b_i^2$$

The same method applies to autoregressive moving average (ARMA) models. However, the result is not simple and needs to be calculated depending on the example used. The expression used for variance is

$$\sigma_y^2 = \frac{\sigma_\xi^2}{2\pi j} \oint_{|z|=1} \frac{b(z^{-1}) b(z)}{a(z^{-1}) a(z)} \frac{dz}{z} \tag{11.38}$$

The autoregressive case has already been discussed with an AR(1) example. It is of course possible to derive the infinite impulse response by long division of the b/a polynomials and summing to infinity the sum of squares of the impulse response coefficients to get the variance of the output.

11.2.6   Contour Integration of Laurent Series

The contour integration of the Laurent series is also important for the study of optimal filters in the frequency domain. It is well known that for a Laurent series, its integral around a closed contour (the z plane) of the form

$$
\frac{1}{2\pi j} \oint_{|z|=1} \left[x_n z^n + x_{n-1} z^{n-1} + \ldots + x_0 + x_1 z^{-1} + x_2 z^{-2} + \ldots + x_n z^{-n} \right] \frac{dz}{z}
$$

$$
= x_0
$$

(11.39)

The proof is given above. In some textbooks the division by z is left out and the solution becomes the x_{-1} term.

11.2.7   Vectors and Random Signals

The output difference equation of an FIR filter drive by zero mean white noise can be written as

$$
y(k) = H(z^{-1}) \xi(k)
$$

(11.40)

Now if

$$
H(z^{-1}) = b_0 + b_1 z^{-1} + b_2 z^{-2} + \ldots + b_n z^{-n}
$$

(11.41)

then the difference equation is

$$
y(k) = b_0 \xi(k) + b_1 \xi(k-1) + b_2 \xi(k-2) + \ldots + b_n \xi(k-n)
$$

(11.42)

Define two vectors. A weight vector

$$
\mathbf{H} = \begin{bmatrix} b_0 \ b_1 \ b_2 \ \ldots \ b_n \end{bmatrix}^T
$$

(11.43)

and a *regressor vector*

$$
\mathbf{X}(k) = \begin{bmatrix} \xi(k) \ \xi(k-1) \ \xi(k-2) \ \ldots \ \xi(k-n) \end{bmatrix}^T
$$

(11.44)

Now re-write (11.40) as

$$
y(k) = \mathbf{H}^T \mathbf{X}(k)
$$
$$
= \mathbf{X}^T(k) \mathbf{H}(k)
$$

(11.45)

Taking expectations of the square to give the variance gives

$$\begin{aligned} E\left[y^2(k)\right] &= \\ E\left[\mathbf{H}^T\mathbf{X}(k)\mathbf{X}^T(k)\mathbf{H}\right] & \\ &= \mathbf{H}^T\mathbf{R}_{xx}(k)\mathbf{H} \end{aligned} \tag{11.46}$$

where

$$\mathbf{R}_{xx}(k) = E\left[\mathbf{X}(k)\mathbf{X}^T(k)\right] \tag{11.47}$$

Is an n-square matrix called a *covariance* matrix. It can be written as

$$\mathbf{R}_{xx}(k) = \mathrm{cov}\left[\mathbf{X}(k),\mathbf{X}^T(k)\right] \tag{11.48}$$

Often in statistics the correlation matrix is shown to be a normalised covariance but in engineering the two terms, correlation and covariance matrices are often used interchangeably. We can also have *cross-covariance* matrix between two random vectors of equal length.

$$\mathbf{R}_{xy}(k) = \mathrm{cov}\left[\mathbf{X}_1(k),\mathbf{X}_2^T(k)\right] \tag{11.49}$$

A random scalar (say u(k)) and a random vector $X(k)$ are related via a *covariance vector*

$$\mathbf{R}_{ux}(k) = E[u(k)\mathbf{X}(k)] \tag{11.50}$$

When a noise vector is white, the covariance is a constant diagonal matrix:

$$\mathrm{cov}\left[\mathbf{X}(k),\mathbf{X}^T(j)\right] = \mathbf{Q}\delta(k-j) \tag{11.51}$$

Here (11.51) uses the Kronecker delta to show that only when k = j are the two vectors correlated. We can usually replace the white covariance matrix with $\mathbf{Q} = \sigma_q^2\mathbf{I}$. Where σ_q^2 is the variance of the noise.

If the random vectors are not white, but correlated, the autocovariance matrix is a special matrix called a Toeplitz matrix. It has the form

$$\mathbf{R}_{xy}(k) = \begin{bmatrix} R_{xx}(0) & R_{xx}(-1) & R_{xx}(-2) & \cdots & R_{xx}(-n) \\ R_{xx}(1) & R_{xx}(0) & R_{xx}(-1) & & R_{xx}(-n+1) \\ R_{xx}(2) & & R_{xx}(0) & & R_{xx}(-n+2) \\ R_{xx}(3) & & & & \\ \cdot & & & \cdot & \cdot \\ \cdot & & & & \cdot \\ R_{xx}(n) & R_{xx}(n-1) & R_{xx}(n-2) & \cdots & R_{xx}(0) \end{bmatrix} \tag{11.52}$$

The individual terms consist of the autocovariance values as calculated from a time-series consisting of the vector data itself. That data will be much longer than the vector since a regressor vector is only the same length as the order of the FIR system. We can calculate the autocovariance values from (10.6) and then construct the matrix, or alternatively we can find the matrix in the following recursive manner.

$$\mathbf{R}_{xy}(k+1) = \mathbf{R}_{xy}(k) + \left(\frac{1}{k+1}\right)\left[\mathbf{X}(k)\mathbf{X}^T(k) - \mathbf{R}_{xy}(k)\right] \tag{11.53}$$

Provided the signal is wide-sense stationary (the mean and variance do not change with time), $\mathbf{X}(k)\mathbf{X}^T(k)$ will converge to a constant matrix. Equation (11.53) is a vector form of recursive variance. We can do the same for covariance vectors. If the data is not wide-sense stationary then (11.53) will not work since it has an infinite memory and will not adapt to changes. Instead we can use the update

$$\mathbf{R}_{xy}(k+1) = \beta\mathbf{R}_{xy}(k) + (1-\beta)\mathbf{X}(k)\mathbf{X}^T(k) \tag{11.54}$$

where $0 < \beta < 1$ is called a *forgetting-factor* and is adjusted so that the matrix tracks any changes from sample to sample. Typically, the forgetting-factor needs to be set close to 1. Equation (11.54) is nothing more than a first-order recursive lowpass filter written in a matrix form. The forgetting factor is the pole of the filter. Digital signal processing is full of algorithms that use *ad-hoc* approaches such as these, but they can be made to work quite successfully with a little care and effort. In the statistics literature an equation such as (11.54) is known as exponential smoothing [3] whereby it is defined with a different parameter $\beta = (1 - \alpha)$ and α is chosen nearer to zero.

Many signal processing algorithms rely on the inverse of a covariance matrix. Normally such an exercise would require $O(n^3)$ operations. Since the matrix is Toeplitz in structure early work has enabled the inverse to be calculated in $O(n^2)$ operations [4, 5]. Another approach is to directly construct the covariance matrix from the raw data and invert it in one operation. This is done using a special feedback control-loop [6].

11.2.8   Differentiation of Vector Products

Vector calculus is a huge field, but we only require some basic results for filter theory. A problem that is commonly met in digital signal processing is differentiating a *scalar* with respect to a *vector*. For example, for a scalar inner product (or dot product) which can be written as:

$$\begin{aligned} y(k) &= \mathbf{H}^T\mathbf{X}(k) \\ &= \mathbf{X}^T(k)\mathbf{H}(k) \end{aligned} \tag{11.55}$$

where the vector $\mathbf{H} = \begin{bmatrix} h_1 \ h_2 \ h_3 \ \dots \ h_n \end{bmatrix}^T$, $\mathbf{X}(k) = \begin{bmatrix} x_1 \ x_2 \ x_3 \ \dots \ x_n \end{bmatrix}^T$ we are required to find the differential of y(k) with respect to the vector $\mathbf{H}$ i.e. $\frac{\partial y(k)}{\partial \mathbf{H}}$.

We require some rules of vector calculus. We write the *gradient* vector *del* as a vector of derivatives with respect to each element in the vector:

$$\nabla_{\mathbf{H}} = \begin{bmatrix} \frac{\partial}{\partial h_1} & \frac{\partial}{\partial h_2} & \frac{\partial}{\partial h_3} & \cdots & \frac{\partial}{\partial h_n} \end{bmatrix} \tag{11.56}$$

Differentiating element by element in the product of the two vectors we get [7]

$$\nabla_{\mathbf{H}}\left[\mathbf{H}^T \mathbf{X}(k) \right] = \mathbf{X}(k) \tag{11.57}$$

$$\nabla_{\mathbf{H}}\left[\mathbf{X}^T(k) \mathbf{H} \right] = \mathbf{X}(k) \tag{11.58}$$

Another common derivative is encountered when $\mathbf{R}$ is an n square matrix:

$$\nabla_{\mathbf{H}}\left[\mathbf{H}^T \mathbf{R} \mathbf{H} \right] = \left[\mathbf{R} + \mathbf{R}^T \right] \mathbf{H} \tag{11.59}$$

If $\mathbf{R}$ is symmetric as is the case for a covariance matrix then $\mathbf{R} = \mathbf{R}^T$ and

$$\begin{aligned} \nabla_{\mathbf{H}}\left[\mathbf{H}^T \mathbf{R} \mathbf{H} \right] &= \left[\mathbf{R} + \mathbf{R}^T \right] \mathbf{H} \\ &= 2 \mathbf{R} \mathbf{H} \end{aligned} \tag{11.60}$$

11.3 The FIR Wiener Filtering Problem

The irony of the heading is that Norbert Wiener, a prodigy in the field of mathematics probably never had anything to do with the work that follows, in that his study was on continuous time filtering. His famous book [8] did lay down the foundation to continuous-time optimal filtering of random signals, but for IIR filters. At the time (1942), the idea of FIR filters probably would not have been of any advantage, since there were very few computers and the likelihood of real-time filtering was more than 50 years later. However, the famous Russian mathematician Kolmogorov did look at the discrete time case independently [9], though again he did not consider FIR filtering. Prior to this there had been some work on autoregressive modelling of systems [10, 11] but this did not concern filtering. The earlier work however is significant because it showed how time-series can be "captured" in the form of autoregressive modelling. This paved the way for the later work of Wiener and contemporaries. Here we will use the basic idea of Wiener but applied to FIR filtering. This has the advantage of stability when applied to many practical problems.

Suppose our time-series (or in the engineering sense a random signal) can be expressed mathematically as white noise passing through an arbitrary LTI Z-transfer function. The filter will be labelled $W(z^{-1})$ and the white noise sequence $\xi(k)$ will have zero mean with variance σ_ξ^2. We have

$$y(k) = W(z^{-1})\xi(k) \tag{11.61}$$

Furthermore, let this signal be contaminated by zero mean, *uncorrelated* white noise $v(k)$ with variance σ_v^2 giving a mixture

$$s(k) = y(k) + v(k) \tag{11.62}$$

It is assumed that the pure signal cannot be measured, but the signal plus noise (11.62) is available. It is important to stress that the measurement additive noise is not correlated with the signal in any way. To begin with we consider the simple case of additive white noise, though the more general case will also be covered later.

Let an optimal filter be defined as an FIR filter with $n + 1$ coefficients or "weights" as they are commonly known in the literature. It can be written either in transfer function form

$$\hat{y}(k) = H(z^{-1})s(k) \tag{11.63}$$

where the hat on top of $y(k)$ denotes an estimate of the signal which is not of course exact.

Equation (10.63) can also be written in vector notation as a moving average

$$\hat{y}(k) = \mathbf{H}^{\mathsf{T}} s(k) \tag{11.64}$$

where the filter weight vector is given be

$$\mathbf{H}^{\mathsf{T}} = \begin{bmatrix} h_0 \ h_1 \ h_2 \ \dots \ h_n \end{bmatrix} \tag{11.65}$$

and the signal plus noise can also be written in its vector regressor form

$$\mathbf{s}^{\mathsf{T}} = \begin{bmatrix} s_k \ s_{k-1} \ s_{k-2} \ \dots \ s_{k-n} \end{bmatrix} \tag{11.66}$$

This is a vector of regressors of the signal plus noise. The signal plus noise model is illustrated in Fig. 11.2.

Create an error between the true signal and the estimate.

$$e(k) = E\big[y(k) - \hat{y}(k)\big] \tag{11.67}$$

We need the expectation operator E(.) in (10.67) since we are dealing with random signals.

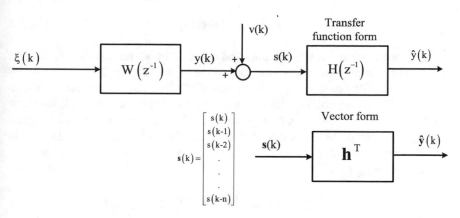

Fig. 11.2   Signal plus noise mixture for FIR Wiener filtering problem

Now define a performance criterion as is done in optimization problems. Choose
the expected value of the error squared.

$$J = E\left[\left(y(k) - \hat{y}(k)\right)^2\right] \tag{11.68}$$

A quadratic criterion makes the problem mathematically tractable, but also gives
results that are sensible from a practical point of view. It is also possible to minimise
the maximum value of the error as well, but this is not considered here. Substitute
for the estimated signal (10.64) and get

$$J = E\left[\left(y(k) - \mathbf{H}^T \mathbf{s}(k)\right)^2\right] \tag{11.69}$$

Write this as

$$J = E\left[\left(y(k) - \mathbf{H}^T \mathbf{s}(k)\right)\left(y(k) - \mathbf{H}^T \mathbf{s}(k)\right)^T\right] \tag{11.70}$$

Expand (11.70)

$$\begin{aligned}
J &= E\left[\left(y(k) - \mathbf{H}^T \mathbf{s}(k)\right)\left(y(k) - \mathbf{H}^T \mathbf{s}(k)\right)^T\right] \\
&= E\left[y^2(k)\right] - E\left[\mathbf{H}^T \mathbf{s}(k)y(k)\right] - E\left[\mathbf{s}^T(k)\mathbf{H}y(k)\right] + E\left[\mathbf{H}^T \mathbf{s}(k)\mathbf{s}^T(k)\mathbf{H}\right]
\end{aligned} \tag{11.71}$$

Assuming a stationary time series, define the auto-covariance matrix of the signal
plus noise

$$\mathbf{R}_{ss} = \mathbf{s}(k)\mathbf{s}^T(k) \tag{11.72}$$

and a cross-covariance vector between the signal plus noise and the true signal as

$$\mathbf{R}_{sy} = E[\mathbf{s}(k)y(k)] \tag{11.73}$$

Often, we leave out the terms auto and cross and just use covariance. Now in (10.72), s(k) can be measured so the covariance matrix is relatively easy to obtain. However, (10.74) gives us a problem since y(k) is the true signal and cannot be measured. The whole point of this exercise is to estimate the signal and if we knew it beforehand there would not be a problem to solve! However, we will later see a solution to this problem. Although not needed in (10.71), we can also define a covariance matrix of the true (desired) signal written in vector regressor form $\mathbf{y}^T(k) = \begin{bmatrix} y_k\ y_{k-1}\ y_{k-2} \cdots y_{k-n} \end{bmatrix}$ as

$$\mathbf{R}_{yy} = E[\mathbf{y}(k)\mathbf{y}^T(k)] \tag{11.74}$$

Also, we know that from the mixture model (11.62)

$$\mathbf{R}_{ss} = \mathbf{R}_{yy} + \mathbf{R}_{vv} \tag{11.75}$$

Now since $v(k)$ and $y(k)$ are uncorrelated, (11.73) is really the correlation between y(k) and y(k) and $\mathbf{R}_{sy}$, since it is a vector, is the first column of the correlation matrix $\mathbf{R}_{yy}$. This follows because in the scalar case for example: $E[\mathbf{s}(k)y(k)] = E[(y(k) + v(k))y(k)] = E[y^2(k)]$. This information can be used later, but continuing from (10.71), substitute the required covariances

$$
\begin{aligned}
J &= E[y^2(k)] - \mathbf{H}^T E[\mathbf{s}(k)y(k)] - E[y(k)\mathbf{s}^T(k)]\mathbf{H} + \mathbf{H}^T E[\mathbf{s}(k)\mathbf{s}^T(k)]\mathbf{H} \\
&= \sigma_y^2 - \mathbf{H}^T \mathbf{R}_{sy} - \mathbf{R}_{sy}^T \mathbf{H} + \mathbf{H}^T \mathbf{R}_{ss}\mathbf{H}
\end{aligned}
\tag{11.76}
$$

Using the results of Eqs. (11.57), (11.58) and (11.60) we differentiate with respect to the weight vector **H** and equate to zero:

$$
\begin{aligned}
\nabla_{\mathbf{H}} J &= -\mathbf{R}_{sy} - \mathbf{R}_{sy} + [\mathbf{R}_{ss} + \mathbf{R}_{ss}^T]\mathbf{H} \\
&= -2\mathbf{R}_{sy} + 2\mathbf{R}_{ss}\mathbf{H} = 0
\end{aligned}
\tag{11.77}
$$

This is the necessary condition for optimality. Solving gives

$$\mathbf{H}_{opt} = \mathbf{R}_{ss}^{-1}\mathbf{R}_{sy} \tag{11.78}$$

where $\mathbf{H}_{opt}$ is the optimal Wiener weight vector. Writing (11.78) as

$$\mathbf{R}_{ss}\mathbf{H}_{opt} = \mathbf{R}_{sy} \tag{11.79}$$

Gives us the so-called Wiener–Hopf set of equations written in matrix form. The original form was in continuous time and an integral equation. They have a similar look to the Yule-Walker equation but in their work, they are estimating the unknown time-series model to model an autoregressive, process whereas here it is the solution to a filtering problem. As stated before, Wiener did not find the solution to this problem but worked in continuous time.

Differentiating (11.77) again, we use the result that for a matrix $\mathbf{A}$ and vector $\mathbf{x}$, $\frac{\partial}{\partial \mathbf{x}} \mathbf{A} \mathbf{x} = \mathbf{A}$. We get

$$\nabla_{\mathbf{H}}^2 J = \mathbf{R}_{ss} \tag{11.80}$$

For white noise the correlation matrix $\mathbf{R}_{vv}$ must be positive definite. But $\mathbf{R}_{ss} = \mathbf{R}_{yy} + \mathbf{R}_{vv}$, and (11.80) will also be positive definite indicating that the optimal solution is also sufficient or is a minimum and not a maximum.

From (11.76), the variance of the error is given by

$$\sigma^2 = \sigma_y^2 - \mathbf{H}^T \mathbf{R}_{sy} - \mathbf{R}_{sy} \mathbf{H} + \mathbf{H}^T \mathbf{R}_{ss} \mathbf{H} \tag{11.81}$$

Substitute the optimal weight vector from (11.78) into (11.81) and we get

$$\sigma_{\min}^2 = \sigma_y^2 - \mathbf{R}_{sy}^T \mathbf{R}_{ss}^{-1} \mathbf{R}_{sy} \tag{11.82}$$

This is the minimum mean-square error or minimum error variance.

11.3.1 FIR Wiener Filter Example with Additive White Noise

Consider a first order AR (all pole) system driven by zero-mean unity variance white noise $\xi(k)$ giving a random signal

$$y(k) = \frac{\sqrt{1 - a^2}}{(1 - az^{-1})} \xi(k), |a| < 1$$

The corrupted signal is

$$s(k) = y(k) + v(k)$$

where $v(k)$ is zero-mean white noise of variance σ_v^2. The optimal FIR Wiener filter is required for three weights, $n = 3$.

The example is picked because it has quite simple autocovariance terms. The autocovariance of y(k) for $\sigma_\xi^2 = 1$ is

$$R_{yy}(k) = a^{|k|}$$

By far the easiest way to very this is the case is to take inverse transforms of the Z-transform spectral density via the method of residues. The spectral density is

$$\Phi_{yy}(z) = \frac{\left(1 - a^2\right)}{\left(1 - az^{-1}\right)(1 - az)} \sigma_{\xi}^2$$

From which its inverse Z-transform is

$$R_{yy}(k) = \frac{\sigma_{\xi}^2}{2\pi j} \oint_{|z|=1} \frac{\left(1 - a^2\right)z^{k-1}}{\left(1 - az^{-1}\right)(1 - az)} dz$$

$$= \frac{\sigma_{\xi}^2}{2\pi j} \oint_{|z|=1} \frac{\left(1 - a^2\right)z^k}{(z - a)(1 - az)} dz$$

Now the integration must be taken anticlockwise around the unit circle. The residues only involve poles within the unit circle. There is only one at z = a provided k is greater or equal to zero. Therefore

$$R_{yy}(k) = \frac{\sigma_{\xi}^2}{2\pi j} \oint_{|z|=1} \frac{\left(1 - a^2\right)z^k}{(z - a)(1 - az)} dz$$

$$= \sigma_{\xi}^2 \frac{\left(1 - a^2\right)z^k}{(1 - az)} \bigg|_{z=a}$$

$$= \sigma_{\xi}^2 a^k, k \geq 0$$

Since $\sigma_{\xi}^2 = 1$ we have the result. Negative k is a little harder to calculate since the z^{-k} term contributes zeros to the expression. The first value for k = −1 is found from the sum of two residues from poles within the unit circle:

$$R_{yy}(-1) = \frac{\sigma_{\xi}^2}{2\pi j} \oint_{|z|=1} \frac{\left(1 - a^2\right)z^{-1}}{(z - a)(1 - az)} dz$$

$$= \sigma_{\xi}^2 \frac{\left(1 - a^2\right)}{(z - a)(1 - az)} \bigg|_{z=0} + \sigma_{\xi}^2 \frac{\left(1 - a^2\right)}{z(1 - az)} \bigg|_{z=a}$$

$$= -\sigma_{\xi}^2 \frac{\left(1 - a^2\right)}{a} + \frac{\sigma_{\xi}^2}{a}$$

Giving

$$R_{yy}(-1) = \sigma_{\xi}^2 a$$

For k = −2 there are three poles and a differentiation is needed resulting in a solution of

$$R_{yy}(-2) = \sigma_\xi^2 a^2.$$

The covariance matrix for $\mathbf{R}_{yy}$ is therefore

$$\mathbf{R}_{yy} = \begin{bmatrix} 1 & a & a^2 \\ a & 1 & a \\ a^2 & a & 1 \end{bmatrix}$$

And the covariance vector is just the first column of $\mathbf{R}_{yy}$, since the noise and signal are uncorrelated.

$$\mathbf{R}_{sy} = \begin{bmatrix} 1 \\ a \\ a^2 \end{bmatrix}$$

The noise covariance is σ_v^2, a scalar, but can be written as a diagonal matrix if the noise is also written as a vector of regressors, hence if we have a vector $\mathbf{v}(k) = \begin{bmatrix} v(k) & v(k-1) & v(k-2) \end{bmatrix}^T$, then

$$\mathbf{R}_{vv} = E\left[\mathbf{v}(k)\mathbf{v}^T(k)\right] = \begin{bmatrix} \sigma_v^2 & 0 & 0 \\ 0 & \sigma_v^2 & 0 \\ 0 & 0 & \sigma_v^2 \end{bmatrix}$$

The covariance of the mixture of signal plus noise is

$$\mathbf{R}_{ss} = \mathbf{R}_{yy} + \mathbf{R}_{vv} = \begin{bmatrix} 1+\sigma_v^2 & a & a^2 \\ a & 1+\sigma_v^2 & a \\ a^2 & a & 1+\sigma_v^2 \end{bmatrix}$$

and the optimal weights are found from

$$\mathbf{H}_{opt} = \mathbf{R}_{ss}^{-1}\mathbf{R}_{sy}$$

$$= \begin{bmatrix} 1+\sigma_v^2 & a & a^2 \\ a & 1+\sigma_v^2 & a \\ a^2 & a & 1+\sigma_v^2 \end{bmatrix}^{-1} \begin{bmatrix} 1 \\ a \\ a^2 \end{bmatrix}$$

The simulation of the signal plus noise, original signal and the optimal estimate is shown in Fig. 11.3. The values used were a $= 0.5$, $\sigma_v^2 = 3$, $\sigma_\xi^2 = 1$.

The signal to noise ratio can be calculated for any random signal and noise with variances σ_x^2, σ_n^2 respectively as:

$$SNR_{db} = 10\log_{10}\left(\frac{\sigma_x^2}{\sigma_n^2}\right) \tag{11.83}$$

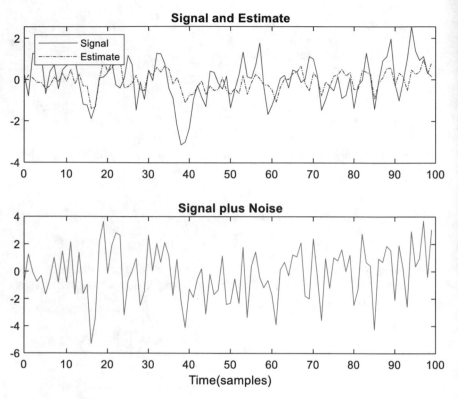

Fig. 11.3 Simple optimal FIR wiener filter

For this example, we find the SNR was found from $SNR_{db} = 10\log_{10}\left(\frac{\sigma_y^2}{\sigma_v^2}\right)$, which is -4.49 dB.

The optimal weight vector was calculated as $\mathbf{H}_{opt} = \begin{bmatrix} 0.2364 & 0.0909 & 0.0364 \end{bmatrix}^T$. the MATLAB code used to generate the above simulation is shown below.

```
MATLAB Code 11.1
%Simple Wiener filter
%First order AR model for signal
%with additive uncorrelated white noise
%Three weights
%Unit variance driving noise
sq=1;
%Additive noise variance
sv=3;
%
a=0.5;
%Covariance matrix of signal plus white noise
Rss=[1+sv a a^2
    a 1+sv a
    a^2 a 1+sv];
%Covariance vector of signal, since noise is uncorrelated
```

```
Rsy=[1 a a^2]';
%Optimal weight vector
hopt=inv(Rss)*Rsy
%
%Simulate the mixture plus filter
Npoints=100;
%Colouring filter for signal
ap=[1 -a];
cp=[1]*sqrt(1-a^2);
%time vector
t=[0:1:Npoints-1]';
%Random Noise length Npoints Variance rn
rn=sqrt(sq)*randn(Npoints,1);
%Filter it to get time-series
%Generate Signal
y=filter(cp,ap,rn);
% Additive uncorrelated noise variance sv
%re-seed to get uncorrelated white noise
seed=1;
rng(seed)
rv=sqrt(sv)*randn(Npoints,1);
%Mixture by adding white noise
s=y+rv;
%filter signal plus noise
yh=filter(hopt,1,s);
subplot(2,1,1)
plot(t,y,'-r',t,yh,'-.b')
legend('Signal','Estimate','Location','northwest')
title('Signal and Estimate')
subplot(2,1,2)
plot(t,s)
set(gcf,'color','w');
title('Signal plus Noise')
set(gcf,'color','w');
xlabel('Time(samples)')
%signal to noise ratio was
snr=10*log10(var(y)/sv)
```

11.3.2 FIR Wiener Filter Improvements

The problem with the above solution is that there are a great many unknowns to be found were this a real problem instead of a contrived theoretical one. To get around this we can try some improvements a stage at a time. First consider an example where we don't know the autocorrelation coefficients and the signal is nonstationary. For speech, the statistics vary with time, but the mean value will remain zero. We will use a short sentence of a human speech signal. We will assume that we can measure the true signal. This is of course unfair since if we could measure it then there would not be a filtering problem. However, it shows us what is at best possible with ideal measurements. Also assume the measurement noise is white and the variance is known.

We can measure the covariance matrix of the speech by using the sample covariance method

$$\mathbf{R}_{yy}(k+1) = \beta \mathbf{R}_{yy}(k) + (1 - \beta)\mathbf{X}(k)\mathbf{X}^T(k)$$

where β is the forgetting factor chosen to be close to unity. The speech must be *segmented* into blocks of data and the filter computed and applied at each segment. This at least tells us if it is possible to track the covariance matrix versus time. Then at each segment we add on the covariance of the measurement noise (as a diagonal matrix) and we then get $\mathbf{R}_{ss} = \mathbf{R}_{yy} + \mathbf{R}_{vv}$. This is sufficient information to calculate the optimal filter. The simulation experiment was carried out with 12 weights. Each block of data was 5000 samples with $\beta = 0.9$. The results are shown in Fig. 11.4. The speech was sampled at 44.1 kHz.

It can be seen that there is a dramatic reduction in signal to noise ratio (SNR). It is difficult to calculate SNR with a nonstationary signal since the SNR varies with time. It can be expressed as a segmented SNR but approximately we can see the noise and speech are off the same level of magnitude. The enhanced speech did not sound as good as it looks on Fig. 11.4. It did remove the noise almost in its entirety during the spaces between words but appeared to cut off the beginning of the words

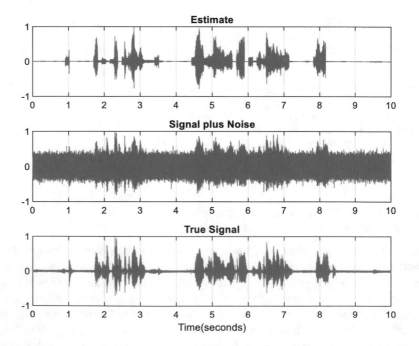

Fig. 11.4 Wiener filter applied to a speech waveform using 12 weights

by small amounts and sounded rather staccato or detached. To make the experiment more realistic the additive noise itself was estimated. Speech is a good type of signal for this since it has areas within it with no signal. If these areas are known (often known as *noise-alone* periods), then the noise covariance can be estimated at these times. The speech signal conveniently has a period of silence at the beginning which is suitable. Furthermore, it was cheating to assume we could measure the true speech signal, so we measure instead the covariance of the mixture of signal plus noise. We then subtract the covariance of the noise to give us the covariance of the signal. Figure 11.5 shows the clean signal, the mixture and the estimated signal.

The filtered speech sounded much more natural than the previous case due to the fact there was some residual noise present. The signal to noise ratio used was larger than previous because in a realistic experiment if the speech power is lower than that of the noise then the areas of noise-alone activity for estimating noise are much harder to find. The noise reduction was calculated to be 18 dB down. The method here of course requires a mechanism to detect when noise alone is present, and this is harder than it first may appear. Such techniques are known as voice activity detectors (VADs). In this experiment only one estimate of the noise covariance was used at the beginning of the speech waveform. This means that the noise must itself be stationary as the filter will only be valid for the period that the statistics stay constant.

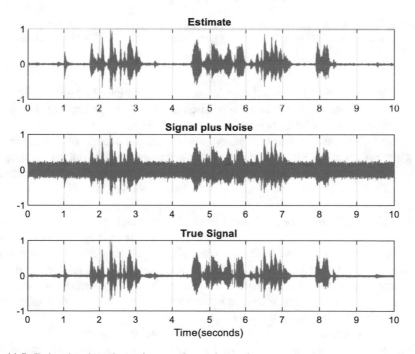

Fig. 11.5 Estimating the noise variance and speech covariance

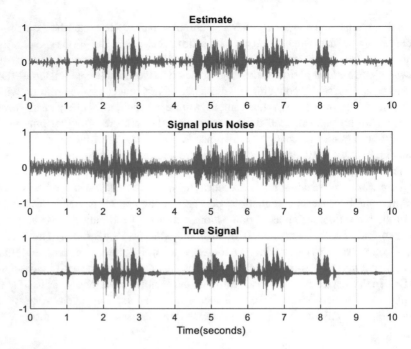

Fig. 11.6 Speech and additive factory noise. Twelve weights were used, and the speech segmented into blocks of length 2000 samples. The forgetting factor was $\beta = 0.999$. Noise reduction was 11.8 dB down on original

Finally, we extend this experiment to coloured (correlated) noise recorded in a factory. The equations do not need to be altered for this case except the covariance matrix of the noise will no longer be diagonal (Fig. 11.6).

The speech sounded better than the original though the quality was not as good as the white noise case. This method also requires that the additive noise be stationary which the factory noise is approximately. Much experimenting was done with the length of a block of data but shorter blocks lead to a pulsating effect on the final filtered speech.

A final experiment was performed by recording voice in a car whilst the car was moving. A silence period of noise alone was left at the beginning to estimate the noise covariance. Figure 11.7 shows the 10 s signal plus engine noise and the signal estimate, whose noise was reduced by approximately 4 dB. This problem is inherently more difficult since there were other noises in the car and the noise was more non-stationary. To make it work better would need more than one estimate of the noise alone period along the sentence.

The steps in the design process can be summarised as follows:

Wiener filter for coloured noise. We have only one file, a corrupted intermittent signal such as a speech waveform. Alternatively, it must be possible to switch off the signal so that the noise alone can be measured.

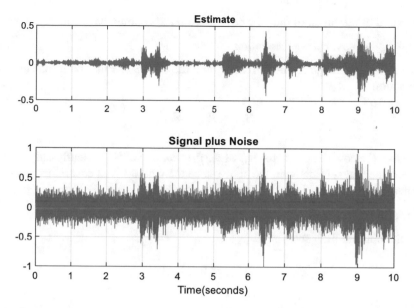

Fig. 11.7 Speech plus engine noise experiment. A noise reduction of 4 dB was achieved using a Wiener filter

Step 1. Initialisation

Divide the waveform into blocks or segments of quasi stationary samples of equal length.

Step 2. Noise covariance estimation

Select the total number of weights m. Before the true signal appears on the waveform estimate recursively the noise alone covariance matrix from smaller vectors of the data $\mathbf{v}(k) = \begin{bmatrix} v(k) & v(k-1) & v(k-2) & \ldots & v(k & m) \end{bmatrix}^T$.

from

$$\mathbf{R}_{vv}(k+1) = \beta\mathbf{R}_{vv}(k) + (1-\beta)\mathbf{v}(k)\mathbf{v}^T(k)$$

giving us the m square noise covariance matrix $\mathbf{R}_{vv}$.

Step 3. Signal plus noise covariance estimation

For each block of data.

Estimate the m square covariance matrix of corrupted speech (signal plus noise) from the smaller vector $\mathbf{s}(k) = \begin{bmatrix} s(k) & s(k-1) & s(k-2) & \ldots & s(k-m) \end{bmatrix}^T$

$$\mathbf{R}_{ss}(k+1) = \beta\mathbf{R}_{ss}(k) + (1-\beta)\mathbf{s}(k)\mathbf{s}^T(k)$$

At the end of the block of data the covariance matrix should have converged.

Step 4. Signal covariance estimation

Since signal and noise are uncorrelated, find the covariance matrix of the signal by subtraction

$$\mathbf{R}_{yy} = \mathbf{R}_{ss} - \mathbf{R}_{vv}$$

Construct the covariance vector $\mathbf{R}_{sy}$ as *the first column* of $\mathbf{R}_{yy}$. Clearly only the first column need be subtracted in the above since the whole signal covariance matrix is not required.

Step 5. Filter the signal for the block of data

Find the optimal filter from

$$\mathbf{H}_{opt} = \mathbf{R}_{ss}^{-1}\mathbf{R}_{sy}$$

and filter the signal plus noise.

Return to step 3 with a new block of data until all data has been processed.

The above method has been named the *correlation subtraction method*. It is very much simplified in that for a real problem we will need to use more than one estimate of the noise covariance, since it is very unlikely to be perfectly stationary. Of course the noise level changes with the speed of the car.

There are other methods which solve this problem, the most notable of which is to *use spectral subtraction*. This was the earliest method and uses the FFT [12, 13]. In its raw form the method tends to introduce an unpleasant phenomenon known as *musical noise*. Another approach is to use the approach above but to use the Periodogram and the FFT to obtain the covariance values. If the periodogram can be estimated, then its inverse FFT is the covariance estimates. Unfortunately, a single FFT is not enough to get good estimates and averaging over numerous FFTs is required using a forgetting factor method like the one above but applied to vectors instead of matrices. Studies on the Wiener filter have shown that the more quality of filtering that is achieved, the quality of the speech signal degrades in proportion [14]. Despite this, Wiener filtering methods appear to give better results than spectral subtraction methods [15]. There are still two problems with this approach, however for real environments.

The first of these requires a reliable and robust speech activity detector (VAD) to find periods where speech exists and therefore regions where it doesn't and there is noise on its own. The VAD problem has been studied over a great many years and a large literature exists. The problem appears at first sight to be quite easy, since a simple energy threshold would suffice if we knew the range of noise that was present. Because of the variety of noise sources and the random nature of noise and signals, it is quite a difficult problem to solve with minimal computation. The problem of speech enhancement when only one sensor measurement of the mixture of signal and noise is available is the hardest of the speech enhancement problems. A good example of a robust solution for this problem is to use feature extraction on the signal plus noise [16]. A good summary of methods is given in [17]. If we are fortunate enough to have more than one sensor, for example a second microphone which picks up the noise on its own (assuming that this is physically possible), or just two microphones in front of the person speaking, then the problem is a little easier, but no less trivial to solve. Agaiby [18] uses two microphones directly in front of the desired speech. The idea is to construct a "zone of activity" for desired speech directly in front of the microphones. If any signal emanates from outside of this zone it is assumed to be noise. The solution works be estimating time-delays between any signal that is present on both microphones. If the time-difference of arrival to each microphone is estimated, then the physical zone can be calculated. Time-delays outside of a defined limit can be rejected and assumed to be noise. The only drawback is that a noise source directly behind the two microphones or in front will be desired speech. A second problem is that in many environments the speech or noise will bounce off the surrounding walls and cause false readings. This is solved to a greater extent by using the *coherence function* which is a measure of correlation versus frequency. A high coherence is assumed to be desired speech since it is known that coherence of reflected (reverberated) signals is reduced at certain frequencies. A lot depends on the application as to whether the more complex VADs.

The second problem that is faced is that the solution for the optimal filter requires a matrix inversion. Fortunately, the matrix is Toeplitz and this problem has been solved some time ago. If inverted using Gaussian elimination methods it would need $O(n^3)$ operations, but with the fast algorithms of Levinson [4] and Durbin [5] the computation is reduced to $O(n^2)$. This computation can be reduced further to $O(8n \log_2^2 n)$ using the generalized Schur algorithm and FFTs [19]. Another algorithm can perform the task with four length n FFTs plus an $O(n)$ term [20]. This is of major importance for real-time applications but not important for offline implementations because the filter length is not long enough to cause significant delays in computing the solution offline.

11.4   The FIR Smoothing and Prediction Problem

In his seminal book, Wiener did not just consider the continuous time filtering problem, he also considered the smoothing and prediction problems too. In discrete time we need a notation to show the difference for an estimate which is smoothed, filtered or predicted. A smoothed estimate is one where there is more data available and the estimate is of a past value of the signal. This in turn introduces a time-delay into the filter. The notation $\hat{y}(k - \ell | k)$ for some integer $\ell \geq 0$ denotes the linear least-squares estimate of the signal y(k) at time $k - \ell$ given observations up to an including time k. For instance, $\hat{y}(k - 5 | k)$ means the estimate is of a signal value which was 5 steps in the past relative to the current observation. For a predictor we have $\hat{y}(k + \ell | k)$ instead which means the linear least-squares estimate of the signal at a future time $k + \ell$ given observations up to and including time k. With more information a smoother will give a better estimate than a filter and with less information a predictor will give a worse estimate than a filter. Often the conditional line | is missed out in some engineering textbooks and confusion arises as to whether a signal is a filtered signal or a one step ahead predicted estimate. Therefore, we need this notation. For instance, $\hat{y}(k + 1|k)$ gives a poorer estimate than $\hat{y}(k|k)$ and $\hat{y}(k - 1|k)$ is better still at the expense of a time-delay of one step.

To obtain the optimal smoother requires a minor modification. The cross-covariance vector for an optimal filter is

$$\mathbf{R}_{sy} = E[\mathbf{s}(k)y(k)]$$

where the vector consists of cross covariance values as follows

$$\mathbf{R}_{sy} = \begin{bmatrix} r_{sy}(0) \\ r_{sy}(1) \\ r_{sy}(2) \\ . \\ . \\ . \\ r_{sy}(n) \end{bmatrix}$$

For a smoother with $\ell = 2$, we need the estimate of a signal which is delayed at time $y(k - \ell)$. This in turn affects the cross-covariance vector of signal with observation and it becomes:

$$\mathbf{R}_{sy} = \begin{bmatrix} r_{sy}(-2) \\ r_{sy}(-1) \\ r_{sy}(0) \\ r_{sy}(1) \\ r_{sy}(2) \\ . \end{bmatrix}$$

The delay in the signal introduces delays in the cross-covariance values. Note that the cross-covariance values are always symmetric so that $r_{sy}(-1) = r_{sy}(1)$, $r_{sy}(-2) = r_{sy}(2)$ and so on. Therefore, to change this vector is a simple task of shifting the array of data downwards by ℓ values and appending the new values at the top. The new values we already have, since the covariance values for negative values and positive ones must be the same. What we have done is to drag optimal weights which are in negative time over to positive time. The filter is always stable because it is FIR.

Consider our previous stationary example with model

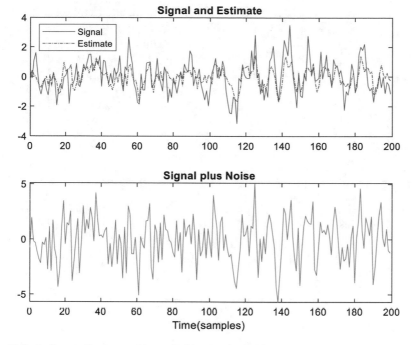

Fig. 11.8 Stationary filtering problem with 30 weights

$$y(k) = \frac{\sqrt{1-a^2}}{\left(1-az^{-1}\right)}\xi(k), |a| < 1$$

The values used were $a = 0.5$, $\sigma_v^2 = 3$, $\sigma_\xi^2 = 1$. Consider the case with 30 weights first and filtering only as shown in Fig. 11.8.

Now repeat the experiment with a smoothing lag of $\ell = 10$.

First, note the delay in the estimate for Fig. 11.9 but also that the estimate looks smoother (hence the name). The smoothing filter impulse response is plotted in Fig. 11.10.

The peak in Fig. 11.10 is where the ordinary optimal filter began. The weights to the left of that are dragged from negative time into positive time. The peak begins at sample 11 since for a zero lag filter the peak begins at time sample 1. Looking back at previous chapters, recall that an ideal brick-wall filter is non causal, but many algorithms make use of the non-causal impulse response by shifting it into positive time like Fig. 11.10.

The plot for mean-square error $E\left[y(k-\ell) - \hat{y}(k-\ell|k)\right]^2$ versus time for the filter and smoother is shown in Fig. 11.11. A longer run-time is shown.

The errors converge recursively since the filter converges with time due to estimation of covariances. The smoother shows a smaller mean-square error (Fig. 11.12).

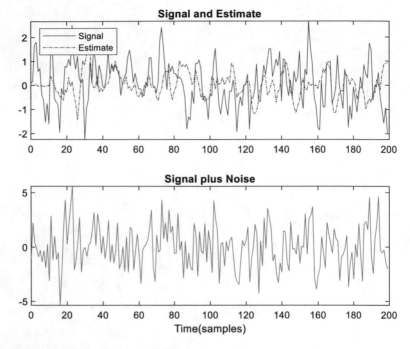

Fig. 11.9 Smoothing estimation problem with $\ell = 10$

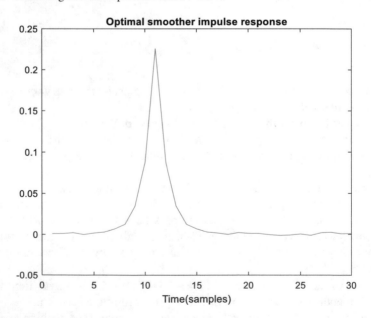

Fig. 11.10 Optimal smoother impulse response

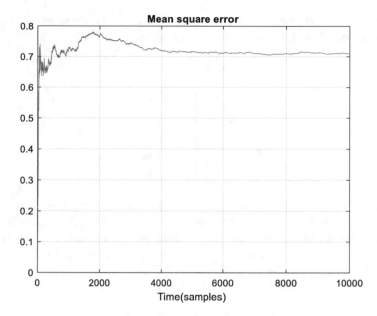

Fig. 11.11   Mean-square error for optimal filter 30 weights

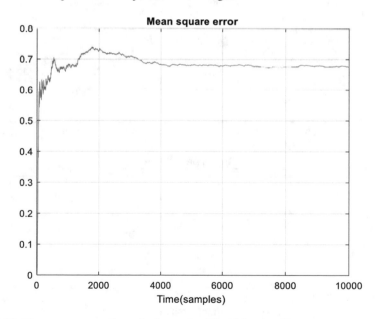

Fig. 11.12   Mean square error for optimal smoother 30 weights $\ell = 10$

For the prediction problem, the cross-covariance vector is shifted upwards. Hence for a 2-step ahead predictor $\hat{y}(k + 2|k)$ we have

$$
\mathbf{R}_{sy} = \begin{bmatrix} r_{sy}(2) \\ r_{sy}(3) \\ r_{sy}(4) \\ . \\ . \\ . \\ . \end{bmatrix}
$$

For this case we have less data available and the estimate is worse than filtering or smoothing. Predictors give inferior results at low SNRs, but at high SNRs they perform adequately and in electrical engineering have been applied to control systems to counter the affect of time-delays. Consider the same example with the values used as a $= 0.5$, $\sigma_v^2 = 0.1$, $\sigma_\xi^2 = 1$ which is a 10 dB SNR.

Note that in Fig. 11.13 the estimate is ahead of time of the signal by 5 samples.

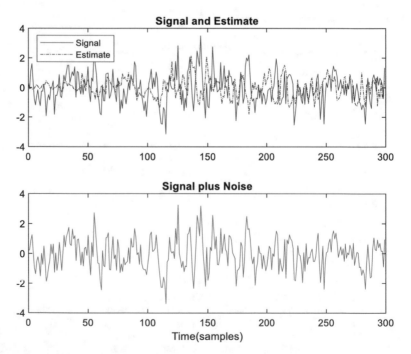

Fig. 11.13 Signal plus noise and predictor output $\hat{y}(k + 5|k)$. 200 weights were used

References

1. R.G. Brown, *Smoothing Forecasting and Prediction of Discrete Time Series* (Englewood Cliffs, Prentice Hall, NJ, 1963)
2. N. Levinson, The Wiener RMS error criterion in filter design and prediction. J. Math. Phys. **25**, 261–278 (1947)
3. J. Durbin, The fitting of time series models. Rev. Inst. Int. Stat **28**, 233–243 (1960)
4. T.J. Moir, Control systems approach to the sample inverse covariance matrix. J. Franklin Inst. **346**, 237–252 (2009)
5. K.B. Petersen, M.S. Pedersen, *The Matrix Cookbook*. Available online at http://matrixcoo kbook.com. (University of Waterloo, Canada, 2012)
6. N. Wiener, *Extrapolation, Interpolation, and Smoothing of Stationary Time Series: With Engineering Applications*, (MIT Press, 1964)
7. A.N. Kolmogorov, Stationary sequences in Hilbert space, in *Selected Works of Mathematics and Its Applications (Soviet Series)*, vol. 26, ed. by A.N. Shiryayev (Springer, Dordrecht, 1992), pp. 228–271
8. G.U. Yule, On a Method of Investigating Periodicities in Disturbed Series, with Special Reference to Wolfer's Sunspot Numbers, Philosophical Transactions of the Royal Society of London. Ser. A **226**, 267–298 (1927)
9. G. Walker, On Periodicity in Series of Related Terms, in *Proceedings of the Royal Society of London, Ser. A,* vol. 131 (1931), pp. 518–532
10. J.S. Lim, A.V. Oppenheim, Enhancement and bandwidth compression of noisy speech. Proc. IEEE **67**(12), 1586–1604 (1979). https://doi.org/10.1109/PROC.1979.11540
11. S. Boll, Suppression of acoustic noise in speech using spectral subtraction. IEEE Trans. Acoust. Speech Signal Process. **27**(2), 113–120 (1979). https://doi.org/10.1109/TASSP.1979.1163209
12. J. Benesty, J. Chen, Y. Huang, S. Doclo, *Study of the Wiener Filter for Noise Reduction, in Speech Enhancement* (Springer, Signals and Communication Technology. Berlin, 2005)
13. S.B Lim, T.Y Chow, J.S. Chang, T.C Tuan, A parametric formulation of the generalized spectral subtraction method, IEEE Trans. Speech Audio Process. **6**(4), 328–337 (1998). https://doi.org/ 10.1109/89.701361
14. B.V. Harsha, A noise robust speech activity detection algorithm, in *Proceedings of 2004 International Symposium on Intelligent Multimedia, Video and Speech Processing*, 20–22 Oct 2004, pp. 322–325. https://doi.org/10.1109/ISIMP.2004.1434065
15. J. Ramirez, J.M. Gorriz, J.C. Segura, Voice Activity Detection. Fundamentals and Speech Recognition System Robustness, in *Robust Speech Recognition and Understanding*, eds by M. Grimm, K. Kroschel (Vienna, Austria, 2007)
16. H. Agaiby, T.J. Moir, A robust word boundary detection algorithm with application to speech recognition, in *Proceedings of 13th International Conference on Digital Signal Processing*, 2–4 July 1997, vol. 2, pp. 753–755. https://doi.org/10.1109/ICDSP.1997.628461
17. G.S. Ammar, W.B. Gragg, Superfast solution of real positive definite Toeplitz systems. SIAM J. Matrix Anal. Appl **9**(1), 61–76 (1988)
18. P.G. Martinsson, V. Rokhlin, M. Tygert, A fast algorithm for the inversion of general Toeplitz matrices. Comput. Math. Appl. **50**, 741–752 (2005)
19. J.F. Barrett, Problems arising from the analysis of randomly disturbed automatic control systems, Ph.D, Engineering Dept., Cambridge, Cambridge, UK, Dec 1958
20. D. Youla, On the factorization of rational matrices. IRE Trans. Inform. Theory **7**(3), 172–189 (1961). https://doi.org/10.1109/TIT.1961.1057636

Chapter 12
IIR Wiener Filtering, Smoothing and Prediction

12.1 Preliminaries and Statement of Problem

The solution to the wiener estimation problem has been addressed in Chap. 11. Although by far the most practical approach due to its guaranteed stability, this was not the original form of the Wiener filter. The original form was to use IIR methods. Consider a filter $H(z^{-1})$ applied to the mixture of a random signal $y(k)$ corrupted by uncorrelated random noise $n(k)$. The mixture process is $s(k)$ where

$$s(k) = y(k) + n(k) \tag{12.1}$$

When the filter is applied to $s(k)$ it will give an estimate

$$\hat{y}(k) = H(z^{-1})s(k) \tag{12.2}$$

At this stage no assumption is made as to what kind of estimate $\hat{y}(k)$ is. It will become either a filtered, smoothed or predicted estimate in what follows. Define the estimation error for filtering

$$e(k) = y(k) - \hat{y}(k) \tag{12.3}$$

The variance of (12.3) can be expressed in the Z-transform frequency domain via Parseval's theorem.

$$E[e^2(k)] = E(y(k) - \hat{y}(k))^2$$
$$= \frac{1}{2\pi j} \oint_{|z=1|} e(z^{-1})e(z)\frac{dz}{z} \tag{12.4}$$

But $e(z^{-1})e(z) = \Phi_{ee}(z)$, the Z-transform spectral density of the error. We can write

© The Author(s), under exclusive license to Springer Nature Switzerland AG 2022
T. J. Moir, *Rudiments of Signal Processing and Systems*,
https://doi.org/10.1007/978-3-030-76947-5_12

$$\sigma_e^2 = \frac{1}{2\pi j} \oint_{|z|=1} \Phi_{ee}(z) \frac{dz}{z}$$

$$= \frac{1}{2\pi j} \oint_{|z|=1} \left[\Phi_{yy} + H\Phi_{ss}H^* - H\Phi_{ys} - \Phi_{sy}H^* \right] \frac{dz}{z} \tag{12.5}$$

In the above we have dropped the z terms to avoid clutter and use the notation for any transfer function F that $F(z^{-1}) = F$, $F(z) = F^*$, where F has all of its poles *inside* the unit circle and the *adjoint* filter F^* has all of its poles *outside* the unit circle. We are required to find the optimal H that minimises (12.5). There are two ways to do this. The first of these is to use the calculus of variations and perturb H by a small arbitrary transfer function. A far simpler method was discovered in the 1950s and uses a method analogous to "completing the square" [1] when applied to finding the minimum of quadratic equations. Forming a square with the error and its conjugate gives us the minimum.

We first need a few other results namely:

$$H\Phi_{yy}H^* + H\Phi_{nn}H^* = H(\Phi_{yy} + \Phi_{nn})H^*$$
$$= H\Phi_{ss}H^* \tag{12.6}$$

and we define

$$\Phi_{ss} = \Phi_{yy} + \Phi_{nn}$$
$$= \Phi_{ss}^+ \Phi_{ss}^- \tag{12.7}$$

We have factorized the spectrum of signal plus noise into the product of two transfer functions:

$$\Phi_{ss} = \Phi_{ss}^+ \Phi_{ss}^- \tag{12.8}$$

Any LTI system can be driven by white noise and model an arbitrary spectrum. For example if white noise $\zeta(k)$ is driven through a transfer function F then it has an output $v(k) = F\zeta(k)$ and it has a spectrum $\Phi_{vv} = FF^*\sigma_\zeta^2$. In (12.8) we are working in reverse. Given the spectrum can we find the transfer function? We ensure Φ_{ss}^+ has all its poles strictly inside the unit circle and its inverse is stable (its zeros also lie strictly within the unit circle). On the other hand, Φ_{ss}^- has all its poles and zeros strictly outside the unit circle. The method is known as *spectral factorization* and was first described by Norbert Wiener. Algorithms to solve such problems appeared in the 1960s and 1970s. Some of the most notable papers in the literature are due to Youla [2], Wilson [3] and Kucera [4]. The poles which are outside the unit circle are mirror images of the ones which are inside. They are reciprocals of each other. A pole at $z = 0.5$ within the transfer function Φ_{ss}^+ will give a pole at $z = 2$ within the transfer function Φ_{ss}^-. Also $\left[\Phi_{ss}^-\right]^* = \Phi_{ss}^+$ and $\left[\Phi_{ss}^+\right]^* = \Phi_{ss}^-$.

Now we write

$$\sigma_e^2 = \frac{1}{2\pi j} \oint_{|z|=1} \left[\Phi_{yy} + H\Phi_{ss}H^* - H\Phi_{ys} - \Phi_{sy}H^* \right] \frac{dz}{z}$$

$$= \frac{1}{2\pi j} \oint_{|z|=1} \left[\left(H\Phi_{ss}^+ - \frac{\Phi_{sy}}{\Phi_{ss}^-} \right) \left(H\Phi_{ss}^+ - \frac{\Phi_{sy}}{\Phi_{ss}^-} \right)^* + \Phi_{yy} - \frac{\Phi_{sy}\Phi_{ys}}{\Phi_{ss}^-\Phi_{ss}^+} \right] \frac{dz}{z}$$

$$= \frac{1}{2\pi j} \oint_{|z|=1} \left[\left(H\Phi_{ss}^+ - \frac{\Phi_{sy}}{\Phi_{ss}^-} \right) \left(H\Phi_{ss}^+ - \frac{\Phi_{sy}}{\Phi_{ss}^-} \right)^* + \Phi_{yy} - \frac{\Phi_{sy}\Phi_{ys}}{\Phi_{ss}} \right] \frac{dz}{z} \quad (12.9)$$

Collecting terms in (12.9)

$$\sigma_e^2 = \frac{1}{2\pi j} \oint_{|z|=1} \left[\left(H\Phi_{ss}^+ - \frac{\Phi_{sy}}{\Phi_{ss}^-} \right) \left(H\Phi_{ss}^+ - \frac{\Phi_{sy}}{\Phi_{ss}^-} \right)^* + \Phi_{yy} \left(1 - \frac{\Phi_{sy}\Phi_{ys}}{\Phi_{yy}\Phi_{ss}} \right) \right] \frac{dz}{z}$$

$$(12.10)$$

Since $\Phi_{ys} = \Phi_{sy}^*$, sometimes this last term in (12.10) is written

$$\Phi_{yy} \left(1 - \frac{\Phi_{sy}\Phi_{sy}^*}{\Phi_{yy}\Phi_{ss}} \right) = \Phi_{yy} \left(1 - C_{sy} \right) \quad (12.11)$$

where

$$C_{sy} = \frac{|\Phi_{sy}|^2}{\Phi_{yy}\Phi_{ss}} \quad (12.12)$$

This is known as the *magnitude-squared coherence* (MSC). The MSC of two signals is like correlation but is a function of frequency as well. Its values always lie between 0 and 1. When two signals are unrelated it has a value of zero. Since it is frequency dependent, two signals could be related at say low frequencies but not at high or vice-versa.

The solution to (12.10) follows immediately as

$$H\Phi_{ss}^+ - \frac{\Phi_{sy}}{\Phi_{ss}^-} = 0 \quad (12.13)$$

From which

$$H_{opt} = \frac{\Phi_{sy}}{\Phi_{ss}} \quad (12.14)$$

Unfortunately, this solution is unstable since the denominator spectrum has poles in the right half plane. It is known as the *noncausal Wiener filter* [5]. It can be written assuming the signal and noise are uncorrelated i.e. $\Phi_{sy} = \Phi_{yy}$ as

$$H_{opt} = \frac{\Phi_{yy}}{\Phi_{yy} + \Phi_{nn}} \tag{12.15}$$

When there is no noise the filter has unity amplitude and when the noise gets larger the magnitude reduces. The minimum mean-square error is found by direct substation of the optimal filter back into (12.10) giving

$$
\begin{aligned}
\sigma_e^2(\min) &= \frac{1}{2\pi j} \oint_{|z|=1} \Phi_{yy}\left(1 - \frac{|\Phi_{yy}|^2}{\Phi_{yy}\Phi_{ss}}\right) \frac{dz}{z} \\
&= \frac{1}{2\pi j} \oint_{|z|=1} \Phi_{yy}\left(1 - \frac{\Phi_{yy}}{\Phi_{ss}}\right) \frac{dz}{z} \\
&= \frac{1}{2\pi j} \oint_{|z|=1} \Phi_{yy}\left(\frac{\Phi_{nn}}{\Phi_{ss}}\right) \frac{dz}{z}
\end{aligned}
\tag{12.16}
$$

Equation (12.15) an be implemented using FFTs, though here we need a transfer function and that would be unstable. To make a causal and hence stable Wiener filter we need a bit of ingenuity. Let $H\Phi_{ss}^+ - \frac{\Phi_{sy}}{\Phi_{ss}^-}$ be split into terms which are causal and noncausal. We use the Wiener notation

$$H\Phi_{ss}^+ - \left\{\frac{\Phi_{sy}}{\Phi_{ss}^-}\right\}_+ = A_+ \tag{12.17}$$

$$-\left\{\frac{\Phi_{sy}}{\Phi_{ss}^-}\right\}_- = A_- \tag{12.18}$$

The special curly bracket notation tells us that if a power-series was made from Eqs. (12.17) and (12.18) they would have the following form

$$A_+ = \alpha_0 + \alpha_1 z^{-1} + \alpha_2 z^{-2} + \alpha_3 z^{-3} + \cdots \tag{12.19}$$

$$A_- = \beta_1 z + \beta_2 z + \beta_3 z^3 \ldots \tag{12.20}$$

Equation (12.10) can then be written as

$$\sigma_e^2 = \frac{1}{2\pi j} \oint_{|z|=1} \left[(A_+ + A_-)(A_+ + A_-)^* + \Phi_{yy}\left(1 - \frac{\Phi_{sy}\Phi_{ys}}{\Phi_{yy}\Phi_{ss}}\right)\right]\frac{dz}{z} \tag{12.21}$$

Only the A terms involve the filter transfer function leaving the second term in (12.21) on its own which we leave. Now by Cauchy's residue theorem, only terms with a dc (z^0) term will be non-zero (since there is a division by z and the coefficient of z^{-1} is the only term that matters).

Now $A_-A_+^*$ and $A_+A_-^*$ have no term in z^0 so their integral is zero. This leaves us with

$$\sigma_e^2 = \frac{1}{2\pi j} \oint_{|z|=1} \left[A_+A_+^* + A_-A_-^* + \Phi_{yy}\left(1 - \frac{\Phi_{sy}\Phi_{ys}}{\Phi_{yy}\Phi_{ss}}\right) \right] \frac{dz}{z} \qquad (12.22)$$

The only term which involves H is the first and therefore the optimal filter that minimises the above integral is when $A_+ = 0$ giving

$$H\Phi_{ss}^+ - \left\{ \frac{\Phi_{sy}}{\Phi_{ss}^-} \right\}_+ = 0$$

Solving for H

$$H_{opt} = \left\{ \frac{\Phi_{sy}}{\Phi_{ss}^-} \right\}_+ \frac{1}{\Phi_{ss}^+} \qquad (12.23)$$

The minimal mean-square error becomes

$$\sigma_e^2(\min) = \frac{1}{2\pi j} \oint_{|z|=1} \left[A_-A_-^* + \Phi_{yy}\left(1 - \frac{\Phi_{sy}\Phi_{ys}}{\Phi_{yy}\Phi_{ss}}\right) \right] \frac{dz}{z}$$

$$\sigma_e^2(\min) = \frac{1}{2\pi j} \oint_{|z|=1} \left[\left\{ \frac{\Phi_{sy}}{\Phi_{ss}^-} \right\}_- \left\{ \frac{\Phi_{sy}}{\Phi_{ss}^-} \right\}_-^* + \Phi_{yy}\left(1 - \frac{\Phi_{sy}\Phi_{ys}}{\Phi_{yy}\Phi_{ss}}\right) \right] \frac{dz}{z} \qquad (12.24)$$

This is clearly larger than for the noncausal optimal filter of (12.16). The celebrated Wiener solution of (12.23) is elegant in the extreme. It is quite general and holds for any stationary signal and noise term. The problem arises as to how to solve such an expression and much of the earlier work in the 1970s involved more elegant solution to this equation that tried to eliminate as far as possible the solving of the {} + brackets. The spectral factorization however remained a problem and although can be solved using Kalman type algorithms it is no less a problem to solve. With modern computing capabilities however many of these problems are becoming routine.

12.2 The IIR Wiener Filter

The solution to the Wiener filtering problem was given by (12.23)

$$H_{opt} = \left\{ \frac{\Phi_{sy}}{\Phi_{ss}^-} \right\}_+ \frac{1}{\Phi_{ss}^+}$$

Define the transfer function which generates the signal spectrum as a ratio of polynomials

$$W(z^{-1}) = \frac{c(z^{-1})}{a(z^{-1})} \tag{12.25}$$

where

$$c(z^{-1}) = c_1 z^{-1} + c_2 z^{-2} + \cdots + c_{nc} z^{-nc} \tag{12.26}$$

$$a(z^{-1}) = 1 + a_1 z^{-1} + a_2 z^{-2} + \cdots + a_{na} z^{-na} \tag{12.27}$$

And it is assumed that $nc \le na$. Let this be driven by zero mean white noise $\xi(k)$ with variance σ_ξ^2.

The desired signal becomes

$$y(k) = \frac{c(z^{-1})}{a(z^{-1})} \xi(k) \tag{12.28}$$

For the time being assume that the additive noise $n(k)$ is zero mean uncorrelated white with variance σ_η^2.

The signal and noise mixture is

$$s(k) = \frac{c(z^{-1})}{a(z^{-1})} \xi(k) + n(k) \tag{12.29}$$

We often drop the z terms in a polynomial argument and write for example $a(z^{-1}) = a$ instead. This removes unnecessary mathematical clutter. The polynomial a is assumed to have all its roots within the unit circle and likewise for c. It is possible to have a c polynomial with zeros outside the unit circle, but we do not consider this case here. Likewise, the *conjugate polynomials* can be described as a^*, c^* have their zeros outside the unit circle.

Therefore, the spectra of the mixture and the signal as expressed in Z-transform format is

$$\Phi_{ss} = \frac{cc^*}{aa^*} \sigma_\xi^2 + \sigma_n^2 \tag{12.30}$$

$$\Phi_{yy} = \frac{cc^*}{aa^*} \sigma_\xi^2 = \Phi_{sy} \tag{12.31}$$

The last equation says that the auto spectrum of y(k) is the same as the cross spectrum between y(k) and n(k). This is simply because the white noise sequences are uncorrelated.

From (12.27)

$$\Phi_{ss} = \Phi_{yy} + \Phi_{nn}$$
$$= \Phi_{ss}^{+}\Phi_{ss}^{-}$$

We have

$$\Phi_{ss} = \frac{cc^*}{aa^*}\sigma_\xi^2 + \sigma_n^2$$
$$= \Phi_{ss}^{+}\Phi_{ss}^{-} \tag{12.32}$$

This is the spectral factorization problem. We must add the two spectra and make it into a third. This third spectrum must generate the same spectrum as the sum of the other two. We add the terms in (12.32) and get

$$\Phi_{ss} = \frac{cc^*}{aa^*}\sigma_\xi^2 + \sigma_\eta^2$$
$$= \frac{dd^*}{aa^*}\sigma_\varepsilon^2 \tag{12.33}$$

With

$$cc^*\sigma_\xi^2 + aa^*\sigma_\eta^2 = dd^*\sigma_\varepsilon^2 \tag{12.34}$$

where

$$d(z^{-1}) = 1 + d_1 z^{-1} + d_2 z^{-2} + \cdots + d_{nd} z^{-nd} \tag{12.35}$$

and $nd = na$ for this white noise case.

We define the transfer function

$$s(k) = \frac{d}{a}\varepsilon(k) \tag{12.36}$$

as the *innovations* model of the mixture process. The white noise term $\varepsilon(k)$ with variance σ_ε^2 is known as the *innovations* white noise sequence. The innovations sequence generates the same spectrum of signal plus noise as the one in (12.33) obtained by the addition of the individual spectra.

The polynomial expression in (12.34) is the spectral factorization problem and d is known as the *spectral factor*. The spectral factor has all its zeros inside the unit circle. We can solve (12.34) for simple cases but for any size of polynomial greater than 2 we need a numerical algorithm as previous discussed. The innovations can be written in terms of the other white noise terms by inverting (12.35). We have

$$\varepsilon(k) = \frac{a}{d}s(k)$$
$$= \frac{a}{d}\left[\frac{c}{a}\xi(k) + n(k)\right]$$
$$= \frac{c}{d}\xi(k) + \frac{a}{d}n(k) \qquad (12.37)$$

The innovations sequence is therefore clearly correlated with the other white noise terms. The transfer function $\frac{a}{d}$ is also a whitening filter for the mixture process. The transfer function of the spectral factor is

$$\Phi_{ss}^{+} = \frac{d}{a}\sigma_{\varepsilon} \qquad (12.38a)$$

and

$$\Phi_{ss}^{-} = \frac{d^*}{a^*}\sigma_{\varepsilon} \qquad (12.38b)$$

Often the spectral factor is shown with the σ_{ε} standard deviation absorbed into the d polynomial but it is better to have it outside and normalize the d polynomial so that $d(0) = 1$.

The term within the $\{\}+$ brackets is

$$\frac{\Phi_{sy}}{\Phi_{ss}^{-}} = \frac{cc^*}{aa^*}\sigma_{\xi}^2\frac{a^*}{d^*}\frac{1}{\sigma_{\varepsilon}}$$
$$= \frac{cc^*\ \sigma_{\xi}^2}{ad^*\ \sigma_{\varepsilon}} \qquad (12.39)$$

If inverse Z-transforms are taken of (12.39) it is a 2-sided sequence. It is messy to calculate manually and requires partial fractions. Consider the following example.

12.2.1   Wiener Filter Example

Consider an example we considered previously.

$$y(k) = \frac{z^{-1}\sqrt{1-a^2}}{\left(1 - az^{-1}\right)}\xi(k), |a| < 1$$

The corrupted signal is

$$s(k) = y(k) + n(k)$$

where n(k) is zero-mean white noise of variance σ_η^2. We have added a one-step delay on the numerator to make our notation correct. The spectrum of y(k) is the same with or without the delay.

Let a = 0.5 and the variance of the driving noise σ_ξ^2 be unity. Make the additive white noise have variance $\sigma_\eta^2 = 3$. Therefore $c(z^{-1}) = z^{-1}\sqrt{1-a^2} = 0.866z^{-1}$, $a(z^{-1}) = 1 - 0.5z^{-1}$. The spectral factorization was performed with the algorithm in [6]. A MATLAB listing is shown below of a function which uses this method.

MATLAB Code 12.1

```
function [ w,w1,ri] = sp_factor( a, c, q,r,step )
%Define a polynomial a and find the Laurent series
%ara* ı cqc*=dr'd*where d is the spectral factor (normalized)
%r' is the innovations variance.
% the star represents conjugate polynomials in z
% make this any length you like but adjust stepsize accordingly.
%uses the method in:
%Moir TJ. A control theoretical approach to the polynomial spectral factorization %problem.
%Circuits, Systems and Signal Processing 2011
na=length(a);
nc=length(c);
%find ara* using convolution of the polynomial with its uncausal counterpart
L1=r*conv(a,fliplr(a))';
%L1 vector produced - this is two-sided Laurent polynomial - has
% negative and positive powers of z within it.
```

```
% do the same for c polynomial
L2=q*conv(c,fliplr(c))';
%L2 vector produced
%Truncate so that only one side is used of the Laurent series (negative powers of z and L0). Start from
%highest negative power and go to zero.
L1=L1(1:na)';
L2=L2(1:nc)';
%check that order of L1 and L2 is the same before adding
nL=max(na,nc);
n=nL;
nmin=min(na,nc);
%There are 3 cases to consider
%case 1
if na>nc

ik=1;
for j=nmin:nL
 La(ik)=L1(j)+L2(ik) ;
 ik=ik+1;
end
  % na is the biggest hence add terms to left of La
  ndif=na-nc;
  we=zeros(ndif,1)';
  for j=1:1:(ndif)
    %create new vector from L1
    we(j)=L1(j);
  end
  % now concatanate LA with we giving [La we]
  L=[we La];
end
%case 2
if nc>na

ik=1;
for j=nmin:nL
 La(ik)=L1(ik)+L2(j) ;
 ik=ik+1;
end
  % nc is the biggest hence add terms to the left of La
  ndif=nc-na;
  we=zeros(ndif,1)';
  for j=1:1:(ndif)
    %create new vector from L2
    we(j)=L2(j);
  end
  % now concatanate LA with we [La we]
  L=[we La];
end
%case 3
```

```
if nc==na
  %just add since both vectors are same dimension
  L=L1+L2;
end
%L is the Laurent series we feed into the loop - the setpoint and the
%output is the spectral factor
%define gain of loop (stepsize)
gain=step;
%Initialise integrator output vector
w=zeros(n,1)';
g=zeros(n,1)';
%initialise error vector
e=zeros(n,1)';
%Main while  loop
go=true;
k=0;
while go
   %increment counter
   k=k+1;
   %Update integrators
w=w+gain*e;
%error first
g=conv(w,fliplr(w));
%truncate in half
F=g(1:n);
%error
e=L-F;
%check norm of error
delt=norm(e);
if delt<1e-6 go=false;
end
end
%Reverse coefficients
w1=fliplr(w);
%remove the first term
co=w1(1);
w=w1/co;
%ri is innovations variance
ri=co^2;
%No of iterations to converge is k
```

The function was used in the following code to solve the spectral factorization problem.

MATLAB Code 12.2

```
%Example program for Spectral Factorization
%define two polynomials
a=[1 -0.5];
c=[0 1]*0.866;
% noise variances
q=1;r=3;
%Define a polynomial a and find the Laurent series
%ara*+cqc*=dr'd*where d is tht spectral factor (normalised)
%r' is the innovations variance.
% the star represents conjugate polynomials in z
% make this any length you like but adjust stepsize accordingly.
%step size - note use default of step=0.1 but can make smaller
% if no convergence then you need to make it smaller
step=0.1;
[ wf,wf1,ri] = sp_factor( a,c, q,r,step );
%Spectral factor - normalized
wf
%innovations variance
ri
% Spectral factor - not normalised
wf1
```

The function returns the value for the normalized spectral factor $d(z^{-1}) = 1 - 0.382z^{-1}$ along with the innovations variance $\sigma_\varepsilon^2 = 3.927$. Now the term within the $\{\} +$ brackets is

$$\left\{ \frac{\Phi_{sy}}{\Phi_{ss}^-} \right\}_+ = \left\{ \frac{cc^*}{aa^*} \sigma_\xi^2 \frac{a^*}{d^*} \frac{1}{\sigma_\varepsilon} \right\}_+$$

$$= \left\{ \frac{cc^*}{ad^*} \right\}_+ \frac{\sigma_\xi^2}{\sigma_\varepsilon}$$

Use partial fractions on the term in brackets

$$\frac{cc^*}{ad^*} = \frac{e}{a} + \frac{f^*}{d^*} \tag{12.40}$$

where e is a polynomial of degree $na - 1$ and f^* is a polynomial in z of degree na. The polynomial e is expressed in *negative* powers of z only and f^* in *positive* powers of z. However, we assume $e(0) = e_0$ and $f^*(0) = 0$. This is so that the term $\frac{f^*}{d^*}$ has no coefficient of z^0 in its power series expansion. We make the term e/a causal and the remainder noncausal, and by forcing $f^*(0) = 0$ this ensures there is no left-over term at the zeroth power of z. Multiplying out (12.40) we get

$$cc^* = ed^* + f^*a \tag{12.41}$$

For our example the order is $n = 1$ and so $a_1 = -0.5$, $c_1^2 = (1 - a_1^2) = 0.75$, $d_1 = -0.382$

$$c_1^2 = e_0 + f_1 a_1$$

$$0 = e_0 d_1 + f_1$$

This can be written in matrix form

$$\begin{bmatrix} 1 & a_1 \\ d_1 & 1 \end{bmatrix} \begin{bmatrix} e_0 \\ f_1 \end{bmatrix} = \begin{bmatrix} 1 - a_1^2 \\ 0 \end{bmatrix}$$

Solving gives $e_0 = 0.927$. We don't care about f_1 since the term involving it will be discarded because it is noncausal. However, its value is $f_1 = 0.354$. The causal term is therefore

$$\left\{ \frac{cc^*}{ad^*} \right\}_+ \frac{\sigma_\xi^2}{\sigma_\varepsilon} = \frac{e_0}{a(z^{-1})} \frac{\sigma_\xi^2}{\sigma_\varepsilon}$$

The Wiener filter is

$$H_{opt} - \left\{ \frac{\Phi_{sy}}{\Phi_{ss}^-} \right\}_+ \frac{1}{\Phi_{ss}^+}$$

and the spectral factor of the mixture process is from (12.38a)

$$\Phi_{ss}^+ = \frac{d}{a} \sigma_\varepsilon$$

The complete filter is

$$H_{opt} = \left\{ \frac{\Phi_{sy}}{\Phi_{ss}^-} \right\}_+ \frac{1}{\Phi_{ss}^+}$$

We use $\left(\frac{\sigma_\xi^2}{\sigma_\varepsilon^2} \right) = 0.2546$ in what follows.

$$\begin{aligned} H_{opt} &= \frac{e_0}{a(z^{-1})} \frac{\sigma_\xi^2 \, a(z^{-1})}{\sigma_\varepsilon^2 \, d(z^{-1})} \\ &= \frac{e_0}{d(z^{-1})} \left(\frac{\sigma_\xi^2}{\sigma_\varepsilon^2} \right) \\ &= \frac{0.236}{1 - 0.382 z^{-1}} \end{aligned}$$

12.2.2   Innovations Form of the Wiener Filter

Previously, in finding the solution to the Wiener filter IIR problem we needed to solve simultaneous equations from (12.41). The higher order the signal transfer function will mean there are more equations to solve. There will be 2n equations for a process of order n. For the white noise case there is a way around solving the {} + bracket term. It was pioneered by several authors, the first of which was Barrett [7] and Shaked [8]. Proceed as follows

$$
H_{opt} = \left\{ \frac{\Phi_{sy}}{\Phi_{ss}^{-}} \right\}_{+} \frac{1}{\Phi_{ss}^{+}} = \left\{ \frac{cc^{*}}{aa^{*}} \sigma_{\xi}^{2} \frac{a^{*}}{d^{*}} \right\}_{+} \frac{1}{\sigma_{\varepsilon}^{2}} \frac{a}{d}
$$

(12.42)

Now substitute from (12.33)

$$
\frac{cc^{*}}{aa^{*}} \sigma_{\xi}^{2} = \frac{dd^{*}}{aa^{*}} \sigma_{\varepsilon}^{2} - \sigma_{\eta}^{2}
$$

(12.43)

into (12.42)

$$
\begin{aligned}
H_{opt} &= \left\{ \left(\frac{dd^{*}}{aa^{*}} \sigma_{\varepsilon}^{2} - \sigma_{\eta}^{2} \right) \frac{a^{*}}{d^{*}} \right\}_{+} \frac{1}{\sigma_{\varepsilon}^{2}} \frac{a}{d} \\
&= \left\{ \frac{d}{a} \sigma_{\varepsilon}^{2} - \frac{a^{*}}{d^{*}} \sigma_{\eta}^{2} \right\}_{+} \frac{1}{\sigma_{\varepsilon}^{2}} \frac{a}{d} \\
&= \left\{ \frac{d}{a} \right\}_{+} \frac{a}{d} - \left\{ \frac{a^{*}}{d^{*}} \right\}_{+} \frac{\sigma_{\eta}^{2}}{\sigma_{\varepsilon}^{2}} \frac{a}{d}
\end{aligned}
$$

(12.44)

The first term in brackets is causal and therefore

$$
\left\{ \frac{d}{a} \right\}_{+} = \frac{d}{a}
$$

(12.45)

The second term can be expressed as a convergent power series in positive powers of z beginning at 1

$$
\left\{ \frac{a^{*}}{d^{*}} \right\}_{+} = \left\{ 1 + p_{1} z + p_{2} z^{2} + ... \right\}_{+} = 1
$$

(12.46)

Combing we get

$$
H_{opt} = 1 - \frac{\sigma_{\eta}^{2}}{\sigma_{\varepsilon}^{2}} \frac{a}{d}
$$

(12.47)

Clearly (12.47) avoids entirely the need to solve a set of equations in the partial fraction expansion. For this example, $\sigma_\varepsilon^2 = 3.927$, $\sigma_\eta^2 = 3$ and $\frac{\sigma_\eta^2}{\sigma_\varepsilon^2} = 0.76394$. Substituting these values gives us

$$H_{\text{opt}} = 1 - \frac{0.764\left(1 - 0.5z^{-1}\right)}{\left(1 - 0.382z^{-1}\right)}$$

Simplifying the above gives us

$$H_{\text{opt}} = \frac{0.236}{1 - 0.382z^{-1}}$$

Which the same answer as using the direct method. The polynomial form of (12.47) was first reported by Hagander and Wittenmark [9]. The earlier work used transfer functions instead of polynomials. From their work we can also write the filtered signal by multiplying (12.47) by s(k) where from (12.35)

$$s(k) = \frac{d}{a}\varepsilon(k)$$

Note that the innovations are also defined from the one-step ahead predicted estimate of the signal $\hat{y}(k|k - 1)$ as follows [10]:

$$\varepsilon(k) = s(k) - \hat{y}(k|k \text{ - } 1)$$

By definition

$$\hat{y}(k|k) = H_{\text{opt}}s(k) \tag{12.48}$$

Resulting in

$$\hat{y}(k|k) = s(k) - \left(\frac{\sigma_\eta^2}{\sigma_\varepsilon^2}\right)\varepsilon(k) \tag{12.49}$$

This is an innovations form and quite convenient for implementation since the innovations white noise sequence can be found by whitening the mixture:

$$\varepsilon(k) = \frac{a}{d}s(k) \tag{12.50}$$

It is important to distinguish the difference between $\hat{y}(k|k \text{ - } 1)$ and $\hat{y}(k|k)$. The former is a one-step ahead predicted estimate and the latter is our instantaneous estimate which has a lower mean-square error. The two are often confused in the literature. Here we describe $\hat{y}(k|k \text{ - } 1)$ as a one-step ahead *predicted estimate* and not a filtered estimate (Fig. 12.1).

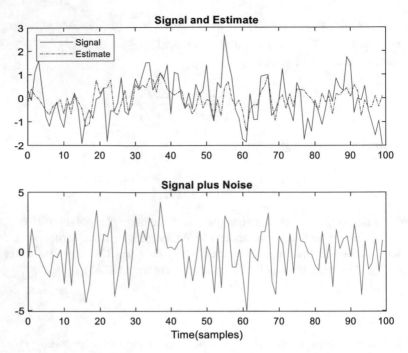

Fig. 12.1 Mixture of signal plus noise, true signal and Wiener filter estimate for additive white noise

The filter of (12.49) was simulated using the MATLAB code below.

MATLAB Code 12.3

```matlab
%IIR Wiener Filter
%using innovations method
clear
close all
Npoints=100;
a=[1 -0.5];
c=[1]*sqrt(1-0.5^2);
%Unit variance driving noise
sq=1;
%Additive noise
sv=3;
t=[0:1:Npoints-1]';
%Random Noise length Npoints Variance rn
rn=sqrt(sq)*randn(Npoints,1);
%Filter it to get time-series
%Generate Signal
y=filter(c,a,rn);
% Additive uncorrelated noise variance sv
seed=1;
rng(seed)
rv=sqrt(sv)*randn(Npoints,1);
s=y+rv;
%%%%
%Innovations model obtained from spectral factor
d=[1 -0.382];
%Innovations variance from spectral factor algorithm too
sinn=3.927;
yinn=filter(a,d,s);
%yinn is now white noise
%we have whitened the mixture process with (a/d).s(k)
%filter signal plus noise
%innovations form of Wiener filter
yh=s-yinn*(sv/sinn);
%%%
subplot(2,1,1)
plot(t,y,'-r',t,yh,'-.b')
legend('Signal','Estimate','Location','northwest')
title('Signal and Estimate')
subplot(2,1,2)
plot(t,s)
set(gcf,'color','w');
title('Signal plus Noise')
set(gcf,'color','w');
xlabel('Time(samples)')
```

12.2.3   The Smoothing and Prediction Problems

Figure. 12.2 illustrates the basic idea behind filtering, smoothing and prediction. In
the case of filtering the delay $\ell = 0$, for smoothing this is a positive integer and for
prediction ℓ is negative.

We have already considered the filtering case. For smoothing the lag ℓ is a positive
integer. We do not need to go back to scratch for this problem, we can continue from
the generic equation for the Wiener filter. The only difference is that a time-delay is
introduced into the cross-spectrum. Hence

$$
\begin{aligned}
H_{opt}^{s} &= \left\{ z^{-\ell} \frac{\Phi_{yy}}{\Phi_{ss}^{-}} \right\}_{+} \frac{1}{\Phi_{ss}^{+}} \\
&= \left\{ z^{-\ell} \frac{cc^{*}}{aa^{*}} \sigma_{\xi}^{2} \frac{a^{*}}{d^{*}} \right\}_{+} \frac{1}{\sigma_{\varepsilon}^{2}} \frac{a}{d}
\end{aligned}
\tag{12.51}
$$

where H_{opt}^{s} is the smoother transfer function. We could again use partial fractions, but
it leads to more unknown to solve for. Instead, we go straight to the second method
already discussed in the previous section and substitute for $\frac{cc^{*}}{aa^{*}} \sigma_{\xi}^{2}$. We get

$$
\begin{aligned}
H_{opt}^{s} &= \left\{ z^{-\ell} \left(\frac{dd^{*}}{aa^{*}} \sigma_{\varepsilon}^{2} - \sigma_{\eta}^{2} \right) \frac{a^{*}}{d^{*}} \right\}_{+} \frac{1}{\sigma_{\varepsilon}^{2}} \frac{a}{d} \\
&= \left\{ z^{-\ell} \frac{d}{a} \sigma_{\varepsilon}^{2} - z^{-\ell} \frac{a^{*}}{d^{*}} \sigma_{\eta}^{2} \right\}_{+} \frac{1}{\sigma_{\varepsilon}^{2}} \frac{a}{d} \\
&= \left\{ z^{-\ell} \frac{d}{a} \right\}_{+} \frac{a}{d} - \left\{ z^{-\ell} \frac{a^{*}}{d^{*}} \right\}_{+} \frac{\sigma_{\eta}^{2}}{\sigma_{\varepsilon}^{2}} \frac{a}{d}
\end{aligned}
\tag{12.52}
$$

The first term in the RHS of (12.52) is causal. The delay just shifts the impulse
response more to the right. The second term is not, and we go back to a power series
expansion.

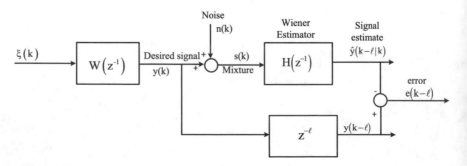

Fig. 12.2 The general estimation problem

$$H_{opt}^s = z^{-\ell} - \left\{ z^{-\ell} \frac{a^*}{d^*} \right\}_+ \frac{\sigma_\eta^2}{\sigma_\varepsilon^2} \frac{a}{d}$$

$$= z^{-\ell} - z^{-\ell} P_\ell(z) \frac{\sigma_\eta^2}{\sigma_\varepsilon^2} \frac{a}{d} \tag{12.53}$$

where

$$\left\{ z^{-\ell} \frac{a^*}{d^*} \right\}_+ = \left\{ z^{-\ell} (1 + p_1 z + p_2 z^2 + \cdots + p_\ell z^\ell + p_{\ell+1} z^{\ell+1} + \cdots) \right\}_+$$

$$= z^{-\ell} (1 + p_1 z + p_2 z^2 + \cdots + p_\ell z^\ell)$$

$$= z^{-\ell} P(z) \tag{12.54}$$

Only the first ℓ terms are causal, so we truncate the series at that coefficient. The reason for this is because they all have negative powers of z. Terms with positive powers of z are noncausal and discarded.

From (12.53) we multiply by s(k) and get the smoothed signal in terms of the innovations sequence $\varepsilon(k)$

$$\hat{y}(k - \ell|k) = s(k - \ell) - \frac{\sigma_\eta^2}{\sigma_\varepsilon^2} \sum_{i=0}^{\ell} p_i \varepsilon(k - \ell + i) \tag{12.55}$$

It is interesting that by using the earlier filtering result

$$\hat{y}(k|k) = s(k) - \left(\frac{\sigma_\eta^2}{\sigma_\varepsilon^2} \right) \varepsilon(k)$$

We re-arrange and write

$$s(k) = \hat{y}(k|k) + \left(\frac{\sigma_\eta^2}{\sigma_\varepsilon^2} \right) \varepsilon(k) \tag{12.56}$$

Then substitute (12.56) into (12.55) giving

$$\hat{y}(k - \ell|k) = \hat{y}(k - \ell|k - \ell) + \left(\frac{\sigma_\eta^2}{\sigma_\varepsilon^2} \right) \varepsilon(k - \ell) - \frac{\sigma_\eta^2}{\sigma_\varepsilon^2} \sum_{i=0}^{\ell} p_i \varepsilon(k - \ell + i)$$

$$= \hat{y}(k - \ell|k - \ell) - \frac{\sigma_\eta^2}{\sigma_\varepsilon^2} \sum_{i=1}^{\ell} p_i \varepsilon(k - \ell + i) \tag{12.57}$$

This equation expresses a smoothed estimate in terms of a filtered estimate. It is like it is subtracting an error from the filtered signal to reduce it further in value. The

extra computation that the smoother needs is in the power series expansion (12.54).
For the same example as the Wiener filter we can calculate with long division.

$$\frac{a^*}{d^*} = \frac{1 - 0.5z}{1 - 0.382z}$$
$$= 1 - 0.118z - 0.0451z^2 - 0.0172z^3 - 0.0066z^4 - 0.0025z^5 + \cdots \quad (12.58)$$

The terms get smaller with larger delay. In this case the delay was chosen as
$\ell = 5$. Simulating the smoother gave the results as shown in Fig. 12.3. There are
several ways to find the power series including a recursive method or upper triangular
Toeplitz matrices.

The estimate is delayed but has a better appearance than the filtered signal of
Fig. 12.1.

The prediction problem is solved by making the time-delay a time advance and
removing the causal parts of the Wiener filter.

$$H_{opt}^p = \left\{ z^\ell \left(\frac{dd^*}{aa^*} \sigma_\varepsilon^2 - \sigma_\eta^2 \right) \frac{a^*}{d^*} \right\}_+ \frac{1}{\sigma_\varepsilon^2} \frac{a}{d}$$
$$= \left\{ z^\ell \frac{d}{a} \sigma_\varepsilon^2 - z^\ell \frac{a^*}{d^*} \sigma_\eta^2 \right\}_+ \frac{1}{\sigma_\varepsilon^2} \frac{a}{d}$$

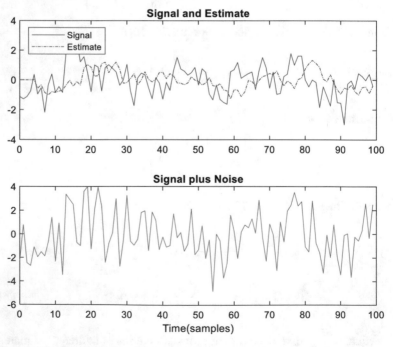

Fig. 12.3 Performance of the optimal smoother for $\ell = 5$

$$= \left\{ z^{\ell} \frac{d}{a} \right\}_+ \frac{a}{d} - \left\{ z^{\ell} \frac{a^*}{d^*} \right\}_+ \frac{\sigma_\eta^2}{\sigma_\varepsilon^2} \frac{a}{d} \tag{12.59}$$

H_{opt}^p is the transfer function of the optimal predictor. The second term in the RHS of (12.59) is noncausal. The predictor now becomes

$$H_{opt}^p = \left\{ z^{\ell} \left(\frac{dd^*}{aa^*} \sigma_\varepsilon^2 - \sigma_\eta^2 \right) \frac{a^*}{d^*} \right\}_+ \frac{1}{\sigma_\varepsilon^2} \frac{a}{d}$$

$$H_{opt}^p = \left\{ z^{\ell} \frac{d}{a} \right\}_+ \frac{a}{d} \tag{12.60}$$

Expanding the term in brackets with a convergent power series in negative powers of z we get

$$z^{\ell} \frac{d}{a} = \left(f_0 + f_1 z^{-1} + f_2 z^{-2} + \cdots \right) z^{\ell} \tag{12.61}$$

Only the terms with negative powers of z are causal. These will depend on ℓ. Expanding gives

$$z^{\ell} \frac{d}{a} = \left(f_0 z^{\ell} + f_1 z^{\ell-1} + f_2 z^{\ell-2} + \cdots f_{\ell} + f_{\ell+1} z^{-1} + f_{\ell+2} z^{-2} + \cdots \right) \tag{12.62}$$

with $f_0 = 1$.

The causal terms start at f_ℓ onwards and the lower terms are noncausal. Write the first $\ell - 1$ terms and the remainder as

$$z^{\ell} \frac{d}{a} = z^{\ell} f + \frac{g}{a} \tag{12.63}$$

where f is a polynomial of order $\ell - 1$ with $f(0) = 1$

$$f(z^{-1}) = f_0 + f_1 z^{-2} + f_2 z^{-2} + \cdots + f_{\ell-1} z^{-(\ell-1)} \tag{12.64}$$

and the polynomial g is of order $n - 1$.

$$g(z^{-1}) = g_0 + g_1 z^{-2} + g_2 z^{-2} + \cdots + g_{n-1} z^{-(n-1)} \tag{12.65}$$

Find the polynomial equation from (12.63) by multiplying by $az^{-\ell}$.

$$d = af + gz^{-\ell} \tag{12.66}$$

The solution to (12.66) can be found by comparison of powers of z on each side of the equation. This gives the polynomial g and the causal part of (12.62) is now

$$\left\{ z^\ell \frac{d}{a} \right\}_+ = \frac{g}{a} \qquad . \tag{12.67}$$

The predictor is found from (12.60)

$$H^p_{opt} = \frac{g}{a} \frac{a}{d}$$

$$= \frac{g}{d} \tag{12.68}$$

This type of predictor was first derived by Wittenmark [11].

A predicted estimated can also be written in innovations form by using $s(k) = \frac{d}{a}\varepsilon(k)$ applied to (12.68)

$$\hat{y}(k + \ell|k) = \frac{g}{d}\varepsilon(k) \tag{12.69}$$

Of special interest is the one step ahead predictor when $\ell = 1$. Examining (12.64)

$$d = af_0 + gz^{-1}$$

The polynomial f since it is of order zero becomes $f = f_0 = 1$. Solving for g gives

$$g = z(d - a)$$

The one-step ahead predictor is

$$\hat{y}(k + 1|k) = \frac{z(d - a)}{d}\varepsilon(k) \tag{12.70}$$

This is of the form

$$\hat{y}(k + 1|k) = \frac{(d_1 - a_1) + (d_2 - a_2)z^{-1} + \cdots + (d_n - a_n)z^{-(n-1)}}{d(z^{-1})}\varepsilon(k) \tag{12.71}$$

The various forms of Wiener filter for the white noise case are given a uniform treatment in reference [12].

To finish the simple example, we derive a 2-step predictor. The g and f polynomials are

$$g = g_0$$

$$f = 1 + f_1 z^{-1}$$

Substitute into the polynomial equation

$$d = af + gz^{-\ell}$$

and obtain

$$\left(1 + d_1 z^{-1}\right) = \left(1 + a_1 z^{-1}\right)\left(1 + f_1 z^{-2}\right) + g_0 z^{-2}$$

Solving results in

$$g_0 = -a_1(d_1 - a_1)$$
$$f_1 = d_1 - a_1$$

The predictor is

$$H_{opt}^p = \frac{g}{d}$$
$$= \frac{g_0}{1} + d_1 z^{-1}$$

A simulation is shown for this example in Fig. 12.4 for $\sigma_\eta^2 = 0.1$.

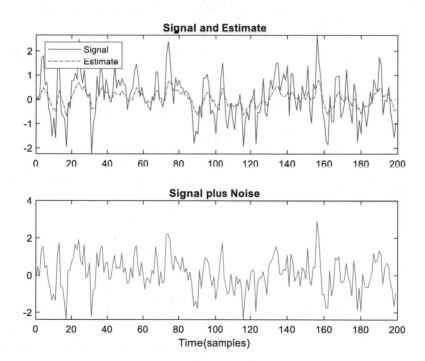

Fig. 12.4 A 2-step ahead predictor with additive white-noise

12.3 Optimal Wiener IIR Filter, Smoother and Predictor for Additive Coloured Noise

This is an extension of the white-noise case. The signal and noise mixture is shown in Fig. 12.5.

The signal is described by

$$s(k) = W\xi(k) + W_n\eta(k) \tag{12.72}$$

There is only additive coloured noise shaped by the transfer function W_n. It is expressed as

$$n(k) = W_n\eta(k) \tag{12.73}$$

It is assumed that any extra white noise has already been absorbed into the noise transfer function via the addition of spectra. Multiple noise sources can be added, and spectral factorization is performed to add them and describe them as a single transfer function W_n driven by the zero-mean innovations white noise $\eta(k)$ with variance σ_η^2. This noise is uncorrelated with the signal. Hence the model of Fig. 12.5 is quite general. The desired signal is $y(k)$ and we need to find the optimal filter H_{opt} which will give us the estimate. Hence

$$\hat{y}(k|k) = H_{opt}s(k) \tag{12.74}$$

is the filtered signal.

To find the filter, start with the general expression for the Wiener filter (12.23)

$$H_{opt} = \left\{ \frac{\Phi_{sy}}{\Phi_{ss}^-} \right\}_+ \frac{1}{\Phi_{ss}^+}$$

Substituting spectra gives

Fig. 12.5 Signal plus coloured noise problem

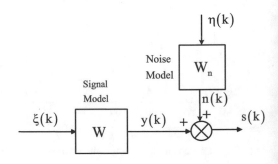

$$\Phi_{sy} = \Phi_{yy}$$
$$= WW^* \sigma_\xi^2 \tag{12.75}$$

$$\Phi_{ss} = WW^* \sigma_\xi^2 + W_n W_n^* \sigma_\eta^2 \tag{12.76}$$

This last spectrum of the mixture can be factorized. Let the stable minimum phase spectral factor transfer function be F with driving white innovations noise $\varepsilon(k)$ of variance σ_ε^2. We have

$$\Phi_{ss} = WW^* \sigma_\xi^2 + W_n W_n^* \sigma_\eta^2$$
$$= FF^* \sigma_\varepsilon^2 \tag{12.77}$$

Substitute into the generic Wiener solution

$$H_{opt} = \left\{ \frac{WW^* \sigma_\xi^2}{F^*} \right\}_+ \frac{1}{F \sigma_\varepsilon^2} \tag{12.78}$$

This is harder than it looks, since the signal and noise generating transfer functions W, W_n will not share the same poles and zeros. We use upper case letter for polynomials here as well as transfer functions. Let the two transfer functions which define the signal and noise be written as a ratio of polynomials

$$W = \frac{C}{A} \tag{12.79a}$$

$$W_n = \frac{P}{Q} \tag{12.79b}$$

where the orders of the C, A, P and Q polynomials are nc, na, np and nq respectively. It is assumed that the numerator polynomials have degree at least the same or less than their corresponding denominators:$nc \leq na, np \leq nq$. All of the polynomials are assumed to have their zeros within the unit circle and it is assumed that $A(0) = Q(0) = 1$.

The spectral factor of the mixture is found from (12.77) by substitution

$$\Phi_{ss} = \frac{CC^*}{AA^*} \sigma_\xi^2 + \frac{PP^*}{QQ^*} \sigma_\eta^2$$
$$= FF^* \sigma_\varepsilon^2 = \frac{DD^*}{RR^*} \sigma_\varepsilon^2 \tag{12.80}$$

where $R = AQ$ and $F = \frac{D}{R}$. In polynomial form the spectral factorization becomes

$$DD^*\sigma_\varepsilon^2 = CQC^*Q^*\sigma_\xi^2 + APA^*P^*\sigma_\eta^2 \tag{12.81}$$

The degree of D will depend on the degree of the other polynomials in (12.81). It will be nd = max(nc + nq, na + np) and we make sure that $D(0) = 1$ and absorb any constant term into the innovations variance σ_ε^2. To solve (12.81) we use the previous algorithm but must first form new polynomials CQ and AQ by multiplication.

From (12.75)

$$\Phi_{yy} = \frac{CC^*}{AA^*}\sigma_\xi^2 \tag{12.82}$$

Now substitute the polynomial expressions into (12.78)

$$H_{opt} = \left\{\frac{CC^*}{AA^*}\sigma_\xi^2 \frac{R^*}{D^*}\frac{1}{\sigma_\varepsilon^2}\right\}_+ \frac{R}{D} \tag{12.83}$$

This is the solution but we have a significant effort on separating the { } + brackets. Now substitute from (12.80) into (12.83) and consider the term in { } + brackets only [13–16]

$$\left\{\frac{CC^*}{AA^*}\frac{R^*}{D^*}\frac{\sigma_\xi^2}{\sigma_\varepsilon^2}\right\}_+$$

$$= \left\{\left[\frac{DD^*}{RR^*}\sigma_\varepsilon^2 - \frac{PP^*}{QQ^*}\sigma_\eta^2\right]\frac{R^*}{D^*\sigma_\varepsilon^2}\right\}_+$$

$$= \left\{\left[\frac{D}{R} - \frac{PP^*}{QQ^*}\frac{R^*}{D^*}\left(\frac{\sigma_\eta^2}{\sigma_\varepsilon^2}\right)\right]\right\}_+ \tag{12.84}$$

Now $\frac{D}{R}$ is causal and so we leave it. Use partial fractions on

$$\frac{CC^*}{AA^*}\frac{R^*}{D^*}\frac{\sigma_\xi^2}{\sigma_\varepsilon^2} = \frac{CC^*}{A}\frac{Q^*}{D^*}\frac{\sigma_\xi^2}{\sigma_\varepsilon^2}$$

$$\frac{CC^*}{A}\frac{Q^*}{D^*}\frac{\sigma_\xi^2}{\sigma_\varepsilon^2} = \frac{G}{A} + \frac{E^*}{D^*} \tag{12.85}$$

where $E^*(0) = 0$ and the degree of the E polynomial is nd. The RHS of (12.85) is now divided into a causal and noncausal transfer function. This doesn't solve the problem entirely. Returning to Eq. (12.84), applying partial fractions a second time

$$\frac{PP^*}{QQ^*}\frac{R^*}{D^*}\left(\frac{\sigma_\eta^2}{\sigma_\varepsilon^2}\right) = \frac{PP^*}{Q}\frac{A^*}{D^*}\left(\frac{\sigma_\eta^2}{\sigma_\varepsilon^2}\right)$$

$$= \frac{T}{Q} + \frac{L^*}{D^*} \tag{12.86}$$

where $L^*(0) = 0$ and the degree of L is nd. This is also divided into causal and noncausal transfer functions. Combining the causal terms, we must have that

$$\frac{G}{A} = \frac{D}{R} - \frac{T}{Q} \tag{12.87}$$

The degrees of G and T must be $na - 1$ and $nq - 1$ respectively.

$$G(z^{-1}) = g_0 + g_1 z^{-1} + \cdots + g_{na-1} z^{-(na-1)}$$

$$T(z^{-1}) = t_0 + t_1 z^{-1} + \cdots + t_{nq-1} z^{-(nq-1)}$$

Expanding (12.87) since $R = AQ$

$$GQ = D - TA$$

Or

$$D = TA + GQ \tag{12.88}$$

This is a polynomial *Diophantine equation* popularized by Kurera in the late 1970s [4]. It has an infinite number of solutions, but there is one minimal degree solution which we seek. The solution can be found in a number of ways. The Euclidian algorithm can be used as applied to polynomials by repeatedly taking the remainder of the division of polynomials. The second approach is just brute strength by multiplying out and solving the unknown polynomial coefficients. It is important to note that the Diophantine will have no solution if A and Q have common factors. The two polynomials must be relatively prime.

The optimal filter transfer function is now given as

$$H_{opt} = \frac{G}{A} \frac{R}{D}$$

$$= \frac{GQ}{D} \tag{12.89}$$

If required, the filter can also be put in innovations format since

$$s(k) = \frac{D}{R} \varepsilon(k) \tag{12.90}$$

Then from (12.74) and using (12.89)

$$\hat{y}(k|k) = \frac{G}{A}\varepsilon(k) \tag{12.91}$$

From the above results we have the following equality

$$\frac{D}{R} - \frac{PP^*}{QQ^*}\frac{R^*}{D^*}\left(\frac{\sigma_\eta^2}{\sigma_\varepsilon^2}\right) = \frac{D}{R} - \frac{T}{Q}\frac{L^*}{D^*}$$

$$= \frac{G}{A} + \frac{E^*}{D^*} \tag{12.92}$$

Looking at the noncausal terms, by comparison we must have that $E^* = -L^*$. From (12.85) and (12.86) we therefore have two Diophantine equations

$$CC^*Q^*\left(\frac{\sigma_\varepsilon^2}{\sigma_\varepsilon^2}\right) = GD^* + AE^* \tag{12.93}$$

$$PP^*A^*\left(\frac{\sigma_\eta^2}{\sigma_\varepsilon^2}\right) = TD^* - QE^* \tag{12.94}$$

These are called *Bilateral Diophantine equations*. For this case we can combine them and get the *Implied Diophantine equation* in (12.88)

$$D = TA + GQ$$

Consider an example where signal and noise models are

$$\frac{C}{A} = \frac{z^{-1}\sqrt{1-a_1^2}}{(1-a_1z^{-1})}, \frac{P}{Q} = \frac{z^{-1}\sqrt{1-q_1^2}}{(1-q_1z^{-1})}$$

with $a_1 = 0.5, q_1 = 0.9$. These noise shaping filters are scaled so that for unit variance white driving noise at the input they give unit variance correlated noise at the output. The noise sources are assume uncorrelated.

Then $\frac{C}{A} = \frac{0.866z^{-1}}{(1-0.5z^{-1})}, \frac{P}{Q} = \frac{0.4359z^{-1}}{(1-0.9z^{-1})}$.

Define the noise variances as $\sigma_\xi^2 = 1, \sigma_\eta^2 = 3$ resulting in an SNR of -4.77 dB. Forming the polynomial spectral factor yields.

$\sigma_\varepsilon^2 = 1.4218, d = 1 - 0.675z^{-1}$ and $\left(\frac{\sigma_\varepsilon^2}{\sigma_\varepsilon^2}\right) = 0.7033$. The polynomial G has order $na - 1 = 0$. Then and we can either solve the implied Diophantine Eq. (12.88) or the first Bilateral one (12.93). Either solution gives $g_0 = 0.438$. We also get $e_1 = -0.1790$ but this is not needed in this solution. The filter is then $H_{opt} = \frac{GQ}{D} = \frac{0.438 - 0.3942z^{-1}}{1-0.675z^{-1}}$. The mixture, desired signal and its estimate are shown in Fig. 12.6.

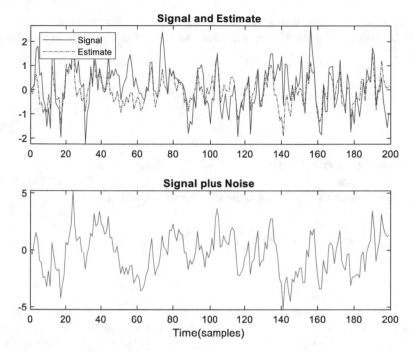

Fig. 12.6 Wiener filter for coloured noise example

12.3.1 Optimal Coloured Noise Smoother

We can proceed in a similar manner for the smoothing problem. From (12.85)

$$\frac{CC^*}{A}\frac{Q^*}{D^*}\frac{\sigma_\xi^2}{\sigma_\varepsilon^2} = \frac{G}{A} + \frac{E^*}{D^*}$$

Introduce a delay $z^{-\ell}$ and take the causal part

$$\left\{\frac{CC^*}{A}\frac{Q^*}{D^*}\frac{\sigma_\xi^2}{\sigma_\varepsilon^2}z^{-\ell}\right\}_+ = \left\{z^{-\ell}\frac{G}{A}\right\}_+ + \left\{z^{-\ell}\frac{E^*}{D^*}\right\}_+ \qquad (12.95)$$

Now $z^{-\ell}\frac{G}{A}$ is causal but $z^{-\ell}\frac{E^*}{D^*}$ has causal and noncausal terms. By long division we can write $\left\{z^{-\ell}\frac{E^*}{D^*}\right\}_+ = \left\{N + \frac{\overline{E}^*}{D^*}\right\}_+ = N$ where N is a polynomial of degree $\ell - 1$. Also $\left\{N + \frac{\overline{E}^*}{D^*}\right\}_- = \frac{\overline{E}^*}{D^*}$ where $\overline{E}$ is a polynomial of degree nd but its coefficients are different to that of E unless the smoothing lag is zero. Therefore, when the causal polynomial N is added to G/A the order of G increases and we must allow for this.

The order of G must change to $na + \ell - 1$ and is greater than that of A. When the delay is zero the order of G reverts back to $na - 1$. The optimal smoother is then from (12.83)

$$H^s_{opt} = \left[\frac{G}{A}\middle|\frac{R}{D}\right]$$

$$= \frac{QG}{D} \tag{12.96}$$

The numerator order is $na + nq + \ell - 1$. The denominator order is nd and it is always stable since the spectral factor defines its poles. To find the solution to the smoothing problem we must solve the Diophantine equation

$$CC^*Q^*\left(\frac{\sigma^2_\xi}{\sigma^2_\varepsilon}\right)z^{-\ell} = GD^* + A\overline{E}^* \tag{12.97}$$

for polynomials $\overline{E}^*$ and G.

Continuing with our example, let the delay be $\ell = 5$. Here $na = nq = 1$ so that the degree of G is $na + \ell - 1 = 5$. The Diophantine equation has 7 linear equations with 7 unknowns. These are 6 coefficients of the G polynomial and one for the $\overline{E}^*$ polynomial. The zeroth coefficient. The dimension of $\overline{E}^*$ is 1 with $\overline{E}^*(0) = \overline{e}_0 = 0$. So $G = g_0 + g_1 z^{-1} + g_2 z^{-2} + g_3 z^{-3} + g_4 z^{-4} + g_5 z^{-5}$ and $\overline{E}^* = \overline{e}_1 z$

We write the Diophantine equation by comparison of powers of z in (12.97)

$$\begin{bmatrix} 1 & d_1 & 0 & 0 & 0 & 0 & a_1 \\ d_1 & 0 & 0 & 0 & 0 & 0 & 1 \\ 0 & 1 & d_1 & 0 & 0 & 0 & 0 \\ 0 & 0 & 1 & d_1 & 0 & 0 & 0 \\ 0 & 0 & 0 & 1 & d_1 & 0 & 0 \\ 0 & 0 & 0 & 0 & 1 & d_1 & 0 \\ 0 & 0 & 0 & 0 & 0 & 1 & 0 \end{bmatrix}\begin{bmatrix} g_0 \\ g_1 \\ g_2 \\ g_3 \\ g_4 \\ g_5 \\ e_1 \end{bmatrix} = \begin{bmatrix} 0 \\ 0 \\ 0 \\ 0 \\ 0 \\ c_1^2 q_1\left(\sigma^2_\xi/\sigma^2_\varepsilon\right) \\ c_1^2\left(\sigma^2_\xi/\sigma^2_\varepsilon\right) \end{bmatrix} \tag{12.98}$$

where $c_1 = \sqrt{1 - a_1^2}$.

Substituting values gives the smoother transfer function after multiplication by the Q polynomial.

$$H^s_{opt} = \frac{QG}{D}$$

$$= \frac{-0.0372 - 0.003z^{-1} - 0.0212z^{-2} - 0.0314z^{-3} - 0.0465z^{-4} + 0.6336z^{-5} - 0.4743z^{-6}}{1 - 0.675z^{-1}}$$

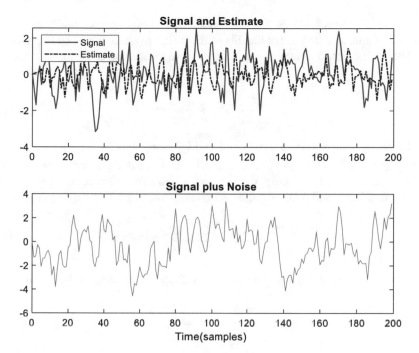

Fig. 12.7 Optimal smoother with lag of 5 steps for coloured additive noise

The numerator order is $na + nq + \ell - 1 = 6$. Figure 12.7 shows the performance of the optimal smoother.

12.3.2   *Optimal Coloured Noise Predictor*

From (12.85)

$$\frac{CC^*}{A}\frac{Q^*}{D^*}\frac{\sigma_\xi^2}{\sigma_\varepsilon^2} = \frac{G}{A} + \frac{E^*}{D^*}$$

Multiply by z^ℓ and get

$$\frac{CC^*}{A}\frac{Q^*}{D^*}\frac{\sigma_\xi^2}{\sigma_\varepsilon^2}z^\ell = \frac{Gz^\ell}{A} + \frac{E^*z^\ell}{D^*} \tag{12.99}$$

whereas smoothing takes the basic filter impulse response and shifts it to the right to make it more like the noncausal filter, prediction does the opposite and shifts the impulse response to the left. This is very similar to the white noise solution.

Equation (12.99) on its RHS has two terms. The second term is totally noncausal
and will disappear with the application of the positive power of z. The first term on
the RHS is composed of two terms. It needs to be divided up using long division as
follows

$$\frac{Gz^\ell}{A} = z^\ell F + \frac{G_p}{A} \tag{12.100}$$

The polynomial F is of degree $\ell - 1$ and G_p is a polynomial of degree $na - 1$.
Clearly $z^\ell F$ is noncausal and $\frac{G_p}{A}$ is causal. The problem is we do not know what the
G polynomial is. In the white noise case, we knew this information. This time we
must solve a polynomial Diophantine equation. From (12.99) and using (12.100)

$$\frac{CC^*}{A}\frac{Q^*}{D^*}\frac{\sigma_\xi^2}{\sigma_\varepsilon^2}z^\ell = \frac{G_p}{A} + \frac{E^* z^\ell}{D^*} + z^\ell F \tag{12.101}$$

We can combine the last two terms in Eq. (12.101) since they are both noncausal.

$$\frac{E^* z^\ell}{D^*} + z^\ell F = \frac{\overline{E}^*}{D} \tag{12.102}$$

where $\overline{E}^*$ has dimension $\ell + nd$ and $\overline{E}^*(0) = 0$. The Diophantine equation follows
from (12.99)

$$CC^* Q^* \frac{\sigma_\xi^2}{\sigma_\varepsilon^2} z^\ell = G_p D^* + A\overline{E}^*, \ell \geq 1 \tag{12.103}$$

Compare with (12.97) for smoothing. The symmetry is clear that only the delay
changes from negative to positive and the dimensions of the polynomials change.
For prediction the noncausal part is larger and the causal part smaller. The opposite
is true for smoothing.

The predictor transfer function becomes

$$H_{opt}^p = \frac{QG_p}{D} \tag{12.104}$$

For our example let the prediction be only one step or $\ell = 1$ since predictors
do not perform well for large prediction steps. They are the opposite of smoothers.
Then the polynomials are $\overline{E}^* = \bar{e}_1 z + \bar{e}_2 z^2$ and $G_p = g_0$. Comparing coefficients
of positive powers of z in the Diophantine Eq. (12.103) gives the set of three linear
equations

$$\begin{bmatrix} 1 & a_1 & 0 \\ d_1 & 1 & a_1 \\ 0 & 0 & 1 \end{bmatrix} \begin{bmatrix} g_0 \\ \bar{e}_1 \\ \bar{e}_2 \end{bmatrix} = \begin{bmatrix} 0 \\ c_1^2\left(\sigma_\xi^2/\sigma_\varepsilon^2\right) \\ c_1^2 q_1\left(\sigma_\xi^2/\sigma_\varepsilon^2\right) \end{bmatrix} \tag{12.105}$$

Solving gives $g_0 = 0.219$, $\bar{e}_1 = 0.438$, $\bar{e}_2 = -0.4747$. We don't need the $\bar{E}$ terms in the solution since they are noncausal.

The optimal predictor transfer function is from (12.104)

$$\begin{aligned} H_{opt}^p &= \frac{QG_p}{D} \\ &= \frac{0.219 - 0.197z^{-1}}{1 - 0.675z^{-1}} \end{aligned} \tag{12.106}$$

Its performance is shown in Fig. 12.8.

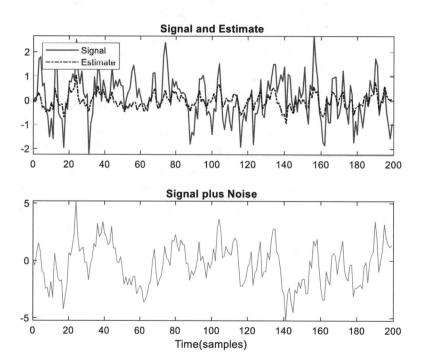

Fig. 12.8 Optimal one step predictor for coloured noise

References

1. J.F. Barrett, Problems arising from the analysis of randomly disturbed automatic control systems, Ph.D, Engineering Department, Cambridge, UK, Dec 1958
2. D. Youla, On the factorization of rational matrices. IRE Trans. Inf. Theor. **7**(3), 172–189 (1961). https://doi.org/10.1109/TIT.1961.1057636
3. G. Wilson, Factorization of the covariance generating function of a pure moving average process. SIAM J. Numer. Anal **6**(1), 1–7 (1969)
4. V. Kucera, *Discrete Linear Control, the Polynomial Equation Approach* (Wiley, NY, 1979)
5. H.W. Bode, C.E. Shannon, A simplified derivation of linear least square smoothing and prediction theory. Proc. IRE **38**(4), 417–425 (1950). https://doi.org/10.1109/JRPROC.1950.231821
6. T.J. Moir, A control theoretical approach to the polynomial spectral-factorization problem. Circ. Syst. Signal Process. **30**(5), 987–998 (2011). https://doi.org/10.1007/s00034-011-9270-4
7. J.F. Barrett, Construction of Wiener filters using the return-difference matrix. Int J. Control **26**(5), 797–803 (1977)
8. U. Shaked, A transfer function approach to the linear discrete stationary filtering and the steady-state discrete optimal control problems. Int J. Control **29**(2), 279–291 (1979)
9. P. Hagander, B. Wittenmark, A self-tuning filter for fixed lag smoothing. IEEE Trans. Inf. Theor. **IT-23**(3), 377–384 (1977)
10. T.J. Moir, M.J. Grimble, Optimal self-tuning filtering, prediction, and smoothing for discrete multivariable processes. IEEE Trans. Autom. Control **29**(2), 128–137 (1984). https://doi.org/10.1109/TAC.1984.1103464
11. B. Wittenmark, A self-tuning predictor. IEEE Trans. Autom. Control **19**(6), 848–851 (1974). https://doi.org/10.1109/TAC.1974.1100734
12. J.F. Barrett, T.J. Moir, A unified approach to multivariable discrete-time filtering based on the Wiener theory. Kybernetika **23**(3), 177–197 (1987)
13. T.J. Moir, D.R. Campbell, H.S. Dabis, A polynomial approach to optimal and adaptive filtering with application to speech. IEEE Trans. Signal Process. **39**(5), 1221–1224 (1991). https://doi.org/10.1109/78.80976
14. H.S. Dabis, T.J. Moir, A unified approach to optimal estimation using diophantine equations. Int. J. Control. **57**(3), 577–598 (1993). (Online). Available: https://doi.org/10.1080/00207179308934408
15. M.J. Grimble, Polynomial systems approach to optimal linear filtering and prediction. I. J. Control. **41**(6), 1545–1564 (1985). https://doi.org/10.1080/0020718508961214
16. M.J. Grimble, Polynomial matrix solution to the discrete fixed-lag smoothing problem. Kybernetika **27**(3), 190–201 (1991)

Chapter 13
FIR Wiener Filters Using Lower Triangular Toeplitz Matrices

13.1 Preliminaries

Chapter 10 has covered the basic theory of lower triangular Toeplitz (LTT) matrices as applied to LTI systems. This chapter will explore their use in optimal FIR filtering. The previous chapter covers the theory of polynomial filtering, smoothing and prediction but suffers from the problem of having to solve polynomial equations known as Diophantine equations. Although there are several methods available to solve such equations, they generally require the solution of a set of linear equations and as such this requires a matrix inversion. The additive white noise case is the exception where the solution is much simpler and only requires a single spectral factorization. Furthermore, such solutions are IIR and if applied to experimental data could lead to stability problems. FIR Wiener filters can also be found using the direct method of Chapter 11, but that method requires the inversion of a full Toeplitz matrix. An interesting approach was used recently which is somewhere between the two methods and gives an FIR solution, yet the model of the noise processes can be IIR or FIR. That method uses LTT matrices and it is the topic of this chapter [1, 2].

Recall the basic idea of a LTT system representation. We can represent any polynomial and hence system in the following way. If a polynomial

$$a(z^{-1}) = 1 + a_1 z^{-1} + a_2 z^{-2} + \cdots + a_n z^{-n} \tag{13.1}$$

then an equivalent LTT matrix is defined in bold thus

$$\mathbf{A} = \begin{bmatrix} 1 & 0 & 0 & 0 & 0 & 0 & 0 \\ a_1 & 1 & 0 & 0 & 0 & 0 & 0 \\ a_2 & a_1 & 1 & 0 & 0 & 0 & 0 \\ . & a_2 & a_1 & 1 & 0 & 0 & 0 \\ . & . & a_2 & a_1 & . & 0 & 0 \\ . & . & . & a_2 & . & . & 0 \\ a_n & a_{n-1} & . & . & . & a_1 & 1 \end{bmatrix} \tag{13.2}$$

© The Author(s), under exclusive license to Springer Nature Switzerland AG 2022 367
T. J. Moir, *Rudiments of Signal Processing and Systems*,
https://doi.org/10.1007/978-3-030-76947-5_13

Reading down the first column or in *reverse along the bottom row* is the polynomial which represents the matrix. When inverting such matrices, the diagonal elements define the eigenvalues and hence the matrix is always non-singular unless the zeroth coefficient of the polynomial is defined as zero. This only becomes a problem with system when they have a time-delay and time delays cannot be inverted in any case since they become noncausal and turn into pure time advances. When such a matrix is inverted the power series (Maclaurin series) expansion of $1/a$ is given by the new matrix. We require the polynomial to have all of its roots within the unit circle otherwise the inverse gives rise to a divergent power-series. For inversion to make sense we require matrices that are much larger than the polynomial. This is essentially zero stuffing at the end of the polynomial. This makes way for expansion when inverted otherwise inaccuracies arise.

The adjoint or noncausal polynomial is given by

$$a^*(z) = 1 + a_1 z + a_2 z^2 + \cdots + a_n z^n \tag{13.3}$$

which is represented by the upper triangular Toeplitz matrix (UTT)

$$\mathbf{A}^* = \begin{bmatrix} 1 & a_1 & a_2 & a_3 & . & . & a_n \\ 0 & 1 & a_1 & a_2 & . & . & a_{n-1} \\ 0 & 0 & 1 & a_1 & a_2 & . & . \\ . & 0 & 0 & 1 & a_1 & a_2 & . \\ . & . & . & 0 & . & a_1 & a_2 \\ . & . & . & . & . & . & a_1 \\ 0 & 0 & . & . & . & 0 & 1 \end{bmatrix} \tag{13.4}$$

This involves only the transform of the original causal matrix. To do more complex work, consider the polynomial expression

$$x = abc/d \tag{13.5}$$

Now (13.5), if expressed as an infinite series would require polynomial operations involving convolution and polynomial division.

This can be computed in Toeplitz form by using

$$\mathbf{X} = \mathbf{ABC}^{\mathrm{T}}\mathbf{D}^{-1} \tag{13.6}$$

Which is simple matrix manipulation.

Note that *the product of two lower(upper) triangular matrices is a lower(upper) triangular matrix.*

A Laurent series which has both positive and negative powers of z will be expressed as a Toeplitz matrix which has upper and lower triangular terms, the separation from causal to noncausal being the leading diagonal. Anything below and including the leading diagonal is causal and anything above is noncausal. This is covered in detail in Chap. 10. We can therefore re-visit the Wiener theory of polynomial filtering and express in the Toeplitz form instead.

13.2 LTT Model Description

We consider a generic description first described by Ahlen and Sternad [3] and is shown in Fig. 13.1.

There are two uncorrelated white noise sources. The first of these white noise sources $\xi(k)$ with variance σ_ξ^2 drives the signal model $W_1(z^{-1})$ and gives rise to the random signal $u(k)$. The second white noise source is $\varepsilon(k)$ has variance σ_ε^2 and it drives through a transfer function $W_n(z^{-1})$ creating additive coloured noise. The transfer function $W_s(z^{-1})$ represents a communication channel or arbitrary transfer function which models a number of problems. When the transfer function $W_s(z^{-1})$ is set to 1, then the model becomes the standard Wiener filtering problem for coloured noise. When $W_s(z^{-1})$ has a defined transfer function, then we estimate the desired signal $u(k)$ which is at its input. This is known as the *deconvolution* problem. Deconvolution arises in several real systems including geophysical signal processing, image processing and channel equalization. Perhaps the most obvious of these is the channel equalization problem where data in a system passes through the environment to get to the receiver. In so doing it becomes phase and amplitude changed in such a way that the received signal is partially degraded. On top of this is additive noise to make the problem even harder. It is clearly not just a matter of inverting the transfer function of the channel (assuming it is known), because the additive noise will get amplified. For example, if the channel transfer function is lowpass in nature, then by cascading a highpass filter to reverse the effect will amplify high frequencies. We require the optimal solution which minimizes the mean-square error. A secondary problem is that the channel transfer function can have zeros outside the unit circle, so its inverse is unstable. In fact, this problem has already been discussed in Chap. 10, but for the noise-free case only.

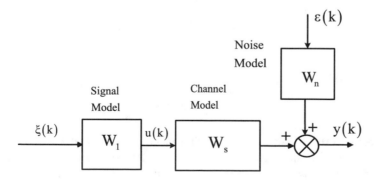

Fig. 13.1 Generic model of signal plus noise

We can write the three transfer functions in polynomial form.

$$W_1(z^{-1}) = \frac{C(z^{-1})}{D(z^{-1})} \tag{13.7}$$

$$W_n(z^{-1}) = \frac{M(z^{-1})}{N(z^{-1})} \tag{13.8}$$

$$W_s(z^{-1}) = \frac{z^{-d}B(z^{-1})}{A(z^{-1})} \tag{13.9}$$

The roots of polynomials, A, D and N are all assumed to lie within or on the unit circle of the z-plane. The d term in (13.9) represents an integer time-delay which the signal may incur whilst passing through. Each of the polynomials can be represented as LTT matrices of appropriate dimension. Let the order of all the matrices be $m + 1$ which is assumed to be significantly bigger than the order of the polynomials i.e. $m >> \max(na, nm, nd, nc, nn, nb + d)$.

Now write each polynomial in LTT form [4], adding zeros for coefficients when the order of the polynomial is exceeded. For example

$$\mathbf{D} = \begin{bmatrix} 1 & 0 & 0 & 0 & 0 & 0 & 0 \\ d_1 & 1 & 0 & 0 & 0 & 0 & 0 \\ d_2 & d_1 & 1 & 0 & 0 & 0 & 0 \\ . & d_2 & d_1 & 1 & 0 & 0 & 0 \\ . & . & d_2 & d_1 & . & 0 & 0 \\ . & . & . & d_2 & . & . & 0 \\ d_m & d_m & . & . & . & d_1 & 1 \end{bmatrix} \tag{13.10}$$

$$\mathbf{C} = \begin{bmatrix} 1 & 0 & 0 & 0 & 0 & 0 & 0 \\ c_1 & 1 & 0 & 0 & 0 & 0 & 0 \\ c_2 & c_1 & 1 & 0 & 0 & 0 & 0 \\ . & c_2 & c_1 & 1 & 0 & 0 & 0 \\ . & . & c_2 & c_1 & . & 0 & 0 \\ . & . & . & c_2 & . & . & 0 \\ c_m & c_m & . & . & . & c_1 & 1 \end{bmatrix} \tag{13.11}$$

And let another LTT matrix be

$$\mathbf{W}_1 = \mathbf{D}^{-1}\mathbf{C} \tag{13.12}$$

The order of these matrices is m which is much greater than the individual orders of the polynomials, and so there are many zeros in these matrices. For example, for a first order a polynomial $a(z^{-1}) = 1 + a_1 z^{-1}$, let $m + 1 = 7$ and then

$$A = \begin{bmatrix} 1 & 0 & 0 & 0 & 0 & 0 & 0 \\ a_1 & 1 & 0 & 0 & 0 & 0 & 0 \\ 0 & a_1 & 1 & 0 & 0 & 0 & 0 \\ 0 & 0 & a_1 & 1 & 0 & 0 & 0 \\ 0 & 0 & 0 & a_1 & 1 & 0 & 0 \\ 0 & 0 & 0 & 0 & a_1 & 1 & 0 \\ 0 & 0 & 0 & 0 & 0 & a_1 & 1 \end{bmatrix} \tag{13.13}$$

Similarly let

$$\mathbf{W_s} = \mathbf{A}^{-1}\mathbf{B} \tag{13.14}$$

and

$$\mathbf{W_n} = \mathbf{N}^{-1}\mathbf{M} \tag{13.15}$$

These matrices are all convolution matrices and not z-transform matrices. They contain the impulse response information.

Now write the white noise terms as vectors with bold notation.

$$\boldsymbol{\xi}^T(k) = \begin{bmatrix} \xi(k) & \xi(k-1) & \xi(k-2) \dots \xi(k-m) \end{bmatrix} \tag{13.16}$$

$$\boldsymbol{\varepsilon}^T(k) = \begin{bmatrix} \varepsilon(k) & \varepsilon(k-1) & \varepsilon(k-2) \dots \varepsilon(k-m) \end{bmatrix}$$

So, the covariance matrices of these noise vectors are both diagonal matrices

$$E\big[\boldsymbol{\xi}(k)\boldsymbol{\xi}^T(j)\big] = \sigma_{\xi}^2 \mathbf{I}\delta(k-j) \tag{13.17}$$

$$E\big[\boldsymbol{\varepsilon}(k)\boldsymbol{\varepsilon}^T(j)\big] = \sigma_{\varepsilon}^2 \mathbf{I}\delta(k-j) \tag{13.18}$$

And the signal model can now be written as vectors

$$\mathbf{y}(k) = \mathbf{W_s}\mathbf{u}(k) + \mathbf{W_n}\boldsymbol{\varepsilon}(k) \tag{13.19}$$

where the desired signal is generated from

$$\mathbf{u}(k) = \mathbf{W_1}\boldsymbol{\xi}(k) \tag{13.20}$$

The delay d is assumed to be absorbed into the polynomial b implicitly using leading zeros. For example, if $b(z^{-1}) = b_0 + b_1 z^{-1}$ and $d = 2$, then we write b as $b(z^{-1}) = 0 + 0z^{-1} + b_0 z^{-2} + b_1 z^{-3}$. This means that $\mathbf{B}$ becomes a singular LTT matrix. It is fortunate that it never has to be inverted, but perhaps not surprising since inverting a pure delay is a pure time-advance and noncausal anyway. We could

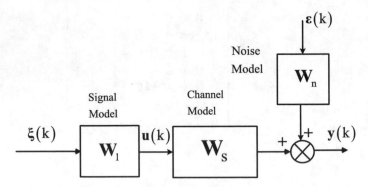

Fig. 13.2 LTT matrix form of deconvolution or estimation problem

have used a special delay matrix to model the delay but there appears to be no great advantage. The LTT model is shown in Fig. 13.2.

13.3   The Estimation Problem

Dropping the (k) argument in signals where necessary to avoid mathematical clutter, the LTT model is now

$$\mathbf{y} = \mathbf{W}_1 \mathbf{u} + \mathbf{W}_n \boldsymbol{\varepsilon} \tag{13.21}$$

and

$$\mathbf{u} = \mathbf{W}_1 \boldsymbol{\xi} \tag{13.22}$$

Define the LTT form of the optimal estimator as

$$\hat{\mathbf{u}}(k - \ell | k) = \mathbf{W}_f \mathbf{y} \tag{13.23}$$

where $\mathbf{W}_f$ is the optimal estimator in LTT form. *It is not a function of z* and the filter coefficients must be taken from it along the bottom row in reverse order. When $\ell = 0$ we have the deconvolution filtering problem otherwise when ℓ is a positive integer we have the deconvolution smoothing problem. As a special case, when $\mathbf{W}_1 = 1$ the problem becomes an ordinary wiener filtering and smoothing problem in coloured noise.

13.3.1 Estimation Error

Minimise the following scalar estimation error J, which is the trace of the error spectral Toeplitz matrix $\mathbf{\Phi}_{ee}$. Note that for this case we have a matrix for spectral density since we are dealing with vector quantities, but it is not a function of z.

$$J = \text{trace}[\mathbf{\Phi}_{ee}] \tag{13.24}$$

The trace is the sum of the leading diagonal elements of a matrix and is used because for a vector $\mathbf{e}$ we can say

$$E[\mathbf{e}^T\mathbf{e}] = \text{traceE}[\mathbf{e}\mathbf{e}^T] \tag{13.25}$$

and the term on the RHS of (13.25) is the Toeplitz form of the error power spectrum. The term on the LHS of (13.25) is the sum of expected value of errors which will be minimised. So minimizing the trace of the power spectrum of the error is the same as minimizing the mean-square error. Often, we write for shortness, *tr* instead of *trace*.

$$J = \text{trE}[\mathbf{u} - \hat{\mathbf{u}}][\mathbf{u}-\hat{\mathbf{u}}]^T \tag{13.26}$$

With a little more algebra and noting that the two noise vectors are un-correlated, we can show

$$\mathbf{\Phi}_{ee} = \mathbf{W}_1\mathbf{W}_1^T\sigma_\xi^2 - \mathbf{W}_1\mathbf{W}_1^T\mathbf{W}_s^T\mathbf{W}_f^T\sigma_\varepsilon^2 - \mathbf{W}_f\mathbf{W}_s\mathbf{W}_1\mathbf{W}_1^T\sigma_\varepsilon^2 + \mathbf{W}_f\mathbf{\Lambda}\mathbf{\Lambda}^T\mathbf{W}_f^T \tag{13.27}$$

where we define the spectral factorization of the measurement vector $\mathbf{y}$ from its Toeplitz spectral density matrix

$$\mathbf{\Phi}_{yy} = \mathbf{\Lambda}\mathbf{\Lambda}^T = \mathbf{W}_s\mathbf{W}_1\mathbf{W}_1^T\mathbf{W}_s^T\sigma_\xi^2 + \mathbf{W}_n\mathbf{W}_n^T\sigma_\varepsilon^2 \tag{13.28}$$

Note that the spectral matrices $\mathbf{\Phi}_{yy}$ and $\mathbf{\Phi}_{ee}$ should not be confused with either their scalar equivalent or a multivariable spectral density. *These matrices have no coefficients of z within them.* They are special square matrices known as *Sylvester* matrices.

13.3.2 Spectral Factorization of Toeplitz Matrices

For spectral factorization to work with this method we need an algorithm that can factorize a spectral matrix like (13.28) into lower and upper triangular matrices of the Toeplitz variety. The lower triangular part is then the causal spectral factor and the upper is noncausal. Fortunately, such a method has existed for some time in the form of the Cholesky factorization method. The method factorizes a Sylvester matrix $S = \Phi_{yy}$ (the name in the literature) into the form

$$S = LD_eL^T \tag{13.29}$$

The matrix D_e is a diagonal matrix and L has 1 s as the leading diagonal.

To understand what is happening more clearly, consider a simpler problem. If we started with coloured noise and additive white noise as the following mixture

$$y(k) = \frac{c}{a}\xi(k) + v(k) \tag{13.30}$$

where $\xi(k)$ has variance q and $v(k)$(assumed uncorrelated from $\xi(k)$) has variance r, then we would end up with an equivalent innovation model

$$y(k) = \frac{d}{a}\varepsilon(k) \tag{13.31}$$

where d satisfies the spectral factorization

$$a(z^{-1})ra(z) + c(z^{-1})qc(z) = d(z^{-1})r_\varepsilon d(z) \tag{13.32}$$

Usually we write the above in shorthand notation as

$$ara^* + cqc^* = dr_\varepsilon d^* \tag{13.33}$$

Here $\varepsilon(k)$ has variance r_ε and $d(0) = 1$. Polynomials with the * represent adjoint polynomials whose roots all lie outside of the unit circle. If we assume the degree of the a polynomial is na and it is larger than or equal to that of nc, then when multiplied out the $ara^* + cqc^*$ $dr_\varepsilon d^*$ forms a symmetric Laurent polynomial

$$s = s_{-na}z^{na} + \cdots + s_{-2}z^2 + s_{-1}z + s_0 + s_1z^{-1} + s_2z^{-2} + \cdots + s_{na}z^{-na} \tag{13.34}$$

We have a similar problem with LTT matrices. Instead of (13.33) we have instead

$$WQW^T + R = S \tag{13.35}$$

where adjoints are represented with the transpose instead making them upper triangular Toeplitz matrices. Q and R are symmetric covariance matrices, in this case diagonal since the source is white noise. The Laurent matrix is the Sylvester matrix S

$$\mathbf{S} = \begin{bmatrix} s_0 & s_1 & \cdots & s_m & 0 & \cdots & & 0 \\ s_{-1} & s_0 & s_1 & \cdot & \cdot & & & \cdot \\ & \cdot & s_{-1} & \cdot & \cdot & & & \cdot \\ & \cdot & & \cdot & \cdot & \cdot & & \\ & \cdot & & & \cdot & \cdot & & 0 \\ s_{-m} & \cdot & & & & s_m & \cdot \\ 0 & s_{-m} & & & & & \cdot & \cdot \\ & \cdot & \cdot & \cdot & & \cdot & \cdot & \cdot \\ & \cdot & & \cdot & \cdot & & \cdot & \cdot & \cdot \\ & \cdot & & & \cdot & \cdot & & \cdot & s_1 \\ 0 & & \cdots & 0 & s_{-m} & \cdots & s_{-1} & s_0 \end{bmatrix} \tag{13.36}$$

We factorize this using Cholesky factorization

$$\mathbf{S} = \mathbf{\Lambda}\mathbf{\Lambda}^{\mathbf{T}} = \mathbf{L}\mathbf{D_e}\mathbf{L}^{\mathbf{T}} \tag{13.37}$$

where $\mathbf{L}$ is lower triangular and $\mathbf{L}^{\mathbf{T}}$ is its adjoint upper triangular matrix with $\mathbf{D_e}$ a diagonal matrix of the innovations covariance. Ordinary Cholesky factorization requires $\mathbf{O}\left(\frac{m^3}{3}\right)$ operations, however fast methods using FFTs now exist to perform this factorization. The spectral factor polynomial can be read off the bottom row of the matrix as per our definition in (13.2). Although it is not by any means compulsory to use the Cholesky method, it seems natural since the solution comes in lower triangular Toeplitz format. Spectral factorization as applied to Toeplitz matrices was first explored by Pousson [5]. The so-called LDL factorization method is available as a tool on MATLAB.

13.3.3 Optimal Filter, Smoother or Predictor

We must differentiate the scalar cost function

$$\mathbf{J} = tr\left[\mathbf{W_1}\mathbf{W_1^T}\sigma_\xi^2 - \mathbf{W_1}\mathbf{W_1^T}\mathbf{W_s^T}\mathbf{W_f^T}\sigma_\varepsilon^2 - \mathbf{W_f}\mathbf{W_s}\mathbf{W_1}\mathbf{W_1^T}\sigma_\varepsilon^2 + \mathbf{W_f}\mathbf{\Lambda}\mathbf{\Lambda}^{\mathbf{T}}\mathbf{W_f^T}\right] \tag{13.38}$$

with respect to the Toeplitz matrix $\mathbf{W_f}$.

Using standard derivatives and properties of matrices under the trace operator [6] we can form the derivative. Note also that LTT matrices can be interchangeable since they are commutative and also matrices under the trace operator are also interchangeable i.e. for any two square matrices $\mathbf{A}$ and $\mathbf{B}$ we have $tr[\mathbf{AB}] = tr[\mathbf{BA}]$. Also $\frac{\partial}{\partial\mathbf{B}}tr[\mathbf{BA}] = \mathbf{A^T}, \frac{\partial}{\partial\mathbf{B}}tr[\mathbf{AB^T}] = \mathbf{A}, \frac{\partial}{\partial\mathbf{B}}tr[\mathbf{BAB^T}] = \mathbf{BA^T} + \mathbf{BA}$. Also note that $\mathbf{W_1}\mathbf{W_1^T} = [\mathbf{W_1}\mathbf{W_1^T}]^T$ and $\mathbf{\Lambda}\mathbf{\Lambda}^{\mathbf{T}} = [\mathbf{\Lambda}\mathbf{\Lambda}^{\mathbf{T}}]^T$.

Differentiating we get

$$\frac{\partial J}{\partial \mathbf{W_f}} = -2\mathbf{W}_1\mathbf{W}_1^T\mathbf{W}_s^T\sigma_\xi^2 + 2\mathbf{W_f}\mathbf{\Lambda}\mathbf{\Lambda}^T = 0 \tag{13.39}$$

Giving

$$\mathbf{W_f}\mathbf{\Lambda}\mathbf{\Lambda}^T = \mathbf{W}_1\mathbf{W}_1^T\mathbf{W}_s^T\sigma_\xi^2 \tag{13.40}$$

Now obtain $\mathbf{W_f}$ on its own by the usual Wiener method except here we use LTT matrices.

$$\mathbf{W_f} = \left[\mathbf{W}_1\mathbf{W}_1^T\mathbf{W}_s^T\mathbf{\Lambda}^{-T}\right]_+ \sigma_\xi^2\mathbf{\Lambda}^{-1} \tag{13.41}$$

Alternatively, by using (13.37) this can also be written in the form

$$\mathbf{W_f} = \left[\mathbf{W}_1\mathbf{W}_1^T\mathbf{W}_s^T\mathbf{L}^{-T}\right]_+ \mathbf{L}^{-1}\left(\frac{\sigma_\xi}{\sigma_e}\right)^2 \tag{13.42}$$

This assumes that the diagonal matrix in the LDL factorization is given by $\mathbf{D_e} = \sigma_e^2\mathbf{I}$. Equation (13.42) *is not a Z-transform* matrix of any kind but is merely a square matrix of real numbers when the terms are substituted.

The brackets are easily solved here using Property 3 in Chap. 10 which gives us the causal solution. This involves reducing all elements within the [.] + bracket to zero above the leading diagonal. For example, for an m dimensional Toeplitz matrix $\mathbf{F} = [\mathbf{H}]_+$ composed of $f_{i,j}$, i, j = 1, 2..m individual elements, then its elements become $f_{i,j} = 0$, i > j after removal of the causal brackets and is converted into LTT format.

The Wiener smoothing and prediction problems are closely related to the filtering problem and only differ by the terms collected in the asymmetric Laurent series within the $[.]_+$ brackets [7]. So (13.42) is the same for filtering, smoothing or prediction except the interpretation of which elements must be retained is different within $\left[\mathbf{W}_1\mathbf{W}_1^T\mathbf{W}_s^T\mathbf{L}^{-T}\right]_+$. For example, a time-delay within these brackets results in a smoothing filter and a time-advance a predictor, though no functions of z can be used here since the matrices themselves are not Z-transforms. They are essentially impulse response convolution matrices. This means that the elements of $\left[\mathbf{W}_1\mathbf{W}_1^T\mathbf{W}_s^T\mathbf{L}^{-T}\right]_+$ are essentially shifted to the left for smoothing or right for prediction. For filtering they are left unchanged. For example, for the full Toeplitz matrix within the $[.]_+$ brackets

$$\mathbf{H} = \begin{bmatrix} h_0 & h_{-1} & h_{-2} & h_{-3} \\ h_1 & h_0 & h_{-1} & h_{-2} \\ h_2 & h_1 & h_0 & h_{-1} \\ h_3 & h_2 & h_1 & h_0 \end{bmatrix} = \mathbf{H_+} + \mathbf{H_-}$$

For deconvolution filtering (no delay) corresponding to an estimate $\hat{\mathbf{u}}(k|k)$ we choose for causality

$$\mathbf{H}_+ = \begin{bmatrix} h_0 & 0 & 0 & 0 \\ h_1 & h_0 & 0 & 0 \\ h_2 & h_1 & h_0 & 0 \\ h_3 & h_2 & h_1 & h_0 \end{bmatrix}$$

For deconvolution smoothing (delay of unity) $\hat{\mathbf{u}}(k-1|k)$ we choose

$$\mathbf{H}_+ = \begin{bmatrix} h_{-1} & 0 & 0 & 0 \\ h_0 & h_{-1} & 0 & 0 \\ h_1 & h_0 & h_{-1} & 0 \\ h_2 & h_1 & h_0 & h_{-1} \end{bmatrix}$$

For deconvolution one-step ahead prediction $\hat{\mathbf{u}}(k+1|k)$ we choose

$$\mathbf{H}_+ = \begin{bmatrix} h_1 & 0 & 0 & 0 \\ h_2 & h_1 & 0 & 0 \\ h_3 & h_2 & h_1 & 0 \\ h_4 & h_3 & h_2 & h_1 \end{bmatrix}$$

The sufficient condition is easily verified by differentiating (13.39) with respect to $\mathbf{W}_f$ which by matrix calculus [8] gives

$$\frac{\partial^2 J}{\partial \mathbf{W}_f^2} = \mathbf{\Lambda}\mathbf{\Lambda}^T \otimes \mathbf{I}_m > 0 \tag{13.43}$$

where $\otimes$ is the Kronecker product of matrices. This equation always gives a positive definite result since $\mathbf{S} = \mathbf{\Lambda}\mathbf{\Lambda}^T$ is a Sylvester matrix and is always full rank.

13.3.3.1 Illustrative Deconvolution Examples

Consider example 1 used in [3]. The work requires a single Diophantine equation and a spectral factorization to get the optimal deconvolution smoother. The system is non-minimum phase with one-step delay and has the form

$$y(k) = z^{-d} \frac{B(z^{-1})}{A(z^{-1})} u(k) + \frac{M(z^{-1})}{N(z^{-1})} \varepsilon(k)$$

With input model

$$u(k) = \frac{C(z^{-1})}{D(z^{-1})} \xi(k)$$

where as, $A(z^{-1}) = 1 - 0.5z^{-1}$, $C(z^{-1}) = 1 - 0.5z^{-1}$, $D(z^{-1}) = 1 - 0.9z^{-1}$. The disturbance polynomials $M = N = 1$. The smoothing lag $\ell = 3$. The noise variances are given as $\sigma_\xi^2 = 1$, $\sigma_\varepsilon^2 = 0.1$.

First select an order of $m = 9$ weights for the optimal smoother and calculate the LTT matrices $\mathbf{W}_s, \mathbf{W}_1, \mathbf{W}_n$. We then use the spectral factorization (13.37) and the filter calculation (13.42) to obtain the FIR smoother polynomial

$$w_f(z) = -0.1382 + 0.499z^{-1} - 0.169z^{-2} - 0.0406z^{-3} + 0.018z^{-4} + \cdots$$

The solution given in [3] is the IIR filter

$$w_f(z) = \frac{-0.1382 + 0.4386z^{-1} + 0.0507z^{-3} - 0.1152z^{-3}}{1 + 0.44z^{-1}}$$

It is easily verified by long division that the two filters are identical.

Consider example 2 used in [3]. The work requires a single Diophantine equation and a spectral factorization to get the optimal deconvolution smoother. What makes this problem difficult is the fact that the signal is integrated white-noise and hence has a pole on the unit circle.

$$y(k) = z^{-d}\frac{B(z^{-1})}{A(z^{-1})}u(k) + \frac{M(z^{-1})}{N(z^{-1})}\varepsilon(k)$$

With input model

$$u(k) = \frac{C(z^{-1})}{D(z^{-1})}\xi(k)$$

where the signal and channel models are given by.

$$B(z^{-1}) = 1 \, , \, A(z^{-1}) = 1 - z^{-1} C(z^{-1}) = 1, D(z^{-1}) = 1 - 0.6z^{-1}$$

and the disturbance is given by $M(z^{-1}) = 1$, $N(z^{-1}) = 1 - 1.9z^{-1} + 0.9425z^{-2}$. The smoothing lag $\ell = 0$. The noise variances are given as $\sigma_\xi^2 = 10$, $\sigma_\varepsilon^2 = 1$.

The Toeplitz method required $m = 64$ FIR coefficients to achieve the comparison of Fig. 13.3. This is a large order for spectral factorization but not a high order for an FIR filter. The spectral factorization can of course be done in polynomial form instead and the $\mathbf{L}$ matrix constructed from it. However, a better approach is in the next section.

13.4   The Noncausal Toeplitz Filtering Solution

Using (13.40) and substitute the Sylvester matrix $\mathbf{S} = \boldsymbol{\Lambda}\boldsymbol{\Lambda}^{\mathbf{T}}$ gives

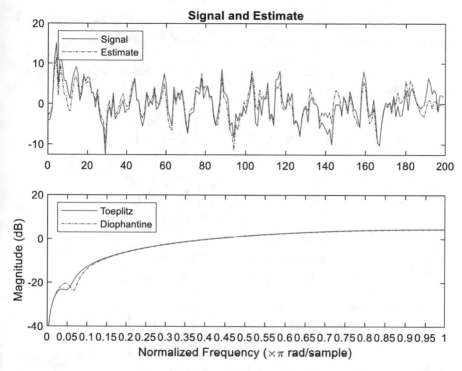

Fig. 13.3 (top) True signal and its estimate, (bottom) comparison of frequency responses for deconvolution filters

$$\mathbf{W_f S} = \mathbf{W}_1 \mathbf{W}_1^T \mathbf{W}_s^T \sigma_\xi^2 \tag{13.44}$$

If we solve for the noncausal solution it becomes

$$\mathbf{W_f} = \mathbf{W}_1 [\mathbf{W}_s \mathbf{W}_1]^T \mathbf{S}^{-1} \sigma_\xi^2 \tag{13.45}$$

Although this solution can be interpreted as noncausal because it is an FIR filter it can never be unstable. This means it is perfectly suitable as a filter. We also know from previous chapters that the minimum mean-square error is smaller for the noncausal case. Eq. (13.45) has three parts to it in the solution. It has a LTT matrix $\mathbf{W}_1$ which multiplies an upper TT matrix $[\mathbf{W}_s \mathbf{W}_1]^T$ and this in turn multiplies the symmetric Sylvester matrix. It is of course then scaled by the variance of the driving noise. Examining a similar problem consider two minimum phase polynomials in z

$$a(z^{-1}) = a_0 + a_1 z^{-2} + a_2 z^{-2}$$
$$b(z^{-1}) = b_0 + b_1 z^{-2} + b_2 z^{-2}$$

Then multiplying ab* gives

$$a(z^{-1})b(z) = (a_0 + a_1 z^{-1} + a_2 z^{-2})(b_0 + b_1 z + b_2 z^2)$$
$$= (a_0 b_0 + a_1 b_1 + a_2 b_2) + (a_1 b_0 + a_2 b_1)z^{-1} + a_2 b_0 z^{-2} + (a_0 b_1 + a_1 b_2)z + a_0 b_2 z^2$$

A Laurent series of finite length. Now represent each polynomial as a triangular Toeplitz matrix

$$\mathbf{A} = \begin{bmatrix} a_0 & 0 & 0 \\ a_1 & a_0 & 0 \\ a_2 & a_1 & a_0 \end{bmatrix}, \mathbf{B}^T = \begin{bmatrix} b_0 & b_1 & b_2 \\ 0 & b_0 & b_1 \\ 0 & 0 & b_0 \end{bmatrix}$$

$$\mathbf{AB}^T = \begin{bmatrix} a_0 & 0 & 0 \\ a_1 & a_0 & 0 \\ a_2 & a_1 & a_0 \end{bmatrix} \begin{bmatrix} b_0 & b_1 & b_2 \\ 0 & b_0 & b_1 \\ 0 & 0 & b_0 \end{bmatrix}$$

$$= \begin{bmatrix} a_0 b_0 & a_0 b_1 & a_0 b_2 \\ a_1 b_0 & a_0 b_0 + a_1 b_1 & a_0 b_1 + a_1 b_2 \\ a_2 b_0 & a_1 b_0 + a_2 b_1 & a_0 b_0 + a_1 b_1 + a_2 b_2 \end{bmatrix}$$

Now examine the product of the two matrices closely. Read from right to left the bottom row and we have the causal part of the Laurent series. Read the last row from bottom to top and see we get the noncausal part of the Laurent series. Of course, we need to exclude the z^0 part from the noncausal part of the series since it begins at positive powers of z upwards. This is illustrated in Fig. 13.4.

From (13.45) to get the two-sided impulse response all we must do is interpret the solution matrix properly since all the information is already there.

Example of noncausal Toeplitz Wiener filter.

This is the same example considered in Chap. 11. The corrupted signal is given by

Laurent series

$$a(z^{-1})b(z) = (a_0 b_0 + a_1 b_1 + a_2 b_2) + (a_1 b_0 + a_2 b_1)z^{-1} + a_2 b_0 z^{-2} + (a_0 b_1 + a_1 b_2)z + a_0 b_2 z^2$$

$$\mathbf{AB}^T = \begin{bmatrix} a_0 & 0 & 0 \\ a_1 & a_0 & 0 \\ a_2 & a_1 & a_0 \end{bmatrix} \begin{bmatrix} b_0 & b_1 & b_2 \\ 0 & b_0 & b_1 \\ 0 & 0 & b_0 \end{bmatrix}$$

$$= \begin{bmatrix} a_0 b_0 & a_0 b_1 & a_0 b_2 \\ a_1 b_0 & a_0 b_0 + a_1 b_1 & a_0 b_1 + a_1 b_2 \\ a_2 b_0 & a_1 b_0 + a_2 b_1 & a_0 b_0 + a_1 b_1 + a_2 b_2 \end{bmatrix} \begin{matrix} z^2 \\ z \\ z^0 \end{matrix} \quad \uparrow \text{noncausal}$$

$$\begin{matrix} z^{-2} & z^{-1} & z^0 \end{matrix}$$

$\longleftarrow$

causal

Fig. 13.4 Shows how two triangular Toeplitz matrices $\mathbf{AB}^T$ give rise to a Laurent series

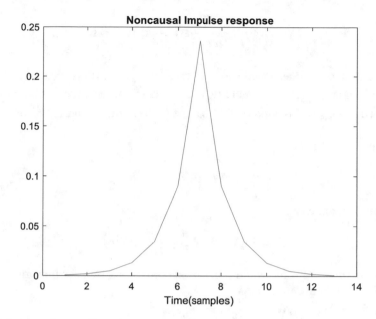

Fig. 13.5 Two-sided impulse response of optimal filter, additive white noise

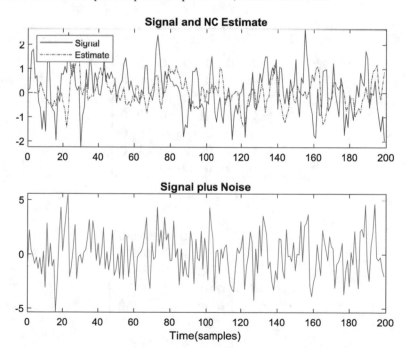

Fig. 13.6 Signal, signal plus noise and the estimated signal

$$y(k) = \frac{\sqrt{1-\alpha^2}}{\left(1-\alpha z^{-1}\right)}\xi(k)+\varepsilon(k)|\alpha| < 1$$

where $\xi(k)$ is zero-mean white noise of variance σ_ξ^2 and $\varepsilon(k)$ is uncorrelated white noise with variance σ_ε^2. The optimal FIR Wiener filter for three weights was previously calculated as $H_{opt} = \begin{bmatrix} 0.2364\ 0.0907\ 0.0364 \end{bmatrix}^T$. This was found for $\sigma_\xi^2 = 1, \sigma_\varepsilon^2 = 3, \alpha = 0.5$.

The noncausal filter is

$$\mathbf{W_f} = \mathbf{W}_1[\mathbf{W_s}\mathbf{W}_1]^T\mathbf{S}^{-1}\sigma_\xi^2$$

where $\mathbf{S}$ is found from

$$\mathbf{S} = \mathbf{W_s}\mathbf{W}_1\mathbf{W}_1^T\mathbf{W_s}^T\sigma_\xi^2 + \mathbf{W_n}\mathbf{W_n}^T\sigma_\varepsilon^2$$

In polynomial form the white noise has no colouring transfer function and the channel is unity transfer function.

Here $\mathbf{W_n} = \mathbf{W_s} = 1, \mathbf{W}_1 = \frac{\sqrt{1-\alpha^2}}{1-\alpha z^{-1}} = \frac{0.866}{1-0.5z^{-1}}$.

In Toeplitz form, assuming $m = 7$ for the matrix orders. $\mathbf{W_n} = \mathbf{W_s} = \mathbf{I}_7$ and

$$\mathbf{W}_1 = \mathbf{D}^{-1}\mathbf{C}$$

$$\mathbf{D} = \begin{bmatrix} 1 & 0 & 0 & 0 & 0 & 0 & 0 \\ -\alpha & 1 & 0 & 0 & 0 & 0 & 0 \\ 0 & -\alpha & 1 & 0 & 0 & 0 & 0 \\ 0 & 0 & -\alpha & 1 & 0 & 0 & 0 \\ 0 & 0 & 0 & -\alpha & 1 & 0 & 0 \\ 0 & 0 & 0 & 0 & -\alpha & 1 & 0 \\ 0 & 0 & 0 & 0 & 0 & -\alpha & 1 \end{bmatrix}$$

$$\mathbf{C} = 0.866\mathbf{I}_7$$

$$\mathbf{W}_1 = 0.866\begin{bmatrix} 1 & 0 & 0 & 0 & 0 & 0 & 0 \\ \alpha & 1 & 0 & 0 & 0 & 0 & 0 \\ \alpha^2 & \alpha & 1 & 0 & 0 & 0 & 0 \\ \alpha^3 & \alpha^2 & \alpha & 1 & 0 & 0 & 0 \\ \alpha^4 & \alpha^3 & \alpha^2 & \alpha & 1 & 0 & 0 \\ \alpha^5 & \alpha^4 & \alpha^3 & \alpha^2 & \alpha & 1 & 0 \\ \alpha^6 & \alpha^5 & \alpha^4 & \alpha^3 & \alpha^2 & \alpha & 1 \end{bmatrix}$$

Simulation of the problem was done on the following MATLAB program.

MATLAB Code 13.1

```
%Deconvolution filter or smoother
%Uses the Toeplitz noncausal method
close all
clear
%samples for simulation
Np=200;
%Disturbance polynomials M/N
mp=[1];%Numerator
np=[1];%denominator
%Make sure the matrices are large enough for implicit delay
m=7;
tr=zeros(1,m);
tr(1)=np(1);
nul=zeros(1,(m-length(np)));
tc=horzcat(np,nul);
N=toeplitz(tc,tr)
%
tr=zeros(1,m);
tr(1)=mp(1);
nul=zeros(1,(m-length(mp)));
tc=horzcat(mp,nul);
M=toeplitz(tc,tr)

output=('Coloured driving Noise is C/D polynomials')
%driving noise model C/D
c=[0.866];
d=[1 -0.5];
tr=zeros(1,m);
tr(1)=c(1);
nul=zeros(1,(m-length(c)));
tc=horzcat(c,nul);
C=toeplitz(tc,tr)
%
tr=zeros(1,m);
tr(1)=d(1);
```

```
nul=zeros(1,(m-length(d)));
tc=horzcat(d,nul);
D=toeplitz(tc,tr)
%
%Channel dynamics b/a
b=[1];
a=[1];
tr=zeros(1,m);
tr(1)=b(1);
nul=zeros(1,(m-length(b)));
tc=horzcat(b,nul);
B=toeplitz(tc,tr)
%
tr=zeros(1,m);
tr(1)=a(1);
nul=zeros(1,(m-length(a)));
tc=horzcat(a,nul);
A=toeplitz(tc,tr)

% Find FIR versions of transfer functions
W=inv(A)*B;
Wn=inv(N)*M;
W1=inv(D)*C;
%Signal driving variance sq
sq=1;
%Driving white noise variance for coloured noise
sv=3;
R=sv*eye(m);
%Sylvester matrix - signal plus noise spectrum Toeplitz verson
S=sq*W*W1*W1'*W'+Wn*Wn'*sv;

%%%%%%%%%%%%%%%%%%%%%%%%%
%%%%%%%%%%%%%%%%%%%%%%%%%
%FIR filter
t=[0:1:Np-1]';

 %Random Noise  Variance
rn=sqrt(sq)*randn(Np,1);
%Desired Noise - coloured u
u=filter(c,d,rn);
%Now re-seed random noise generator
seed=1;
rng(seed)
```

```
%Filter u via b/a to give s
%Generate Signal
s=filter(b,a,u);
% Additive noise variance sv passes through M/N
%Random Noise  Variance sv
rv=sqrt(sv)*randn(Np,1);
nt=filter(mp,np,rv);
% Signal plus disturbance is y=s+nt
y=s+nt;
fd=[1];

%Noncausal version of filter
Hu=sq*W1*W1'*W'*inv(S)
%Negative powers of z
fc=fliplr(Hu(m,:));
%noncausal part
fu=(fliplr(Hu(:,m)'));
%Delete first element
fu(1)=[];
%now stitch the causal and noncausal together
ft=horzcat(fliplr(fu), fc);
%This is the impulse response - two sided
figure
plot(ft)
set(gcf,'color','w');
xlabel('Time(samples)')
title('Noncausal Impulse response')
set(gcf,'color','w');
figure
%Filter noncausal filter
uhu=filter(ft,fd,y);
subplot(2,1,1)
plot(t,u,'-r',t,uhu,'-.b')
legend('Signal','Estimate','Location','northwest')
title('Signal and NC Estimate')
subplot(2,1,2)
plot(t,y)
title('Signal plus Noise')
set(gcf,'color','w');
xlabel('Time(samples)')
set(gcf,'color','w');
```

The program gave the following solution to the estimation problem.

$$\mathbf{W}_f = \begin{bmatrix} 0.191 & 0.0729 & 0.0279 & 0.0106 & 0.0041 & 0.0016 & 0.0006 \\ 0.0729 & 0.2188 & 0.0836 & 0.0319 & 0.0122 & 0.0047 & 0.0019 \\ 0.0279 & 0.0836 & 0.2229 & 0.0852 & 0.0326 & 0.0125 & 0.0050 \\ 0.0106 & 0.0319 & 0.0852 & 0.2235 & 0.0855 & 0.0329 & 0.0131 \\ 0.0041 & 0.0122 & 0.0326 & 0.0855 & 0.2238 & 0.0861 & 0.0344 \\ 0.0016 & 0.0047 & 0.0125 & 0.0329 & 0.0861 & 0.2254 & 0.0902 \\ 0.0006 & 0.0019 & 0.0050 & 0.0131 & 0.0344 & 0.0902 & 0.2361 \end{bmatrix}$$

Reading The bottom row from right to left we see the first three values correspond approximately with the previous result i.e. 0.2361,0.0902,0.0344…

Reading up the last column gives us the noncausal values. (ignoring the first element which is already included in the bottom row.

The estimate looks like a smoothed signal with the time-delay. This is to be expected since a smoother usually only drags terms from the negative part of the impulse response into the causal region.

References

1. T.J. Moir, Toeplitz matrices for LTI systems, an illustration of their application to Wiener filters and estimators. Int. J. Syst. Sci. **49**(4):800–817 (2018). https://doi.org/10.1080/00207721.2017.1419306
2. U. Grenander, G. Szego, *Toeplitz Forms and Their Applications* (University of California Press, Los Angeles USA, 1958)
3. A. Ahlen, M. Sternad, Optimal deconvolution based on polynomial methods. IEEE Trans Acoust Speech Signal Process **37**(2), 217–226 (1989). https://doi.org/10.1109/29.21684
4. R. Gray, Toeplitz and circulant matrices: a review," in *Foundations and trends in communications and information theory*, vol. 2, no. 3. MA,USA: Now publishers Inc (2006)
5. H.R. Pousson, Systems of toeplitz operators on H². II. Trans. Am. Math. Soc. **133**(2), 527–536 (1968)
6. K.B. Petersen, M.S. Pedersen, *The Matrix Cookbook*. Available online at http://matrixcookbook.com: University of Waterloo,Canada, 2012
7. J.F. Barrett, T.J. Moir, A unified approach to multivariable discrete-time filtering based on the Wiener theory. Kybernetika **23**(3), 177–197 (1987)
8. J.E. Gentle, *Matrix Algebra, Theory, Computation and Applications in Statistics* (Springer, Berlin, 2007)

Chapter 14
Adaptive Filters

14.1 Overview and Motivation

The previous chapters have looked at methods for designing optimal filters based on mathematical modelling of the signal and noise processes. This is either done through statistical properties such as autocovariance functions or Z-transform models. The Z-transform approach leads us to areas such as spectral factorization and Diophantine equations as part of the solution. These can be computationally demanding for real-time applications. Moreover, the Diophantine approach realizes pole-zero filters which could have stability issues in finite precision arithmetic implementations, especially if the process models are inaccurate. The FIR Wiener filter is not troubled by stability issues, but it does require autocovariance information of the signal plus noise and the noise on its own. To a certain extent the FIR approach is already a kind of adaptive filter since we can estimate noise covariances from direct measurements. This does of course require a means of determining what part of the data is signal and what is signal plus noise and this in turn necessitates the use of voice-activity detectors (VADs) in the case where the signal is a speech waveform. Other approaches which have been tried in the past such as spectral subtraction could also be described as a form of adaptive filter. These approaches are *ad-hoc* in nature however and more of an after-thought from the main theory of Wiener filtering. They are not generically designed from the ground upwards, but after the filtering theory is complete the thinking goes into how the solution can be made adaptive or "self-tuning". The first attempts at adaptive filtering were analogue, and after a digital theory was developed the computer technology lagged behind the theory and so hardware versions of adaptive filters had to be developed to get the speed faster [1]. Although attempts were made at IIR adaptive filters [2], the FIR method appears to have withstood the test of time. Adaptive FIR filters were implemented in CMOS and hard-wired in the early days of the theory. The earliest application was an adaptive equalizer for communication channels [3]. It used a small number of FIR weights (as they are commonly known instead of coefficients) and used steepest descent to optimize them. Only later did the inventor become aware [4] of the more general work

© The Author(s), under exclusive license to Springer Nature Switzerland AG 2022 387
T. J. Moir, *Rudiments of Signal Processing and Systems*,
https://doi.org/10.1007/978-3-030-76947-5_14

of Widrow which is the topic of the next section. A few years earlier Widrow and
Hoff had developed a more generic approach to adaptive filters [5] by inventing the
least-mean squares algorithm (LMS). Adaptive equalizers of different forms are used
to this day in cellular communication systems to minimize inter-symbol interference.
Much later the idea of "self-tuning" filters was contrived (which included smoothers
and predictors), with analogy to self-tuning control which was being developed at the
time. This however was another form of IIR adaptive filter [6, 7]. The IIR theory was
later extended to filters which minimized the H infinity norm rather than mean-square
error [8].

14.2   The Least-Mean Squares (LMS) Method

The earliest and yet most common of all of the adaptive filtering approaches is the
LMS method. It is not the fastest method but is by far the simplest and as such
works sparingly on computer resources. It is based on FIR filters only, but this is an
advantage, even though FIR filters may require many weights to give comparable
results to IIR filters of lower order. The comfort of knowing the solution is always
stable outweighs any advantages there are in reduced model orders as used in IIR
adaptive filters. It is best to view the LMS algorithm as an algorithm which can
estimate an unknown system driven by white noise as shown in Fig. 14.1. This is
also known as *system identification.*

The first analysis is like the FIR Wiener filter. Let the LMS output be a scalar
represented by the dot product of two vectors

$$\mathbf{W}^{\mathrm{T}}\mathbf{X}(k) = \mathbf{X}^{\mathrm{T}}(k)\mathbf{W} \tag{14.1}$$

Fig. 14.1 The LMS
Algorithm as a block
diagram used for system
identification

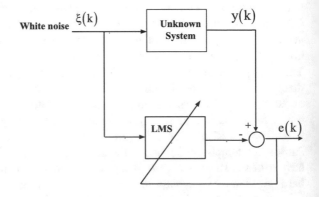

We assume $\mathbf{W} = [w_0 \ w_1 \ w_2 \ \ldots w_n]^T$ is the vector of (n + 1) weights which will be updated at each iteration and $\mathbf{X} = [\xi(k) \ \xi(k-1) \ \xi(k-2) \ \ldots \ \xi(k-n)]^T$ is the vector of regressors which is composed of past values of zero-mean white noise $\xi(k)$ which has variance σ_ξ^2.

Let y(k) be the unknown system output where

$$y(k) = \mathbf{H}\mathbf{X}(k) = \mathbf{X}^T(k)\mathbf{H}$$

and $\mathbf{H}$ is the unknown weight vector to be estimated by the LMS method. That is, we require that $\mathbf{W}_{opt} \to \mathbf{H}$ as the optimal solution.

Now form an error between the unknown system output and the LMS output

$$e(k) = y(k) - \mathbf{W}^T\mathbf{X}(k) \tag{14.2}$$

The expected value of the scalar error squared gives a quadratic (convex) weight space from which we need to find the solution. Convex spaces are known to have only one unique minimum. We multiply by ½ without losing generality for mathematical convenience. This is true for any cost function that is to be minimised.

$$J = \frac{1}{2}E[e^2(k)] = \frac{1}{2}E\left[\left(y(k) - \mathbf{W}^T\mathbf{X}(k)\right)\left(y(k) - \mathbf{W}^T\mathbf{X}(k)\right)^T\right] \tag{14.3}$$

Multiply out like the Wiener filter case and get

$$J = \frac{1}{2}\left\{E[y^2(k)] - E[\mathbf{W}^T\mathbf{X}(k)y(k)] - E[\mathbf{X}^T(k)\mathbf{W}y(k)] + E[\mathbf{W}^T\mathbf{X}(k)\mathbf{X}^T(k)\mathbf{W}]\right\} \tag{14.4}$$

Differentiate wrt the weight vector $\mathbf{W}$ to find the minimum. This is written as taking the gradient of (14.4) term by term

$$\nabla_{\mathbf{W}}J = 0 - E[\mathbf{X}(k)y(k)] + E[\mathbf{X}(k)\mathbf{X}^T(k)]\mathbf{W} \tag{14.5}$$

Define the symmetric covariance matrix of the regressors of dimension (n + 1) square as

$$\mathbf{R} = \mathbf{R}^T = E[\mathbf{X}(k)\mathbf{X}^T(k)] \tag{14.6}$$

And the (n + 1) cross-covariance vector as

$$\mathbf{R}_{xy} = E[\mathbf{X}(k)y(k)] \tag{14.7}$$

Equation (14.5) then becomes

$$\nabla_{\mathbf{W}}J = -\mathbf{R}_{xy} + \mathbf{R}\mathbf{W} = 0 \tag{14.8}$$

Solving, the optimal weight vector is

$$\mathbf{W}_{opt} = \mathbf{R}^{-1}\mathbf{R}_{xy} \qquad (14.9)$$

Note that the gradient from (14.5) an also be written as (omitting the expected value)

$$\nabla_{\mathbf{W}}J = -\mathbf{X}(k)\big(y(k) - \mathbf{X}^{T}(k)\mathbf{W}\big)$$
$$= -\mathbf{X}(k)e(k) \qquad (14.10)$$

The problem is that (14.9) has theoretical values for the cross-covariance vector and covariance matrix and we don't know what they are. The problem can be overcome by using steepest descent (or gradient descent as it is also known as). The method was invented by the famous French mathematician Cauchy who stated that if we have an initial guess for a vector, then the next best guess to minimize a quadratic criterion is if you step in the direction of the negative gradient. This is a bit like finding your way down a hill blindfold step by step by looking for the steepest gradient each time. This way guarantees your trip to the bottom. We need a step size which we label $\mu > 0$. Its upper limit will be discussed later. Steepest descent is

$$\mathbf{W}(k + 1) = \mathbf{W}(k) - \mu\nabla_{\mathbf{W}}J \qquad (14.11)$$

However, since we already know the gradient as Eq. (14.10), on substituting we get

$$\mathbf{W}(k + 1) = \mathbf{W}(k) + \mu\mathbf{X}(k)e(k) \qquad (14.12)$$

This is the celebrated LMS algorithm together with the fact that the error which was already defined is $e(k) = y(k) - \mathbf{W}^{T}\mathbf{X}(k) = y(k) - \mathbf{X}^{T}(k)\mathbf{W}$ and we now restate it but make the weight vector a function of time too because during convergence it is not a constant. Thus, we can write $e(k) = y(k) - \mathbf{X}^{T}(k)\mathbf{W}(k)$. But how do we know this will indeed converge to the desired values we require? One way to proceed is to take expected value of the left and right of Eq. (14.12).

$$E[\mathbf{W}(k + 1)] = E[\mathbf{W}(k)] + \mu E[\mathbf{X}(k)e(k)]$$
$$= E[\mathbf{W}(k)] + \mu E\mathbf{X}(k)\big[\big(y(k) - \mathbf{X}^{T}(k)\mathbf{W}(k)\big)\big]$$
$$= E[\mathbf{W}(k)] - \mu E\big[\mathbf{X}(k)\mathbf{X}^{T}(k)\big]\mathbf{W}(k) + \mu E[\mathbf{X}(k)y(k)]$$

Now write the last line after removing expectations

$$\mathbf{W}(k + 1) = \mathbf{W}(k) - \mu\mathbf{R}\mathbf{W}(k) + \mu\mathbf{R}_{xy} \qquad (14.13)$$

We can write this as

$$\mathbf{W}(k+1) = [\mathbf{I} - \mu\mathbf{R}]\mathbf{W}(k) + \mu\mathbf{R}_{xy} \qquad (14.14)$$

This is a discrete-time state-space description. Recall that we can write it like

$$\mathbf{x}(k+1) = \mathbf{F}\mathbf{x}(k) + \mathbf{G}u(k)$$

And where in this case $\mathbf{G} = \mu\mathbf{R}_{xy}, \mathbf{F} = \mathbf{I} - \mu\mathbf{R}$ with $u(i) = 1$. It has the solution using the convolution summation and the response to initial conditions.

$$\mathbf{x}(k) = \mathbf{F}^k\mathbf{x}(0) + \sum_{i=0}^{k-1} \mathbf{F}^{k-i-1}\mathbf{G}u(i)$$

By comparison to (14.13)

$$\mathbf{W}(k) = \mathbf{F}^k\mathbf{x}(0) + \sum_{i=0}^{k-1} \mathbf{F}^{k-i-1}\mathbf{G} \qquad (14.15)$$

Now provided the eigenvalues of $\mathbf{F}$ have all magnitude less than unity, the initial condition part $\mathbf{F}^k\mathbf{x}(0)$ will die out towards zero. This means that the eigenvalues of

$$\mathbf{F} = \mathbf{I} - \mu\mathbf{R} \qquad (14.16)$$

Must have eigenvalues less than unity. By the *eigenvalue shift theorem* [9, 10], if $\mathbf{R}$ has eigenvalues $\lambda_1, \lambda_2 \ldots \lambda_n$, then $\mathbf{F} = \mathbf{I} - \mu\mathbf{R}$ will have eigenvalues $1 - \mu\lambda_1, 1 - \mu\lambda_2 \ldots 1 - \mu\lambda_n$. Its stability will be determined by the maximum eigenvalue of $\mathbf{I} - \mu\mathbf{R}$. If this eigenvalue is $1 - \mu\lambda_{max}$, then stability becomes apparent provided

$$|1 - \mu\lambda_{max}| < 1 \qquad (14.17)$$

Which is an inequality which means that the step size must lie in the range

$$0 < \mu < \frac{1}{\lambda_{max}} \qquad (14.18)$$

Provided this is satisfied, the first term on the RHS of (14.15) will die out to zero leaving

$$\mathbf{W}(k) = \sum_{i=0}^{k-1} \mathbf{F}^{k-i-1}\mathbf{G}$$

$$= \sum_{i=0}^{k-1} \mathbf{F}^i\mathbf{G} \qquad (14.19)$$

In the above, provided $\mathbf{F}$ has all of its eigenvalues within the unit circle we can express the summation in closed form since it is a convergent matrix geometric series.

$$\sum_{i=0}^{s-1} \mathbf{F}^i = [\mathbf{I} - \mathbf{F}]^{-1}[\mathbf{I} - \mathbf{F}^k] \tag{14.20}$$

Substitute $\mathbf{F} = \mathbf{I} - \mu\mathbf{R}$ into (14.20) and obtain

$$\sum_{i=0}^{k-1} \mathbf{F}^i = \mu^{-1}\mathbf{R}^{-1}[\mathbf{I} - (\mathbf{I} - \mu\mathbf{R})^k] \tag{14.21}$$

Now from (14.19) and substituting $\mathbf{G} = \mu\mathbf{R}_{xy}$

$$\begin{aligned} \mathbf{W}(k) &= \mu^{-1}\mathbf{R}^{-1}[\mathbf{I} - (\mathbf{I} - \mu\mathbf{R})^k]\mu\mathbf{R}_{xy} \\ &= \mathbf{R}^{-1}[\mathbf{I} - (\mathbf{I} - \mu\mathbf{R})^k]\mathbf{R}_{xy} \end{aligned} \tag{14.22}$$

This is only the convolution part of the solution. The entire solution including the response to initial weight conditions $\mathbf{W}(0)$ is from (14.15)

$$\mathbf{W}(k) = (\mathbf{I} - \mu\mathbf{R})^k\mathbf{W}(0) + \mathbf{R}^{-1}[\mathbf{I} - (\mathbf{I} - \mu\mathbf{R})^k]\mathbf{R}_{xy} \tag{14.23}$$

This must converge to $\mathbf{R}^{-1}\mathbf{R}_{xy}$ in steady state, which is the optimal solution.

Example. Suppose the unknown system is a first order AR process (or all-pole model) given by.

$y(k) = \frac{1}{1-0.5z^{-1}}\xi(k)$, where $\xi(k)$ is zero-mean unity variance white noise. A MATLAB program is shown below for this example.

MATLAB Code 14.1

```
%Least-mean squares (LMS) algorithm.
clear
close all
%Number of samples in simulation
Npoints = 1000;
%Number of weights
Nweights = 5;
%Step size
mu = 0.05;
a = [1 -0.5];
c = [1];
%Unit variance driving noise
sq = 1;
%Random Noise length Npoints Variance rn
rn = sqrt(sq)*randn(Npoints,1);
%Filter it to get time-series
%Generate Signal
y = filter(c,a,rn);
```

```
%%%
%Regression vector X;
X = zeros(Nweights,1);
%Weight Vector
W = zeros(Nweights,1);
%Array to store mean square error
mse = zeros(Nweights,1);
%Arrays for weights - store
ws = zeros(Npoints,Nweights);
%Main Loop
%initialise
mer = 0;
for i = 1:1:Npoints
%Shuffle regression vector
for k = Nweights:-1:2
X(k) = X(k-1);
end
X(1) = rn(i);
%Create error signal
e = y(i)-X'*W;
%Update weights
W = W + mu*X*e;
%Store weights
for i1 = 1:Nweights
ws(i,i1) = W(i1);
end
%Mean square error
mer = mer + (1/(i + 1))*(e^2-mer);
mse(i) = mer;
end
%This array v is a plotting index only
v = zeros(Nweights,1);
for j = 1:Nweights
v(j) = j;
end
figure
plot(ws(:,v))
title('Weight estimates')
set(gcf,'color','w');
xlabel('Time(samples)')
figure
plot(mse)
grid
title('Mean square error')
xlabel('Time(samples)')
set(gcf,'color','w');
```

The run length of the simulation was 1000 samples. A step size was chosen as $\mu = 0.05$ Fig. 14.2 shows the evolution of the weights versus time and hence their convergence. Since the LMS algorithm is fitting an FIR (MA) model to an AR model its impulse response is infinite. Therefore, only the first 5 weights were selected.

The mean-square error (mse) was recursively calculated to show the convergence of the error to its minimum. The recursive variance was used. For an error e(k) with zero dc, the recursive variance is given by (Fig. 14.3)

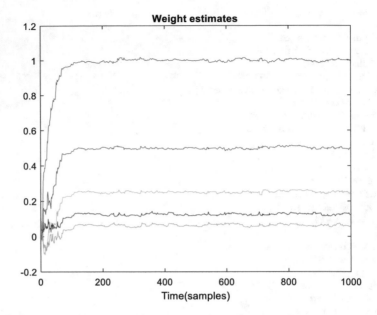

Fig. 14.2 System identification of an IIR system using the LMS algorithm

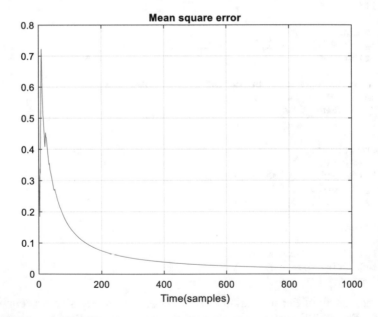

Fig. 14.3 Mean-square error versus time showing convergence of LMS algorithm

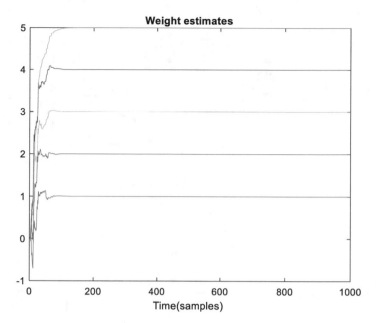

Fig. 14.4 System identification of an FIR system

$$\sigma^2(k+1) = \sigma^2(k) + \frac{1}{k+1}\left[e^2(k) - \sigma^2(k)\right] \tag{14.24}$$

The best approximation to the transfer function can be found by doing a long division $\frac{1}{1-0.5z^{-1}} = 1 - 0.5z^{-1} + 0.25z^{-2} - 0.125z^{-3} + 0.0625z^{-4} + \dots$. The LMS algorithm converges to $W_{opt} = \begin{bmatrix} 0.9941 & 0.4961 & 0.2575 & 0.1323 & 0.0764 \end{bmatrix}^T$.

For an FIR system rather than IIR, the LMS method return results with zero weight error. For example, if $y(k) = \left(1 + z^{-2} + 2z^{-2} + 3z^{-3} + 4z^{-4} + 5z^{-5}\right)\xi(k)$ then the weight estimates are shown in Fig. 14.4.

Convergence is fast, but there is no additive noise. Note that the weight error is defined as $\varepsilon(k) = W(k) - W_{opt}$ and is a vector quantity, not to be confused with the scalar $e(k)$, the error at the output.

14.3 LMS for Wiener Estimators

The LMS algorithm can be used to find an optimal filter, smoother or predictor [11]. The block diagrams are shown in Fig. 14.5. This is still not an adaptive filter as such since we need to know the ideal signal and in practice it is unknown (else there would be no filtering problem needed). It is a useful exercise however and will converge to the optimal FIR estimator.

For the example considered in Chap. 11:

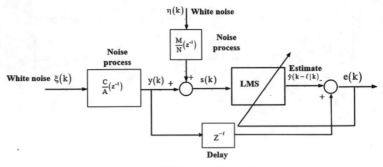

Filter or Smoother

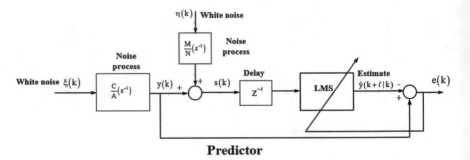

Predictor

Fig. 14.5 Filtering, smoothing or prediction using the LMS algorithm

A system driven by zero-mean unity variance white noise $\xi(k)$ giving a random signal

$$y(k) = \frac{\sqrt{1-a^2}}{\left(1 - az^{-1}\right)}\xi(k), a = 0.5$$

The corrupted signal is

$$s(k) = y(k) + v(k)$$

where $v(k)$ is zero-mean white noise of variance $\sigma_v^2 = 3$. The optimal FIR Wiener filter is required for three weights.

Using covariance methods, the optimal weight was calculated to be $\mathbf{H}_{opt} = \left[0.2364\ 0.0909\ 0.0364\right]^T$. Using LMS, the weights converged to $\mathbf{H}_{opt} = \left[0.2309\ 0.0925\ 0.0374\right]^T$. This was for a long run time of 200,000 samples and a step size of $\mu = 0.5 \times 10^{-6}$. This is very slow convergence, but for the LMS algorithm we must trade off accuracy with speed of response. A small step size gives slow convergence but more accurate results than using a large step size. If the step size is chosen to be too big the algorithm goes unstable. The steepest descent method

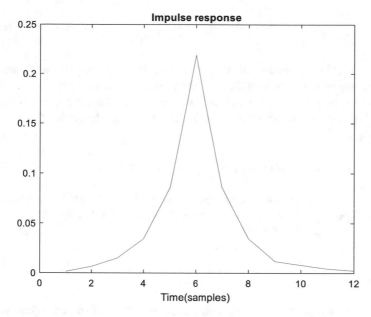

Fig. 14.6 Optimal smoother for a lag of 5 samples and impulse length 12 samples

is known as not being the fastest of the optimization methods but is used extensively in many machine learning algorithms due to its simplicity. The simplicity is particularly appealing for real-time engineering problems. Introducing a time-delay of 6 samples gives us a smoother instead of a filter (see Fig. 14.5). We choose 12 weights overall and the impulse response when converged is shown in Fig. 14.6.

14.4 Choice of Step Size and Normalized LMS

The problem with the ordinary LMS algorithm is the choice of step size. For stationary noise and a simulation, it can be made to look good, but for many practical applications where the noise is nonstationary, the algorithm will go unstable. There are two bounds on the step size which can be used. The first of these has been discussed and is shown in Eq. (14.18) uses the maximum eigenvalue of the $\mathbf{X}$ vector covariance matrix

$$0 < \mu < \frac{1}{\lambda_{\max}}$$

This is only a necessary condition for stability and the correlation matrix itself is unlikely to be known, and even if it was, taking the eigenvalues is a tall order in a real-time system. A better approach is to use convergence in the mean-square, from which we arrive ta yet another bound on the step size.

$$0 < \mu < \frac{1}{(n+1)\sigma_\xi^2} \tag{14.25}$$

where $(n+1)$ is the number of weights and σ_ξ^2 is the driving noise variance. For nonstationary noise the value of the step size needs to reduce as the variance increases. It also needs to increase as the variance reduces to maximize the speed of tracking. As an *ad-hoc* approach the variance can be tracked using a forgetting factor method. For example

$$\sigma_\xi^2(k+1) = \beta\sigma_\xi^2(k) + (1-\beta)\xi^2(k) \tag{14.26}$$

where β is a forgetting factor which needs to be greater than zero but less than unity. Its value depends on how fast the variance changes with time. We can then use the time-varying variance and make

$$0 < \mu(k) < \frac{1}{(n+1)\sigma_\xi^2(k)} \tag{14.27}$$

where the step size $\mu(k)$ now varies with time to keep pace with the changing variance. Inverse variance can be calculated directly using a special algorithm and is based on control theory rather than ad-hoc forgetting factor methods [12]. A better solution than these methods however is to use a slightly modified LMS algorithm known as the *normalized LMS algorithm* [13].

$$\mathbf{W}(k+1) = \mathbf{W}(k) + \frac{\mu}{\delta + \overline{\mu}\|\mathbf{X}(k)\|^2}\mathbf{X}(k)e(k) \tag{14.28}$$

where δ is a small constant to avoid a division by zero (e.g. $\delta = 0.001$) and $\|\mathbf{X}(k)\|^2$ is the norm squared of the regression vector. $\|\mathbf{X}(k)\|^2 = \mathbf{X}^T(k)\mathbf{X}(k)$. This is approximately $\mathbf{X}^T(k)\mathbf{X}(k) \approx (n+1)\sigma_\xi^2$ and hence tracts the variance of the regressor signal as it changes with time. The constant $\overline{\mu}$ is usually taken to be about 0.01 depending on the speed of tracking required. Some trial and error is always needed with LMS type algorithms.

14.5   Adaptive Noise Cancellation

We have already examined the problem of Wiener filtering using LMS, but although it converges to the correct filter, it is of little practical use since it requires the desired signal as an input and if we knew what it was there would be no problem in the first place. Therefore, a slight modification is needed. Rather than using a measure of the signal, why not use the noise. Now the noise will also not be available directly since if it was, we would just subtract it from signal plus noise and the problem would be solved. Even if we could measure it, its magnitude and phase would not be

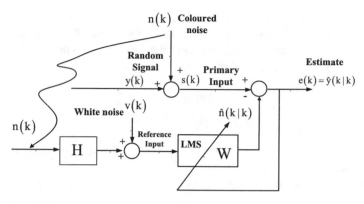

Fig. 14.7   Adaptive noise cancelling problem

matched to the signal, especially if it was an acoustic noise source. However, there are applications where the noise may well be available for measurement but not in ideal form. We assume the noise can be measured as a correlated version of the ideal noise, like as if the noise has passed through some extra filter H before we get to measure it. Such a problem may exist when cancelling noise from a flight cockpit. Putting a sensor near the engine or noise source will produce a related version of the noise that is hear somewhere else. It won't be the same, but a filtered version. In statistical terms it is *correlated* noise. Refer to Fig. 14.7. The top path of the diagram is often referred to as the *Primary* input and the bottom input to the LMS algorithm is referred to as the *Reference* input. It is important that the reference input is free of signal and has noise alone.

Without loss of any generality we can assume the coloured noise n(k) is directly added to the desired signal y(k) and doesn't have to go through another transfer function. In reality it probably will but the problem is still solvable in that case too. The signal plus noise is measurable and given by s(k). The noise is measured as Hn(k) and not n(k), where H is a transfer function. The v(k) term is uncorrelated white noise of variance σ_v^2. The signal and noise are assumed to have arbitrary Z-transform spectral densities Φ_{yy} and Φ_{nn} respectively. There is no additive white noise modelled at the signal, only coloured noise because if it were added it could be combined to give a new colouring transfer function in the form of an innovations model. The spectrum of the noise entering directly the lower or reference part of LMS algorithm is given by

$$\Phi_{xx} = H\Phi_{nn}H^* + \sigma_v^2 \tag{14.29}$$

This too can be factorized into

$$\Phi_{xx} = \Phi_{xx}^+ \Phi_{xx}^- \tag{14.30}$$

It is assumed that the output of the LMS algorithm is a noise estimate and not a signal estimate as for the ordinary Wiener filtering problem. This is then subtracted from signal plus noise to yield the signal estimate which is also the error signal e(k). For analysis consider the error signal

$$e(k) = s(k) - WHn(k) - Wv(k)$$
$$= y(k) + n(k) - WHn(k) - Wv(k) \tag{14.31}$$

In the above error we have dropped the Z-transform argument in the Z-transforms for mathematical clarity.

The expected value of the error squared is

$$E\big[e^2(k)\big] = E\big[y(k) + n(k) - WHn(k) - Wv(k)\big]^2 \tag{14.32}$$

Expanding and remembering that the noise terms are uncorrelated with each other, we then use Parseval's theorem and write

$$E\big[e^2(k)\big] = \frac{1}{2\pi j} \oint\limits_{|z|=1} \Phi_{ee} \frac{dz}{z} \tag{14.33}$$

where

$$\Phi_{ee} = \Phi_{yy} + \Phi_{nn} - \Phi_{nn}HW - \Phi_{nn}H^*W^* + WW^*\big(H\Phi_{nn}H^* + \sigma_v^2\big) \tag{14.34}$$

Using (14.29) in (14.34)

$$\Phi_{ee} = \Phi_{yy} + \Phi_{nn} - \Phi_{nn}HW - \Phi_{nn}H^*W^* + \Phi_{xx}WW^* \tag{14.35}$$

Completing the square

$$E\big[e^2(k)\big] = \frac{1}{2\pi j}\left[\oint\limits_{|z|=1} \Phi_{yy} + \Phi_{nn} + \left(\Phi_{xx}^+ W - \frac{\Phi_{nn}H^*}{\Phi_{xx}^-}\right)\right.$$
$$\left. \left(\Phi_{xx}^+ W - \frac{\Phi_{nn}H^*}{\Phi_{xx}^-}\right)^* - \frac{\Phi_{nn}\Phi_{nn}HH^*}{\Phi_{xx}}\right]\frac{dz}{z} \tag{14.36}$$

This is simplified accordingly

$$E\big[e^2(k)\big] = \frac{1}{2\pi j}\left[\oint\limits_{|z|=1} \Phi_{yy} + \Phi_{nn}\left(1 - \frac{H\Phi_{nn}H^*}{\Phi_{xx}}\right)\right.$$

$$+\left(\Phi_{xx}^{+}W - \frac{\Phi_{nn}H^{*}}{\Phi_{xx}^{-}}\right)\left(\Phi_{xx}^{+}W - \frac{\Phi_{nn}H^{*}}{\Phi_{xx}^{-}}\right)^{*}\right]\frac{dz}{z} \qquad (14.37)$$

The two-sided or noncausal solution for the optimal filter is therefore

$$\Phi_{xx}^{+}W - \frac{\Phi_{nn}H^{*}}{\Phi_{xx}^{-}} = 0$$

or

$$W = \frac{\Phi_{nn}H^{*}}{\Phi_{xx}} \qquad (14.38)$$

where the reference spectrum $\Phi_{xx} = H\Phi_{nn}H^{*} + \sigma_{v}^{2}$.

The minimum mean-square error is given from (14.37) by

$$E\left[e^{2}(k)\right]_{min} = \frac{1}{2\pi j}\oint_{|z|=1} \Phi_{yy} + \Phi_{nn}\left(1 - \frac{H\Phi_{nn}H}{\Phi_{xx}}\right)\frac{dz}{z} \qquad (14.39)$$

For zero measurement noise $\Phi_{xx} = H\Phi_{nn}H^{*}$ and (14.38) reduces to $W = \frac{1}{H}$. This makes sense since the filter would then equalise the H transfer function with its reciprocal giving the ideal noise signal at its output which then gets subtracted from signal plus noise. This would give a minimum mean-square error of $\frac{1}{2\pi j}\oint_{|z|=1}\Phi_{yy}\frac{dz}{z}$ for the ideal case. The two-sided Wiener filter will only be apparent if a delay is introduced in the primary path in Fig. 14.7 and this would make the filter a type of smoother.

14.5.1 Example with Noisy Speech

A recording of around 10 s of clean speech was made sampled at 44.1 kHz. Factory noise was added to this to give a signal plus noise. The SNR was measured to be − 10 dB. The same noise was passed through a transfer function given by $H(z^{-1}) = \frac{1-0.2z^{-1}}{1-z^{-1}+0.6z^{-2}}$ (see Fig. 14.7). Then uncorrelated white noise of variance 0.001 was added and this in turn passed to the input of the normalized LMS algorithm. The result is shown in Fig. 14.8.

The number of weights used was 128 for the LMS algorithm and a delay of 64 was put in the forward path to make a smoothing filter. The estimated impulse response is shown in Fig. 14.9.

The quality of the speech was excellent and much better than the ordinary Wiener filter used in a previous chapter. In fact, two or more channel approaches usually give better results than single channel filtering. This problem of course is quite contrived in that the noise is assumed to arrive with a unity transfer function to the first channel

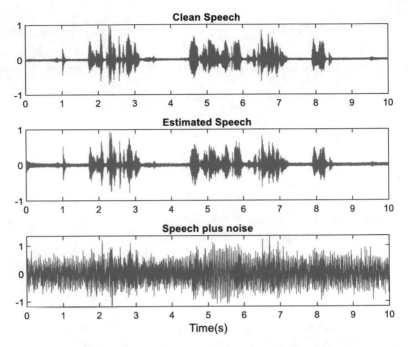

Fig. 14.8 Speech plus noise, estimated speech and clean speech comparison for noise canceller

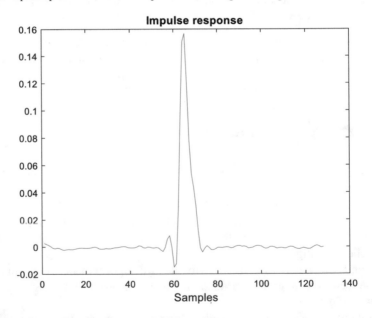

Fig. 14.9 Estimated impulse response of noise-canceller

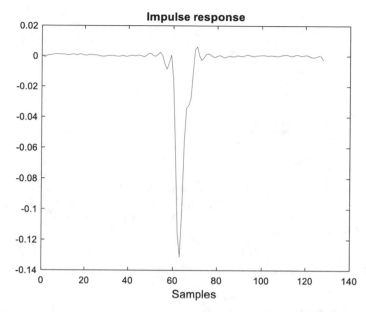

Fig. 14.10 Impulse response when H is nonminimum phase

although this should not make too much difference. In this simulation even if the transfer function H is changed so that its zero is outside the unit circle (a nonminimum phase zero), the canceller delivers just as good results. For $H(z^{-1}) = \frac{1-2z^{-1}}{1-z^{-1}+0.6z^{-2}}$ the impulse response of the filter is shown in Fig. 14.10.

This is an interesting result and perhaps unexpected since most Wiener filters have a positive peak at the centre if there is a delay.

14.6 Adaptive Noise Cancellation, Method 2

The problem with the standard noise cancelling approach is that although it can be made to work in certain environments, in other environments where the noise is not available free of the signal, a residual signal is embedded in the noise alone signal. The result of this is that it cancels the speech as well as the noise. A possible solution to this is that the noise sensor be placed further apart. Doing so will result in larger filter lengths however, assuming it is possible in the first instance. What is needed is an approach that the sensors can be placed together. Such a way is shown in Fig. 14.11.

The signal is assumed to be directly in front of the two sensors whilst the noise is nearby but not in front. Two path transfer functions represent the passage of the noise to each sensor. Additive white noise can also be put onto the diagram, but it is easier to see the operation by initially leaving this out.

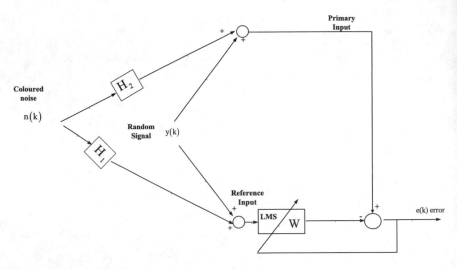

Fig. 14.11 Sensors are placed together instead of apart

When the signal is absent $y(k) = 0$.
The error signal is found from

$$e(k) = H_2 n(k) - W H_1 n(k) \qquad (14.40)$$

The LMS must minimise the expected value of this error. In this simple case it must be zero when

$$W = \frac{H_2}{H_1} \qquad (14.41)$$

This can be a problem if H_2 and H_1 behave as time-delays and the delay for H_1 is greater than that of H_2. This would make W noncausal and therefore the sensors would have to be swapped around or the geometry does not match the physics of the problem. Alternatively, a delay can be present in the upper arm (or Primary input) of Fig. 14.11 which is a good thing in any case making any optimal filter a smoothing filter. During periods of absence of signal we therefore adapt the LMS algorithm and it converges to $\frac{H_2}{H_1}$ or $\frac{z^{-\ell}H_2}{H_1}$ depending whether a delay is present or not.
The weights are then frozen, and the second part of the algorithm becomes when the signal is present.
When the signal and noise are present together.
This pre-supposes the signal reaches each sensor simultaneously since the signal is going to be quite close to the sensors. The error signal now becomes

$$e(k) = y(k) + H_2 n(k) - W H_1 n(k) - W y(k)$$

$$= y(k)\left[1 - \frac{H_2}{H_1}\right] + n(k)[H_2 - WH_1] \qquad (14.42)$$

Substitute (14.41) and the estimated is the error signal

$$e(k) = y(k)\left[1 - \frac{H_2}{H_1}\right] \qquad (14.43)$$

Whilst this is not the signal itself, it is a filtered version of the signal and practically makes little difference in the case of say speech plus noise. If for example both H_2 and H_1 were pure time-delays as in anechoic passage of sound through the air (though practically this is rarely the case except in outdoor environments), then the error will be of the form $e(k) = y(k)\left[1 - z^{-\ell 1}\right]$. The term $\left[1 - z^{-\ell 1}\right]$ is an FIR transfer function of a comb filter and has zeros equally spaced out along the unit circle. In more realistic environments the filter $\left[1 - \frac{H_2}{H_1}\right]$ will be lowpass in nature and in the case of speech make little difference to the quality. When white noise is present, another transfer function will occur, but tests have found little difference to the quality of speech.

If uncorrelated white noise $v(k)$ with variance σ_v^2 is present at the lower reference arm of the noise canceller (the LMS input), then we can show using similar methods to the previous section that the noncausal Wiener filter when signal is absent becomes [14]

$$W_{opt} = \frac{H_2 \Phi_{nn} H_1^*}{\Phi_{xx}} \qquad (14.44)$$

where the reference spectrum is

$$\Phi_{xx} = H_1 \Phi_{nn} H_1^* + \sigma_v^2 \qquad (14.45)$$

which goes to the previous result $\frac{H_2}{H_1}$ with no measurement noise $v(k)$.

14.6.1   Example with Noisy Speech

A similar problem as used in the conventional noise canceller was used here except the signal was assumed to be speech directly in front of two microphones so that the speech arrives at the same time on both. There were 128 weights used and a delay of 64 (half of the filter length) was used in the top path of the noise canceller (in Fig. 14.11) to implement a smoothing filter. The coloured factory noise travelled via two separate transfer functions $H_2 = \frac{1 - 2z^{-1}}{1 - 0.9z^{-1} + 0.4z^{-2}}$ and $H_1 = \frac{1 - 1.5z^{-1}}{1 - 0.7z^{-1} + 0.8z^{-2}}$ which are both nonminimum phase. Additive uncorrelated white noise of variance 0.001 was added to the signal plus filtered noise at the input to the LMS algorithm. The

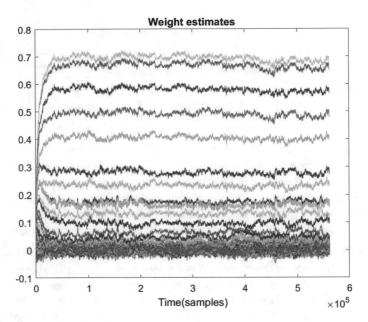

Fig. 14.12   Convergence of LMS weights for noise canceller

LMS algorithm was used on the noise signals on their own and the LMS algorithm converged to the values shown in Fig. 14.12.

The signal plus noise, clean signal and estimated signal is shown in Fig. 14.13. The quality of the enhanced speech was good but not as good as for the ordinary noise canceller. The impulse response is shown in Fig. 14.14. This method is quite a practical approach and has been implemented in real-time, but there are a few problems. The main problem is that it requires a voice activity detector (VAD) to find the noise alone periods. VADs which are robust are quite hard to find and often have difficult in real environments though can appear to work ok in simulation. The second problem is that the noise canceller can require a great many weights to reduce noise by any significant amount. Thirdly, the speech is assumed to be directly in front of the two microphones and arrive simultaneously. This is rarely if ever the case in practice.

14.7   Two Input Beamformer

SNR reductions of 6–12 dB are possible with slight modifications of this algorithm. For example, a two input *beamformer* uses two LMS algorithms. This is an extension of the method used previously. Beamforming is a method to make essentially a super-direction sensor or in the case of speech signal a microphone that is highly directional. Adaptive beamforming [15] is an extension where the beam can be

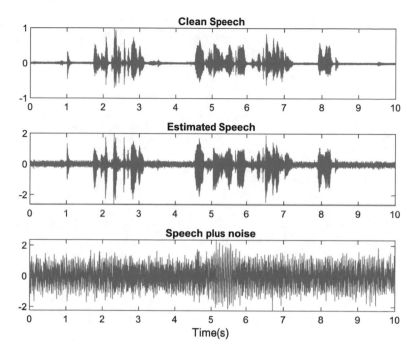

Fig. 14.13 Signal plus noise, estimate and clean signal

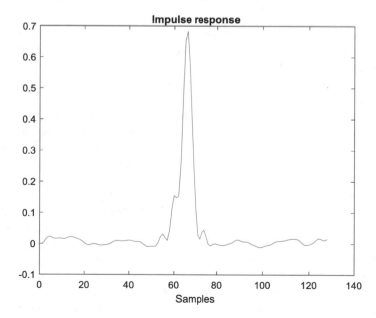

Fig. 14.14 Impulse response of noise-canceller

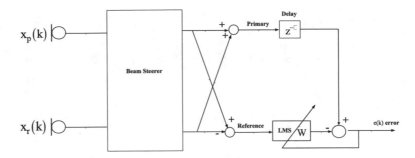

Fig. 14.15 Beamformer for two inputs

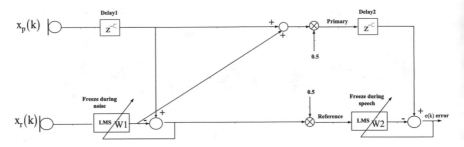

Fig. 14.16 Beamformer that uses two LMS algorithms

steered automatically in a particular direction. The method that follows is not a beamformer in the sense that it has an array of microphones but is technically classed as a beamformer (Fig. 14.15).

The method that is a natural progression from ordinary noise cancelling is the one shown in Fig. 14.16.

The idea is to get the reference signal (the input to the LMS algorithm which has a regressor vector as the input) free of desired signal and have noise only present, but without having to separate the sensors. As such it is closely related to the previous noise canceller with one major difference, it has two instead of one normalized LMS algorithm [16]. The first LMS algorithm works during speech and noise and freezes its values during noise, whilst the second LMS algorithm works during noise and freezes during speech. The basic idea is that the first LMS can align the paths between the signal source and sensors. If this is achieved (we assume the noise source is unchanged) then when subtracted, there will be a noise alone mixture and when added there will be twice the signal. This sum and difference are scaled by a factor of 0.5 (not absolutely necessary in fact but just to keep the algebra correct). The delays in the primary paths ensure that there is no problem with noncausal path differences or nonminimum phase transfer functions. The purpose of the second LMS algorithm at the front end is therefore to get a reference signal to the other LMS algorithm free of signal, and with noise on its own. A robust word boundary algorithm (or VAD) was used with this [17] and implemented in real-time using LabView programming

[18]. The composite system can work well in some environments where the noise source is not too nonstationary and provided the VAD can keep up with the speed of the speech. The type of VAD used works by calculating the time-delay between two microphones and from this information a zone of activity is set up directly in front of the two microphones. Anything out with of the zone is assumed to be noise. The VAD needs to be extremely robust to cope with many real room environments.

When operating in real-time with no voice, but after alignment of the front end LMS algorithm, the other LMS algorithm runs continuously to do noise cancelling and has an interesting effect on the ambient background noise. We can average the spectrum of the noise with no canceller and then switch on the canceller and repeat the procedure. The result is shown in Fig. 14.17 and is quite dramatic. It clearly reduces the noise floor right across the measurable spectrum up to half sampling frequency.

When running with speech as the desired signal directly in front of two microphones and a radio acting as noise, the effect is quite dramatic as can be seen in

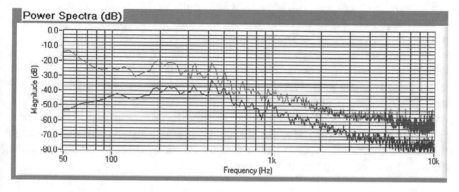

Fig. 14.17 Effect of beamformer running continuously on ambient noise. Top: ambient noise, Bottom: after cancellation

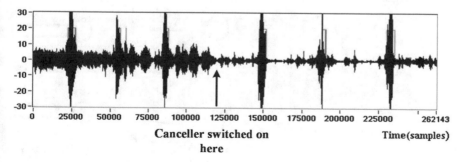

Fig. 14.18 Shows the beamformer when cancelling is switched on. Uses separated words with a radio as the background noise source

Fig. 14.18. It is effective for such problems, but when the speech is continuous and not in isolated words as in Fig. 14.18 the beamformer is far less effective.

14.8   The Symmetric Adaptive Decorrelator (SAD)

Up to now we have discussed the evolution of adaptive noise cancelling and the main problem has always been to get a clean noise signal that is free of signal (this is known as *crosstalk*). The beamformer is one approach, but there have been other attempts to make the noise canceller free from crosstalk. Notable of these approaches was the method that uses two "back to back" LMS type algorithms in a kind of bootstrapping topology. The method was generally known as crosstalk-resistant adaptive noise cancellation (CRANC) and more specifically the method trying to separate two sources from one another is known as the symmetric adaptive decorrelator (SAD) [19–21].

The basic problem with two sources is shown in Fig. 14.19.

Two people are simultaneously talking into two microphones. The problem is to separate the two signals. The two speech signals are labelled $t^1(k)$ and $t^2(ks)$. By the time the signals reach the microphones however, they have passed through two acoustic transfer functions. These acoustic transfer functions can be FIR or IIR but are usually taken to be FIR. In real environments the transfer functions model the impulse response paths to each microphone. They can be quite complex and have long lengths depending on how reverberant the environment is. The received signals are labelled $x^1(k)$ and $x^2(k)$. Because the speech signals each pass through two transfer functions, we must have a transfer function matrix which describes the mixing. This is known as the mixing matrix.

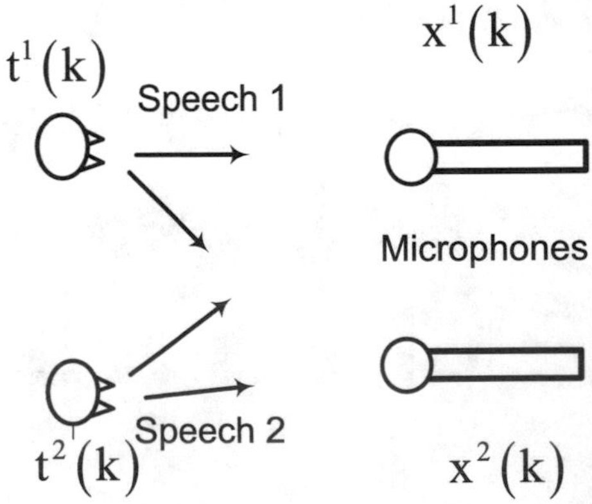

Fig. 14.19 Two speech signals simultaneously into two microphones

Fig. 14.20 Mixing of two
speech signals to the
microphones

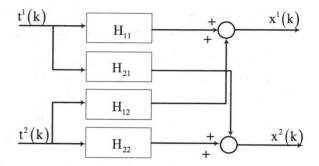

The 2×2 mixing matrix of FIR transfer functions **H** is non-singular with

$$\mathbf{x}(k) = \mathbf{H}(z^{-1})\mathbf{t}(k) \tag{14.46}$$

Or as shown in Fig. 14.20:

$$\begin{bmatrix} x^1(k) \\ x^2(k) \end{bmatrix} = \begin{bmatrix} H_{11}(z^{-1}) & H_{12}(z^{-1}) \\ H_{21}(z^{-1}) & H_{22}(z^{-1}) \end{bmatrix} \begin{bmatrix} t^1(k) \\ t^2(k) \end{bmatrix} \tag{14.47}$$

To proceed, we must further assume that H_{22} and H_{11} are both minimum-phase.
and re-define the inputs according to

$$\begin{bmatrix} x^1(k) \\ x^2(k) \end{bmatrix} = \begin{bmatrix} 1 & G_{12}(z^{-1}) \\ G_{21}(z^{-1}) & 1 \end{bmatrix} \begin{bmatrix} s^1(k) \\ s^2(k) \end{bmatrix} \tag{14.48}$$

where $s^1(k) = H_{11}t^1(k)$ and $s^2(k) = H_{22}t^2(k), G_{12} = H_{12}/H_{22}, G_{21} = H_{21}/H_{11}$ and
we estimate the s terms instead of the t terms. The s signals will only differ by FIR
transfer functions from the t signals. This is illustrated in Fig. 14.21.

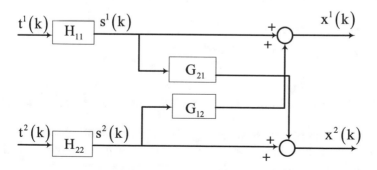

Fig. 14.21 Absorbing some terms and re-defining the input gives a simpler model

We now only need two unknown transfer functions to invert the matrix $\begin{bmatrix} 1 & G_{12}(z^{-1}) \\ G_{21}(z^{-1}) & 1 \end{bmatrix}$ but this is still a difficult problem since both terms $G_{12}(z^{-1})$ and $G_{21}(z^{-1})$ in this transfer function are unknown.

Separation is then found via the weight-vector updates of two *coupled* LMS algorithms:

$$\mathbf{w}^1(k+1) = \mathbf{w}^1(k) + \mu e^1(k)(\mathbf{X}_k^1)^T \tag{14.49a}$$

$$\mathbf{w}^2(k+1) = \mathbf{w}^2(k) + \mu e^2(k)(\mathbf{X}_k^2)^T \tag{14.49b}$$

where

$$e^i(k) = s^i(k) - \left[\mathbf{X}^i(k)\right]^T \mathbf{w}^i(k), i = 1, 2 \tag{14.50}$$

With the regressor vectors having entries which are the errors of the opposite LMS algorithm.

$\mathbf{X}^1(k) \quad = \quad [e^2(k-1), e^2(k-2) \ldots e^2(k-n)]^T \quad$ and $\quad \mathbf{X}^2(k) \quad = \quad [e^1(k-1), e^1(k-2) \ldots e^1(k-n)]^T$.

The two decorrelated random signals are $\hat{s}^i(k) = e^i(k), i = 1, 2$. Assume for strict causality that the two polynomials $G_{12}(0) = G_{21}(0) = 0$. A delay can be introduced into the upper path of the decorrelator in a similar way as for ordinary adaptive noise cancellation [22].

The convergence of this algorithm is dependent on an analysis which ignores certain cross-correlation terms. This means that although the algorithm is simple in structure it is not guaranteed to converge or to converge to the true values. It has been reported that in practice it will converge and if it doesn't the converged filter will be at least stable. However, a later refinement to the algorithm accounts for the extra terms missing in the earlier work and guarantees convergence at the expense of a far more complex algorithm [23].

14.8.1   *Example of Two Mixed Signals*

Consider the recording made in the Salk Institute. The recording made was of a man talking in a normal office room with loud music in the background [24, 25] (Fig. 14.22).

The estimates of the two separated signals are shown in Fig. 14.23.

The quality of the enhanced speech was remarkably good. Some residual music can still be heard in the background but is greatly attenuated. Care must be taken with the step size to avoid instability. The number of weights used was 100 in each LMS algorithm and a time-delay of 50 samples was introduced in the primary path.

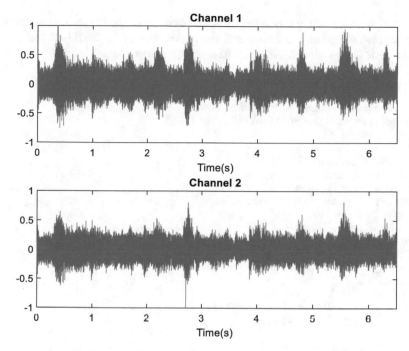

Fig. 14.22 Original speech plus music, two channels

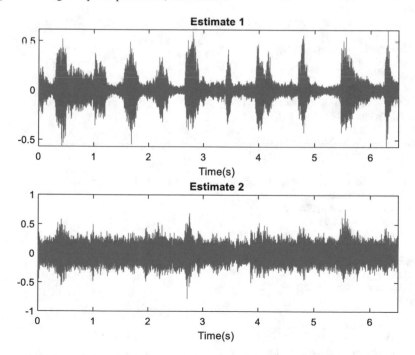

Fig. 14.23 Estimated channels for the separation of continuous speech and music

The basic Wiener filter or smoother used in the earlier chapters can remove more of the noise, but the quality of the speech was not as good as for the SAD method.

The code that produced these results is shown below.

MATLAB Code 14.2

```
%Cross coupled LMS Algorithm
%Symmetric adaptive Decorrelator (SAD)
clear;close all;
tmax = 6.5%seconds
%read clean signals
[s1,Fs] = audioread('r1.wav');
[s2,Fs] = audioread('r2.wav');
%Delay in primary channel 1
Lag = 50;
%Step size
mu = 0.1;
% Normalise the signals
%Code for normalisation on MATLAB Web pages
%Signal 1
if abs(min(s1)) > max(s1)
max_range_value = abs(min(s1));
min_range_value = min(s1);
else
max_range_value = max(s1);
min_range_value = -max(s1);
end
norm_value = 2.* s1./ (max_range_value - min_range_value);
s1 = norm_value;
Npoints1 = length(s1)
%Signal 2
%%%
if abs(min(s2)) > max(s2)
max_range_value = abs(min(s2));
min_range_value = min(s2);
else
max_range_value = max(s2);
min_range_value = -max(s2);
end
norm_value = 2.* s2./ (max_range_value - min_range_value);
s2 = norm_value;
Npoints2 = length(s2)
Nweights = 100;
%%%%%%%%%%
if Npoints1 < Npoints2
Npoints = Npoints1
else
Npoints = Npoints2
end
%Regression vector X1;
X1 = zeros(Nweights,1);
%Weight Vector1
W1 = zeros(Nweights,1);
%Regression vector X2;
```

```
X2 = zeros(Nweights,1);
%Weight Vector2
W2 = zeros(Nweights,1);
%Estimate 1
yhat1 = zeros(Npoints,1);
%Estimate 2
yhat2 = zeros(Npoints,1);
%Time vector
t = [1:1:Npoints]*(1/Fs);
%Main Loop
%Initialise
e1 = 0;
e2 = 0;
for i = 1:1:Npoints
%Shuffle regression vector 1
for k = Nweights:-1:2
X1(k) = X1(k-1);
end
X1(1) = e2;
%Create error signal e1
if (i-Lag) > 0
e1 = s1(i-Lag)-X1'*W1;
else
e1 = 0;
end
%error is signal estimate
yhat1(i) = e1;
%Update weights 1
W1 = W1 + X1*e1*mu;
%Shuffle regression vector 2
for k = Nweights:-1:2
X2(k) = X2(k-1);
end
X2(1) = e1;
%Create error signal e2
e2 = s2(i)-X2'*W2;
%error is signal estimate
yhat2(i) = e2;
%Update weights 1
W2 = W2 + X2*e2*mu;
%done
end
%Now do the plotting etc
figure
subplot(2,1,1)
plot(t,yhat1)
set(gcf,'color','w');
title('Estimate 1')
ax = gca;
ax.XLim = [0 tmax];
set(gcf,'color','w');
xlabel('Time(s)')
%
subplot(2,1,2)
```

```
plot(t,yhat2)
set(gcf,'color','w');
title('Estimate 2')
ax = gca;
ax.XLim = [0 tmax];
set(gcf,'color','w');
xlabel('Time(s)')
%%%%
Figure(2)
subplot(2,1,1);
plot(t,s1)
title('Channel 1');
ax = gca;
ax.XLim = [0 tmax];
xlabel('Time(s)')
subplot(2,1,2);
plot(t,s2)
title('Channel 2');
ax = gca;
ax.XLim = [0 tmax];
set(gcf,'color','w');
xlabel('Time(s)')
%play out estimate 1
sound(yhat1,Fs)
pause
input("")
%play out estimate 2
sound(yhat2,Fs)
```

So this method is by no means the most robust or mathematical precise technique, but it is by far the simplest method and suitable for real-time implementation with modest resources. The LMS algorithm is a simple algorithm but because it is based on steepest descent, the convergence is slow compared with other algorithms. However, other faster algorithms can be substituted using the same cross-coupled SAD method to gain better results at the expensive of more computational overhead [26, 27].

14.9   Independent Component Analysis. (ICA), Some Brief Notes

Often in the scientific literature, a method such as adaptive filtering progresses step by step as what we have just seen, but then in parallel a new method emerges which offers an entirely different approach. Such is the case with ICA as applied to signal processing. As we have seen with the filtering signals there are a great many unknowns. For example, with Wiener filters the signal and noise models and covariances are unknown and must be estimated experimentally or using recursive estimation methods. With adaptive filters we usually assume a noise alone reference signal is available, which makes what appears to be a simple solution quite complex. In certain circumstances all this is possible, and we can proceed. Even then, the quality of the recovered signal is not as good as the theory may suggest. An optimal filter may have estimates that look clean, but in the case of speech signals may not sound good to the human ear. In the mid-1990s a new theory emerged which instead of modelling the signal and noise models, used the probability density functions (PDF) of the signals instead. The method doesn't work on Gaussian signals but exploits the fact that many real-world signals are not Gaussian. Speech for example has a *Laplacian* distribution which has higher cumulants beyond the first two associated with Gaussian signals. An important cumulant which is often used is *Kurtosis*.

For a random variable, its Kurtosis is defined as

$$kurt(y) = E[y^4] - 3$$

If the Kurtosis is negative the signals is said to be *sub-Gaussian*, positive and it is called *super-Gaussian*. When the Kurtosis is zero the signal is Gaussian.

Another piece of information which is true of nearly all such problems is that the random signals are uncorrelated or statistically independent (hence the name). An alternative name in this field was developed: *Blind source separation (BSS)* [28, 29] which was more generic and tended to apply to the mixture of many signals with known PDFs. Adding to this, when this statistical method was received by engineers, the name changed again [30, 31] to *unsupervised adaptive filters*. The older Widrow type noise-cancelling methods were then re-named *supervised adaptive filters*. Here the word supervised implies some form of learning or training algorithm (as in machine learning). This is in fact what the LMS algorithm does as it converges.

The BSS area is quite vast, and many books have been written to date, but the method has to be included for completeness. Without a doubt the method is more general than the previous ad-hoc methods, but they do require significant processing power and part of an engineer's duty is not only to evaluate such algorithms but to trade-off performance with complexity, so that they can be used in real-time. Of course, there are many offline problems where the complexity is not an issue and we can at least evaluate such algorithms and compare with previous methods. As processing power became faster, the ability to implement BSS algorithms became more of a reality.

The original BSS problem was described as a mixing matrix. This of course differs from the mixing transfer function matrix described in the previous section. That problem is relatively easy to solve, but the original method required the inverse of a matrix. A breakthrough in the computation was made by Amari [32] using the Natural Gradient method which reduced the computational complexity and increased the speed of convergence. Amari's breakthrough was to realise that matrix updates of the earlier kind which had a matrix inverse converged faster when the curvature of the space was taken into account. However, this only worked for constant mixing matrices and not transfer-function matrices as is the case for speech mixed in a real environment. The constant matrix only applies to anechoic environments and is a simplification. What was needed was the solution to the *convolution mixing matrix* problem. An extension to the basic idea however was given in [33].

For a vector of m desired signals $\mathbf{s}(k) = \begin{bmatrix} s^1(k) & s^2(k) & s^3(k) & \cdots & s^m(k) \end{bmatrix}^T$ and a vector of n received signals $\mathbf{x}(k) = \begin{bmatrix} x^1(k) & x^2(k) & x^3(k) & \cdots & x^n(k) \end{bmatrix}^T$, the mixing process is given by

$$\mathbf{x}(k) = \sum_{p=-\infty}^{\infty} \mathbf{H}_p \mathbf{s}(k - p) \tag{14.51}$$

where $\mathbf{H}_p$ are nxm matrices forming the matrix convolution (14.51). This is a two-sided matrix impulse response, a bit like the two-sided Wiener filter. To make it work we must of course truncate it suitably.

Then the un-mixer will also be a convolution of similar form to (14.51). For L *weight matrices* (the matrix form of number of vector weights when we had LMS) this will be

$$\mathbf{y}(k) = \sum_{p=0}^{L} \mathbf{W}_p \mathbf{x}(k - p) \tag{14.52}$$

The convolution natural gradient algorithm becomes for non-complex signals[33]

$$\mathbf{W}_p(k + 1) = \mathbf{W}_p(k) + \mu(k) \left[\mathbf{W}_p(k) - \mathbf{f}(\mathbf{y}(k - L)) \mathbf{u}^T(k - p) \right] \tag{14.53}$$

where

$$\mathbf{u}(k) = \sum_{q=0}^{L} \mathbf{W}_{L-q}^T \mathbf{y}(k - q). \tag{14.54}$$

and $\mathbf{f}(.)$ is a vector of nonlinearities dependant on the PDFs of the source signals. Note that the step size $\mu(k)$ is given in the general case of being time-varying to aid convergence, but it can be constant. The choice of the nonlinearities is found from $f_i(y_i) = -d \log(p_i(y_i))/dy_i$ where $p_i(y_i)$ are the individual PDFs of each individual signal. A typical function that works is $f_i(y_i) = \tanh(\gamma y_i)$ for $\gamma > 2$ provided the signals have positive Kurtosis. Mixing Gaussian with non-Gaussian signals may prove a problem and needs a different approach.

The same example from the SALK institute as run in the previous section for the SAD method was run using a much later BSS approach [34]. The results are shown in Fig. 14.24.

The noise is overall removed, but there is still a residual background effect making the quality of the speech unnatural with a slight echo and ringing. The music sounded completely natural, however.

Girolami [35] made an interesting discovery that makes use of the PDFs of the sources, but applied it to the duel LMS (SAD) algorithm of the previous section. It uses the sign of the error instead of the error itself and the theory is based on maximum likelihood parametrization. Hence it is computationally simpler and does result in

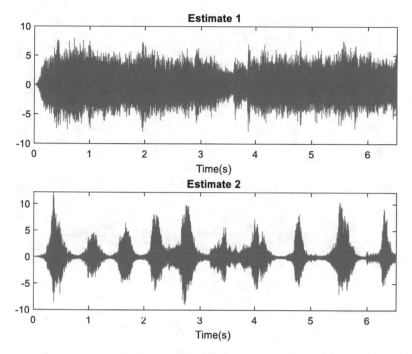

Fig. 14.24 Blind source separation on convolutive mixing of speech and music

better SNRs than ordinary LMS. However, the quality of the speech is not always as good. This is a common problem with all such algorithms as was mentioned in an earlier chapter on Wiener filters. A Wiener filter with known statistics can remove most of the noise from a human speech signal, but the quality of the speech is not always pleasant to the human ear.

$$\mathbf{w}^1(k+1) = \mathbf{w}^1(k) + \mu \frac{sign(e^1(k))}{E\{|e^1(k)|\}} (\mathbf{X}_k^1)^\mathrm{T} \tag{14.55}$$

$$\mathbf{w}^2(k+1) = \mathbf{w}^2(k) + \mu \frac{sign(e^2(k))}{E\{|e^2(k)|\}} (\mathbf{X}_k^2)^\mathrm{T} \tag{14.56}$$

Equations (14.45) and (14.46) show the equations of the modified SAD based on probability theory. Figure 14.25 shows for the same example the separated speech and music.

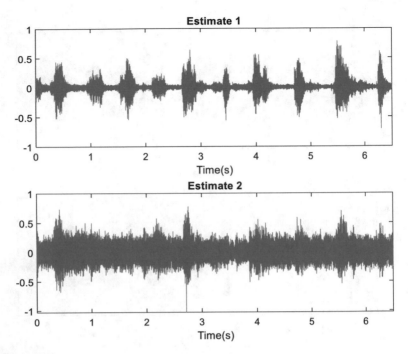

Fig. 14.25   Separated speech and music using the Girolami method

References

1. A. Carusone, D.A. Johns, Analogue adaptive filters: past and present. IEE Proc. Circuits Devices Syst. **147**(1), 82–90 (2000). https://doi.org/10.1049/ip-cds:20000052
2. G.A. Williamson, Adaptive IIR filters, in *Digital signal Processing Handbook*, ed. by V.K. Madisetti, D. Williams (CRC Press LLC, Boca Raton, USA, 1999)
3. R. Lucky, Automatic equalization for digital communications. Bell Syst. Tech. J. **44**(April), 547–588 (1965)
4. R. Lucky, The adaptive equalizer. IEEE Signal Process. Mag. **23**(3), 104–107 (2006)
5. B. Widrow, M. Hoff, Adaptive switching circuits. IRE Wescon Conv. **4**, 96–104 (1960)
6. P. Hagander, B. Wittenmark, A self-tuning filter for fixed lag smoothing. IEEE Trans. Inf. Theor. **IT-23**(3), 377–384 (1977)
7. B. Wittenmark, A self-tuning predictor. IEEE Trans. Autom. Control **19**(6), 848–851 (1974). https://doi.org/10.1109/TAC.1974.1100734
8. M.J. Grimble, H/sub infinity/fixed-lag smoothing filter for scalar systems. IEEE Trans. Signal Process. **39**(9), 1955–1963 (1991). https://doi.org/10.1109/78.134428
9. A. Brauer, Limits for the characteristic roots of a matrix. IV. applications to stochastic matrices. Duke. Math J. **19**, 75–91 (1952)
10. C.Y. Chiang, M.M. Lim, The eigenvalue shift technique and its eigenstructure analysis of a matrix. J. Comput. Appl. Math. **253**(1), 235–248 (2013)
11. S. Haykin, *Adaptive Filter Theory* (Prentice Hall, New Jersey, 1996)
12. T.J. Moir, Automatic variance control and variance estimation loops. Circuits Syst. Signal Process. **20**(1), 1–10, 2001/01/01 2001. https://doi.org/10.1007/BF01204918
13. J. Nagumo, A. Noda, A learning method for system identification. IEEE Trans. Autom. Control **12**(3), 282–287 (1967). https://doi.org/10.1109/TAC.1967.1098599

14. T.J. Moir, H.S. Dabis, Solution of generalised multivariable noise cancelling problem. Electron. Lett. **27**(12), 1060–1061 (1991). https://doi.org/10.1049/el:19910658
15. L. Griffiths, C. Jim, An alternative approach to linearly constrained adaptive beamforming. IEEE Trans. Antennas Propag. **30**(1), 27–34 (1982). https://doi.org/10.1109/TAP.1982.114 2739
16. J.V. Berghe, J. Wouters, An adaptive noise canceller for hearing aids using two nearby microphones. J. Acoust Soc Amer. **103**(6), 3621–3626 (1998)
17. H. Agaiby, T.J. Moir, A robust word boundary detection algorithm with application to speech recognition, in *Proceedings of 13th International Conference on Digital Signal Processing*, 2–4 July 1997 1997, vol. 2, pp. 753–755. https://doi.org/10.1109/ICDSP.1997.628461
18. T. Moir, Real-time acoustic beamformer on a PC. WSEAS Trans. Signal Process. **2**(2), 167–174 (2006)
19. S. Van Gerven, D. Van Compernolle, Signal separation by symmetric adaptive decorrelation: stability, convergence, and uniqueness. IEEE Trans. Signal Process. **43**(7), 1602–1612 (1995). https://doi.org/10.1109/78.398721
20. G. Mirchandani, R.L. Zinser, J.B. Evans, A new adaptive noise cancellation scheme in the presence of crosstalk (speech signals). IEEE Trans. Circuits Syst. II: Analog Digital Signal Process. **39**(10), 681–694 (1992). https://doi.org/10.1109/82.199895
21. R. Zinser, G. Mirchandani, J. Evans, Some experimental and theoretical results using a new adaptive filter structure for noise cancellation in the presence of crosstalk, in *ICASSP '85. IEEE International Conference on Acoustics, Speech, and Signal Processing*, 26–29 April 1985 1985, vol. 10, pp. 1253–1256. https://doi.org/10.1109/ICASSP.1985.1168121
22. S.M. Kou, W.M. Peng, Principle and applications of asymmetric crosstalk-resistant adaptive noise canceler. J. Franklin Inst. **337**(1), 57–71 (2000)
23. P. Scalart, L. Lepauloux, On the convergence behavior of recursive adaptive noise cancellation structure in the presence of crosstalk, in *IET Conference Proceedings*, pp. 11–11. (Online). Available: https://digital-library.theiet.org/content/conferences/10.1049/ic.2010.0230
24. Salk Institute. Blind Source Separation of recorded speech and music signals: Audio examples. Salk Institute. https://cnl.salk.edu/~tewon/Blind/blind_audio.html. Accessed 2020
25. D. Yellin, E. Weinstein, Multichannel signal separation: methods and analysis. IEEE Trans. Signal Process. **44**(1), 106–118 (1996). https://doi.org/10.1109/78.482016
26. M. Gabrea, E. Mandridake, M. Menez, M. Najim, A. Vallauri, Two microphones speech enhancement system based on a double fast recursive least squares (DFRLS) algorithm, in *1996 8th European Signal Processing Conference (EUSIPCO 1996)*, 10–13 Sept 1996, pp. 1–4
27. T.J. Moir, Adaptive crosstalk-resistant noise-cancellation using H infinity filters, in *2019 IEEE International Conference on Signals and Systems (ICSigSys)*, 16–18 July 2019 2019, pp. 24–28. https://doi.org/10.1109/ICSIGSYS.2019.8811051
28. S. Amari, A. Cichocki, H.H. Yang, A new algorithm for blind signal separation. Adv. Neural Information Process. **8**, 757–763 (1996)
29. A. Cichocki, S. Amari, *Adaptive Blind Signal and Image Processing* (John Wiley and Sons, Ltd., Chichester, UK, 2002)
30. S.E. Haykin, *Unsupervised Adaptive Filtering: Blind Deconvolution* (John Wiley and Sons, Inc, 2000)
31. J.M.T. Romano, R.F. de Attux, C.C. Cavalcante, R. Suyama, *Unsupervised Signal Processing, Channel Equalization and Source Separation* (CRC Press, Boca Raton, 2011)
32. S. Amari, Natural gradient works efficiently in learning. Neural Compt. **10**, 251–276 (1998)
33. S. Amari, S. Douglas, A. Cichocki, H.H. Yang, Multichannel blind deconvolution and equalization using the natural gradient, in *First IEEE Signal Processing Workshop on Signal Processing Advances in Wireless Communications*, 16–18 April 1997, pp. 101–104. https://doi.org/10.1109/SPAWC.1997.630083
34. H. Sawada, S. Araki, S. Makino, Underdetermined convolutive blind source separation via frequency bin-wise clustering and permutation alignment. IEEE Trans. Audio Speech Lang. Process. **19**(3), 516–527 (2011). https://doi.org/10.1109/TASL.2010.2051355

35. M. Girolami, Symmetric adaptive maximum likelihood estimation for noise cancellation and signal separation. Electron. Lett. **33**(17), 1437–1438 (1997). https://doi.org/10.1049/el:199 70977

Chapter 15
Other Common Recursive Estimation Methods

15.1 Motivation

So far, we have studied the least-mean squares (LMS) algorithm and seen it applied to adaptive filters. It is still the most popular method used in adaptive filtering dues to its simplicity. A wealth of papers has been written about its convergence and tracking abilities. It is applied almost universally to FIR problems and adaptive filter type methods. It is not easy to keep stable without the modification of the normalized LMS algorithm, and there are a great many variants to add to that too. The problem with LMS is that it is based on steepest descent which is inherently slow convergence. In fact, what is needed is a step-size that is time varying. The step size needs to be large at the start and as the optimal solution is getting nearer, progressively smaller to get accuracy. Speed of adaption or at least speed of convergence is important for adaptive filters since conditions in many real environments can change rapidly with time. We begin with one of the most common alternative methods, recursive least-squares.

15.2 The Recursive Least Squares (RLS) Method

We could derive this method from first principles, but there are a host of books which cover the topic and here we are more interested in the application and advantages. The RLS algorithm was used extensively in the early to late 1960s, not so much by signal processing engineers, but rather control system researchers who were active at the time. The topic was not adaptive filters, but adaptive control. Specifically, it was used on self-tuning regulators [1]. A few years later the convergence of such algorithms was extensively studied [2, 3]. At the time, many thought the RLS algorithm to be a clever extension of least-squares which had been developed around the same time as self-tuning control. It turned out however that a mathematician in 1950 by the name of Plackett [4] has discovered that Gauss had already made the discovery in 1821!

© The Author(s), under exclusive license to Springer Nature Switzerland AG 2022   423
T. J. Moir, *Rudiments of Signal Processing and Systems*,
https://doi.org/10.1007/978-3-030-76947-5_15

The RLS update looks very similar to that of LMS.

$$\mathbf{W}(k+1) = \mathbf{W}(k) + \mathbf{K}(k)e(k) \tag{15.1}$$

With the error update

$$e(k) = y(k) - \mathbf{W}^T\mathbf{X}(k) \tag{15.2}$$

In the control system literature usually θ is more commonly used instead of $\mathbf{W}$ for the weight vector. From (15.1) it is clear that there is no longer a constant step size, but instead we have a vector of gains (not a matrix) $\mathbf{K}(k)$ which is time-varying. For n weights the gain vector is an nx1 vector given by:

$$\mathbf{K}(k) = \frac{\mathbf{P}(k)\mathbf{X}(k)}{1 + \mathbf{X}^T(k)\mathbf{P}(k)\mathbf{X}(k)} \tag{15.3}$$

$\mathbf{P}(k)$ is an n square positive definite error covariance matrix. It in turn is found from

$$\mathbf{P}(k) = \left[\mathbf{I} - \mathbf{K}(k)\mathbf{X}^T(k)\right]\mathbf{P}(k-1) \tag{15.4}$$

We must initialise the algorithm by setting the error covariance matrix to a diagonal matrix. For example $\mathbf{P}(0) = p_0\mathbf{I}$ where p_0 is some small constant such as 0.01. The RLS algorithm is the least-squares solution, but unlike least squares it does not require a matrix inversion. It still requires $O(n^2)$ operations, however. The RLS algorithm has an infinite memory, it remembers all the past history whereas LMS does not. This means that LMS can adapt better to changes whereas RLS can get stuck in the trajectory it is heading towards. There is a partially successful fix to this problem known as a *forgetting factor* λ. Usually the forgetting factor is close to 1. The equations get modified accordingly. The gain vector becomes

$$\mathbf{K}(k) = \frac{\mathbf{P}(k)\mathbf{X}(k)}{\lambda + \mathbf{X}^T(k)\mathbf{P}(k)\mathbf{X}(k)} \tag{15.5}$$

And the error covariance matrix update becomes

$$\mathbf{P}(k) = \frac{1}{\lambda}\left[\mathbf{I} - \mathbf{K}(k)\mathbf{X}^T(k)\right]\mathbf{P}(k-1) \tag{15.6}$$

This solution minimizes the weighted cost function

$$J = \sum_{k=0}^{n}\lambda^{n-k}e^2(k) \tag{15.7}$$

For example, if $\lambda = 0.99$, then for the past 100 samples, it attenuates past errors by the amount $0.99^{100} = 0.366$. The theory and practice with forgetting factors are fraught with problems. A forgetting factor does not provide robust estimation by any means but is a half-way fix to a difficult problem of how to track a time-varying set of weights as fast as possible. It is important to note that the RLS algorithm is *deterministic*. Nowhere within it is any statistical information. One of the great advantages of RLS however is its rate of convergence to constant weights. In real-time applications it is probably better to operate the RLS algorithm in blocks or buffer worth amounts of data. This way the algorithm gets reset as each batch of data comes in and it doesn't have to track continuously at all. In fact the FFT is implemented in a similar manner since it too cannot track (the short-time fast Fourier transform). Block data is processed, and initialization happens at the start of each block.

Large number of weights is impractical to implement in real-time, but the error is very small with RLS and compensates in this respect. More than often in control systems, it is used to estimate pole-zero (ARMA) type models and stability can be compromised if the model changes rapidly with time. Therefore, for pole-zero models the stability must be monitored continuously.

In this book we will use RLS on FIR models only to ensure stability. We re-visit the coupled twin LMS noise canceller (SAD). When using RLS instead of LMS, the likelihood of any real-time implementation is less possible, but the results are dramatically better. Recall in the previous chapter that two cross coupled LMS algorithms were used, each with 100 weights and a delay of 50 (see Fig. 14.23). If we repeat the same experiment but with two RLS algorithms in the SAD configuration we get the results of Fig. 15.1.

Only 10 weights were used with a delay of 5 steps. The results look and sound similar to the LMS method which used 100 weights and a delay of 50. In fact even with 4 weights and a delay of 2 samples only marginally worse results were obtained.

The code that produced Fig. 15.2 is shown below in MATLAB.

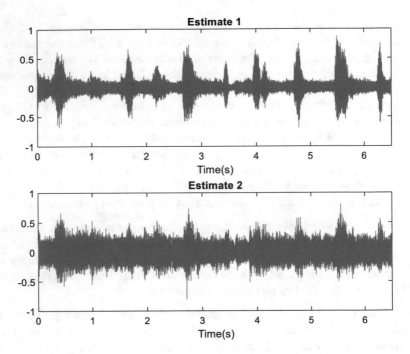

Fig. 15.1 Separation of speech and music using a SAD and two RLS algorithms

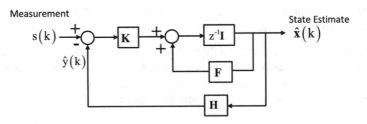

Fig. 15.2 Discrete Kalman filter (one step predictor version)

```
MATLAB Code 15.1
%Cross coupled RLS Algorithm
%Symmetric adaptive Decorrelator (SAD)
%With Forgetting factor
clear;close all;

tmax=6.5%seconds
%read clean signals

[s1,Fs]=audioread('r1.wav');
[s2,Fs]=audioread('r2.wav');

%Delay in primary channel 1
Lag=2;

Nweights=4;
forget=1;



% Normalise the signals
%Code for normalisation on MATLAB Web pages
%Signal 1
if abs(min(s1)) > max(s1)
      max_range_value = abs(min(s1));
      min_range_value = min(s1);
  else
      max_range_value = max(s1);
      min_range_value = -max(s1);
  end
norm_value = 2 .* s1 ./ (max_range_value - min_range_value);
s1=norm_value;

    Npoints1=length(s1)
    %Signal 2
    %%%

    if abs(min(s2)) > max(s2)
      max_range_value = abs(min(s2));
      min_range_value = min(s2);
  else
      max_range_value = max(s2);
      min_range_value = -max(s2);
  end
norm_value = 2 .* s2 ./ (max_range_value - min_range_value);
s2=norm_value;

    Npoints2=length(s2)
```

```
%%%%%%%%%%
if Npoints1 < Npoints2
    Npoints=Npoints1
else
    Npoints=Npoints2
end

%Regression vector X1;
X1=zeros(Nweights,1);
%Weight Vector1
W1=zeros(Nweights,1);
%Regression vector X2;
X2=zeros(Nweights,1);
%Weight Vector2
W2=zeros(Nweights,1);
%Estimate 1
yhat1=zeros(Npoints,1);
%Estimate 2
yhat2=zeros(Npoints,1);
%Gain  Vector1
K1=zeros(Nweights,1);
%Gain  Vector2
K2=zeros(Nweights,1);
%error covariance P1
P1=eye(Nweights)*0.01;
%error covariance P2
P2=eye(Nweights)*0.01;


%Time vector

t=[1:1:Npoints]*(1/Fs);
%Main Loop
%Initialise
e1=0;
e2=0;
for i=1:1:Npoints

    %Shuffle regression vector 1
    for k=Nweights:-1:2
    X1(k)=X1(k-1);
    end
```

```
    X1(1)=e2;
     %Create error signal e1

     if (i-Lag) >0
    e1=s1(i-Lag)-X1'*W1;
     else
         e1=0;

     end

    %error is signal estimate
    yhat1(i)=e1;
    %Update K
    K1=P1*X1/(forget+X1'*P1*X1);
    %Update P
    P1=(1/forget)*(eye(Nweights)-K1*X1'*P1)*P1;

%Update weights 1
W1=W1+e1*K1;

%Shuffle regression vector 2
    for k=Nweights:-1:2
    X2(k)=X2(k-1);
    end

    X2(1)-c1;
     %Create error signal e2
    e2=s2(i)-X2'*W2;

    %error is signal estimate
    yhat2(i)=e2;

    %Update K
    K2=P2*X2/(forget+X2'*P2*X2);
    %Update P
    P2=(1/forget)*(eye(Nweights)-K2*X2'*P2)*P2;


%Update weights 1
W2=W2+e2*K2;
%done
@@@@
end
figure
```

```
subplot(2,1,1)
plot(t,yhat1)
set(gcf,'color','w');
title('Estimate 1')
ax = gca;
ax.XLim = [0 tmax];
set(gcf,'color','w');
xlabel('Time(s)')
%
subplot(2,1,2)
plot(t,yhat2)
set(gcf,'color','w');
title('Estimate 2')
ax = gca;
ax.XLim = [0 tmax];
set(gcf,'color','w');
xlabel('Time(s)')
%%%%

figure(2)
subplot(2,1,1);
plot(t,s1)
title('Channel 1');
ax = gca;
ax.XLim = [0 tmax];
xlabel('Time(s)')

subplot(2,1,2);
plot(t,s2)
title('Channel 2');
ax = gca;
ax.XLim = [0 tmax];
set(gcf,'color','w');
xlabel('Time(s)')
%play out estimate 1
sound(yhat1,Fs)
  pause
input("")
%play out estimate 2
sound(yhat2,Fs)
```

As explained in the previous chapter, the SAD algorithm is not guaranteed to converge to the correct values, but more than often it does.

15.3 The Kalman Filter

The Wiener filter and the theory behind it has a great many uses as has been previously covered. The biggest problem it had was the calculation of the filter itself, and the need for accurate mathematical models of the noise and signal process. It also works best on random signals and noise that are wide sense stationary. The theory does not apply to time-variant models of the signal or noise. Usually the tracking of other algorithms (such as LMS) means that Wiener filters can be used when the statistics are time-variant, but the estimate is not necessarily optimal. In 1960, Kalman came up with a general theory for a new type of filter that does apply to nonstationary models. It will give a lower mean square error than a Wiener filter in such circumstances. The Kalman filter [5] as it came to be known inspired whole industries of interest in the new method, especially in aerospace. At the time, inertial navigation was emerging as the way to keep space vehicles and missiles stable. Rate gyros were used which gave out noisy signals proportional to angular velocity, but angular position was needed for navigation purposes. It would appear at first sight, that ordinary integration is all that would be required to get from velocity to position, but analogue (and even digital) integrators do not work at all well on signals of this nature. The slightest dc offset, and large errors occur in the measurements. Similarly, despite having highly precise mechanical gyroscopes, they would drift out of position over time and introduce large errors. If mankind was to reach the moon for instance it needed accurate ways to measure the position of a vehicle in space. The Kalman filter came along just at the right time. The theory of this original paper by Kalman was in discrete time. He later wrote another paper with a colleague to solve the continuous-time problem and this is known as the Kalman-Bucy filter. It is interesting that he chose discrete time, since computers at the time were in their infancy and analogue systems were dominant until at least the mid-1970s. The theory was a general theory of not just filtering but also control. He showed that filtering and control (as in closed-loop state-feedback control) are theoretically entwined and are one of the same problem. In this book we consider the filtering aspects only and so there will be no control-system input modelled in the Kalman filter. The Kalman filter went on to find application in all manner of industries from image processing to robotics. It is still a heavily researched topic to this day.

Kalman, unlike his predecessor Wiener was not so much interested in filtering a random signal from noise. Because he was predominantly interested in control systems, he tackled the problem of estimation of state variables from noisy measurements of the output of a system. It was like an *observer* that takes account of the statistical properties of the noise. An observer is an algorithm that reconstructs the states of a system from measurements of the output [6]. It does this by having a mathematical model of the system within it. Kalman's model was based on state-variables. For n state variables such a model becomes

$$\mathbf{x}(k+1) = \mathbf{F}\mathbf{x}(k) + \mathbf{D}\xi(k) \tag{15.8}$$

where the state $\mathbf{x}(k)$ is a column vector consisting of possible real-world quantities such as position, velocity and acceleration. The random signal defined as

$$y(k) = \mathbf{H}\mathbf{x}(k) \tag{15.9}$$

The random signal is here defined as scalar, but the theory accounts for systems with multiple outputs. These are called *multivariable systems* in the control literature. The noise that is added is white and so the measured signal is

$$s(k) = \mathbf{H}\mathbf{x}(k) + v(k) \tag{15.10}$$

Both $\xi(k)$ and $v(k)$ are zero-mean uncorrelated white noise with variances σ_ξ^2 and σ_v^2 respectively. The noise sequence $\xi(k)$ is known as *process noise* and $v(k)$ *measurement noise*. In the more general case of a multivariable system, the output is a vector and the measurement noise is a vector of white noise terms that has a covariance matrix. Even in this simpler form above, often the vector $\mathbf{D}$ in the state-space model (15.8) is omitted altogether and a vector of white noise sources used instead. For example, we can instead of (15.8) write

$$\mathbf{x}(k+1) = \mathbf{F}\mathbf{x}(k) + \boldsymbol{\xi}(k) \tag{15.11}$$

where $\boldsymbol{\xi}(k)$ is a vector of white noise sources with covariance matrix $\mathbf{Q} > 0$

$$\begin{aligned} \mathbf{Q} &= \mathrm{cov}\big[\boldsymbol{\xi}(k)\big] \\ &= \mathrm{E}\big[\boldsymbol{\xi}(k)\boldsymbol{\xi}^{\mathrm{T}}(k)\big] \end{aligned} \tag{15.12}$$

This change in definition does not affect the generality of the solution since we can define

$$\boldsymbol{\xi}(k) = \mathbf{D}\xi(k) \tag{15.13}$$

and taking expectations we arrive at

$$\mathbf{Q} = \mathbf{D}\mathbf{D}^{\mathrm{T}}\sigma_\xi^2 \tag{15.14}$$

We can therefore either use a vector of white noise terms or a scalar white noise with a $\mathbf{D}$ vector instead.

The transfer function of the signal generating model is found by taking Z-transforms of (15.8). We find

$$y(z) = \mathbf{H}(z\mathbf{I} - \mathbf{F})^{-1}\mathbf{D}\xi(z) \tag{15.15}$$

The Kalman filter itself looks a bit like an observer and has the form

$$\hat{x}(k+1) = F\hat{x}(k) + Ke(k) \qquad (15.16)$$

with error

$$e(k) = s(k) - \hat{y}(k) \qquad (15.17)$$

The hat over a random variable indicates an estimate rather than an exact value. Then $\hat{x}(k)$ and $\hat{y}(k)$ are the state and signal estimates respectively. The signal estimate is

$$\hat{y}(k) = H\hat{x}(k) \qquad (15.18)$$

where H is a row vector of dimension n. The K variable in (15.16) is the *Kalman gain vector* of dimension n (which can be a matrix in the more general multivariable case). This vector is found by a recursion. First define the n square positive definite *aposteriori* state error covariance matrix

$$P(k|k) = \text{cov}\big[x(k) - \hat{x}(k)\big] \qquad (15.19a)$$

And the *apriori* error-covariance matrix

$$P(k|k-1) = \text{cov}\big[x(k) - \hat{x}(k|k-1)\big] \qquad (15.19b)$$

As with the RLS algorithm it needs an initial value and usually is defined at time zero as

$$P(0|-1) = p_0 I \qquad (15.20)$$

Then the following two equations must be iterated until convergent or for as long as necessary to keep track of the estimate

$$K(k) = FP(k|k-1)H(k)\big[\sigma_v^2 + H(k)P(k|k-1)H^T(k)\big]^{-1} \qquad (15.21)$$

$$P(k|k) = FP(k|k-1)F^T + DD^T\sigma_\xi^2 - K(k)HP(k|k-1)F^T \qquad (15.22)$$

This is the simplest form of the Kalman filter and if studied in detail [7] we find that the estimate it gives is a one-step ahead predicted estimate $\hat{x}(k|k-1)$ and $\hat{y}(k|k-1)$. It is possible to get the instantaneous estimate or even a smoothed estimate, but usually the basic Kalman filter is the one quoted in the literature. Likewise, the model can be extended so that it filters coloured noise instead of white noise. This is done by augmenting the state (adding new states to the existing state variables) which model the coloured noise. In (15.22) we can substitute $\mathbf{Q}=\mathbf{DD}^T\sigma_\xi^2$ if necessary and get a slightly modified form. Equation (15.22) is known as a matrix *Riccati* equation. We do not need to solve it, instead we simply iterate it until it converges. If the model is time-variant (for example, suppose we have $\mathbf{F}(k)$ instead of $\mathbf{F}$), then the equations still apply by inserting the time-variant matrices. Likewise, for nonstationary noise. This is easy to say, but in practice it is unlikely an accurate model will be available if the process is time variant. If the noise terms are stationary and the system is LTI then the Kalman filter will converge in steady state to a Wiener filter. The block diagram of a discrete Kalman filter is shown in Fig. 15.2.

15.3.1 Kalman Filter Example

Consider a random process in state variable format

$$\mathbf{x}(k+1) = \mathbf{Fx}(k) + \mathbf{D}\xi(k)$$
$$y(k) = \mathbf{Hx}(k)$$

With measurement

$$s(k) = \mathbf{Hx}(k) + v(k)$$

where the variance of the process noise $\sigma_\xi^2 = 1$ and the measurement noise is uncorrelated with variance $\sigma_v^2 = 5$.

Assume $\mathbf{F} = \begin{bmatrix} 0 & 1 \\ -a_2 & -a_1 \end{bmatrix}$, $\mathbf{H} = \begin{bmatrix} 1 & 1 \end{bmatrix}$, $\mathbf{D} = \begin{bmatrix} 0 & 1 \end{bmatrix}^T$. This defines the scalar system $y(k) = \frac{z+1}{z^2+a_1 z+a_2}\xi(k)$ with measurement $s(k) = y(k) + v(k)$. Suppose $a_1 = -1.5$, $a_2 = 0.8$ and we require the state estimates. The two random states and their estimates are shown in Fig. 15.3.

The MATLAB code is shown below. MATLAB has its own function for calculating the gain vector but here we use the ones in the Eqs. (15.21, 15.22) instead.

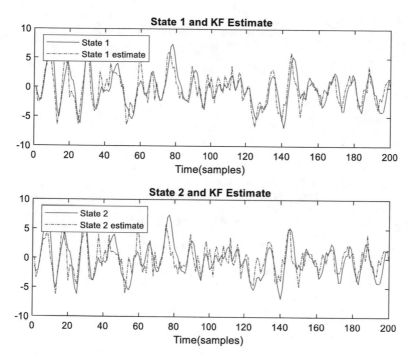

Fig. 15.3 States and state estimates for Kalman filter example. The states are one step ahead predicted

```matlab
MATLAB Code 15.2
% Finds Kalman Filter and shows the state estimates
x0=[0;0]; % Initial conditions on state vector
xh0=[0;0]; % Init conditions for discrete time KF
Npoints = 200; % number of time steps
y=zeros(Npoints,1);
s=zeros(Npoints,1);
x1=zeros(Npoints,1);
x2=zeros(Npoints,1);
xh1=zeros(Npoints,1);
xh2=zeros(Npoints,1);
t=[1:1:200];
% Define the system
F = [0, 1; -0.8 1.5]; % system matrix
D = [0 1]'; % system process noise vector
H = [1 1]; % observation matrix
Q = D*D'*1; %Process noise covariance is unity

R=5 %Measurement noise variance - scalar
% Initialize the covariance matrix
P = eye(2)*0.1 % Covariance matrix for initial state error
%Generate Random Noise Variance =1
zeta=randn(Npoints,1);
%re-seed noise generator
rng(2);
% Additive uncorrelated white noise
rv=sqrt(10)*randn(Npoints,1);
% Loop through and perform the Kalman filter equations recursively
x=x0;
xh=xh0;
for k = 1:Npoints
% State equations of system
x=F*x+D*zeta(k);
y(k)=H*x;
s(k)=y(k)+rv(k);
% Store states for plotting later
x1(k)=x(1);
x2(k)=x(2);
%Update Kalman gain vector : one step ahead version
K=F*P*H'/(H*P*H'+R);
% Update the covariance from  Riccati equation
P = F*P*F' + Q-K*H*P*F';
%Kalman Filter
xh=F*xh+K*[s(k)-H*xh];
%Store estimated states for plotting
xh1(k)=xh(1);
xh2(k)=xh(2);
end
%Plot results
figure
subplot(2,1,1)
plot(t,x1,'-r',t,xh1,'-.b')
legend('State 1','State 1 estimate','Location','northwest')
```

```
title('State 1 and KF Estimate')
xlabel({'Time(samples)'})
subplot(2,1,2)
plot(t,x2,'-r',t,xh2,'-.b')
legend('State 2','State 2 estimate','Location','northwest')
title('State 2 and KF Estimate')
xlabel({'Time(samples)'})
ax = gca;
set(gcf,'color','w');
```

Now consider the same example but the parameter in the $\mathbf{F}$ matrix a_2 is time-variant according to $\mathbf{F}(k) = \begin{bmatrix} 0 & 1 \\ -a_2 & 1 + 0.5\sin\left(2\pi\left(\frac{f}{fs}\right)k\right) \end{bmatrix}$. Here $\left(\frac{f}{fs}\right)$ represents a normalised frequency and is set to $\left(\frac{f}{fs}\right) = 0.02$. This would be a difficult problem with a Wiener filter, but assuming the changing parameter is known, can easily be included into a Kalman filter.

In Fig. 15.4 the Kalman filter can still track the states even though the system is time-variant. This is like having a nonstationary signal. However, it would be a rare occurrence if such information was readily available for many real-world problems.

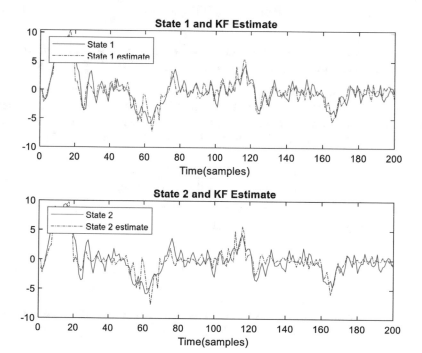

Fig. 15.4 Kalman gain when a parameter is time-variant

15.4   The Kalman Filter Used in System Identification

The Kalman filter has many applications in the area of stochastic filtering. Another use which is less common is to use the Kalman filter as an estimator to identify unknown systems. The RLS approach has much faster convergence than that of the LMS, but it is deterministic and does not track very well if the parameters vary with time. If we model an unknown FIR systems parameter (or weights) as follows [8]

$$\mathbf{W}(k + 1) = \mathbf{W}(k) + \boldsymbol{\xi}(k) \tag{15.23}$$

then this is nothing more than integrated white noise or a random walk. This is a good model of weights that change with time. It is still a state space model with $\mathbf{W}(k) = [w_0(k), w_1(k) \ldots w_n(k)]^T$ as the state vector. We have $\mathbf{F} = \mathbf{I}$ by comparison with the ordinary state space model. The vector $\mathbf{D}$ is no longer there, having been replaced by a vector $\boldsymbol{\xi}(k)$ of white noise sequences with covariance matrix $\mathbf{Q}$. We can define the output as

$$s(k) = \mathbf{x}^T(k)\mathbf{W}(k) + v(k) \tag{15.24}$$

From which we identify $\mathbf{x}^T(k) = \mathbf{H}$. If the system input is $u(k)$ then $\mathbf{x}(k) = [u(k), u(k-1) \ldots u(k-n)]^T$ is the vector of regressors just like LMS or RLS.

The measurement noise covariance is σ_v^2. Of course, with no process noise we can also consider a constant weights model $\mathbf{W}(k+1) = \mathbf{W}(k)$, but leave the measurement noise to model a noisy measurement of the system output. The Kalman filter equations then become very similar to RLS. The Kalman gain vector is from (15.21)

$$\mathbf{K}(k) = \frac{\mathbf{P}(k|k-1)\mathbf{X}(k)}{\sigma_v^2 + \mathbf{X}^T(k)\mathbf{P}(k|k-1)\mathbf{X}(k)} \tag{15.25}$$

and $\mathbf{P}(k|k-1)$ is the n square positive definite error covariance matrix found from the Riccati equation below

$$\mathbf{P}(k|k) = \left[\mathbf{I} - \mathbf{K}(k)\mathbf{X}^T(k)\right]\mathbf{P}(k|k-1) + \mathbf{Q} \tag{15.26}$$

For constant weights we set $\mathbf{Q} = 0$. When $\sigma_v^2 = 1$ the two algorithms are identical assuming the RLS forgetting factor is 1.

15.4.1 Illustrative Example

Consider unity variance white noise being passed through an FIR system with constant impulse response $\{1, 2, 3, 4, 5, 6\}$. Additive uncorrelated white noise is added to the measured output with variance $\sigma_v^2 = 0.001$. Figure 15.5 shows the RLS estimates and Fig. 15.6 shows the Kalman filter estimates. It is assumed that the Kalman filter knows the variance of the additive noise.

The Kalman estimates converge faster than RLS. The RLS is a deterministic algorithm and has no information about noise statistics. The Kalman filter does have information on noise statistics but often in real problems estimating the noise covariance accurately can be problematic.

15.5 The LMS Algorithm with Optimal Gains

The LMS algorithm or its variants is the most extensively used of the recursive algorithms. The main problem with it however is that it has slow convergence, and to speed up convergence mean that the step size must be increased and this in turn can lead to instability. The step size is usually constant, but there are variants where the step size is varied with time. These are usually ad-hoc in that it is known that starting from a large step size and slowly reducing it as the algorithm converges leads to

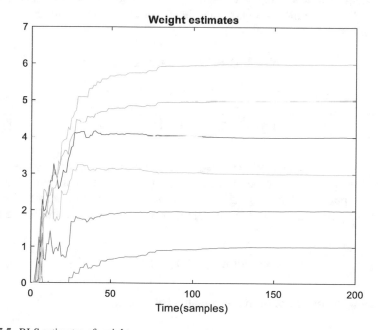

Fig. 15.5 RLS estimates of weights

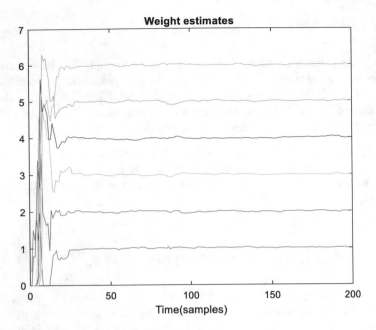

Fig. 15.6 Kalman filter estimate of weights

good results. It is of great interest then to find out just what is the best gain whether it be constant or time varying. We can do this by re-writing the LMS algorithm in a different form. We must also consider the effect of additive white measurement noise v(k). Instead of the usual

$$\hat{\mathbf{W}}(k + 1) = \hat{\mathbf{W}}(k) + \mu \mathbf{X}(k)e(k)$$

We write instead

$$\hat{\mathbf{W}}(k|k) = \hat{\mathbf{W}}(k|k - 1) + \mathbf{G}(k)\mathbf{X}(k)e(k) \tag{15.27}$$

We define $\mathbf{G}(k)$ as square matrix of time-varying gains of dimension n, where n is the number of weights. Usually the ˆ notation is not used in LMS, but here it is needed only for the purpose of distinguishing an estimate from the true vector. The error stays the same (though we could write it at time k − 1 it makes no difference to the result)

$$e(k) = y(k) - \mathbf{X}^{T}(k)\hat{\mathbf{W}}(k|k - 1) \tag{15.28}$$

The problem becomes one of minimizing the cost function

$$J = E\left\|\mathbf{W}(k) - \hat{\mathbf{W}}(k|k)\right\|_2^2 \tag{15.29}$$

with respect to the gain matrix. The cost function is the Euclidian norm squares of the vector weight-error. Here $\mathbf{W}(k)$ represents the true weight vector and $\hat{\mathbf{W}}(k|k)$ its estimate. We use the usual approach and define an error-covariance matrix and then minimise its trace with respect to the gain matrix. Define a positive-definite symmetric *aposteriori* error covariance matrix of the weights as $\mathbf{P}(k|k)$ where

$$\mathbf{P}(k|k) = \text{cov}\left[\mathbf{W}(k) - \hat{\mathbf{W}}(k|k)\right] \tag{15.30}$$

In the above equation $\hat{\mathbf{W}}(k|k)$ is assumed to be the LMS estimate, that is it is assumed to be an instantaneous and not a one-step predicted estimate. Substitute (15.27) and (15.28) into (15.30)

$$\mathbf{P}(k|k) = \text{cov}\left[\mathbf{W}(k) - \left(\hat{\mathbf{W}}(k|k-1) + \mathbf{G}(k)\mathbf{X}(k)\left(y(k) - \mathbf{X}^T(k)\hat{\mathbf{W}}(k|k-1)\right)\right)\right] \tag{15.31}$$

When uncorrelated white noise $v(k)$ of variance σ_v^2 is added to the measurement we have

$$y(k) = \mathbf{X}^T(k)\mathbf{W}(k|k-1) + v(k) \tag{15.32}$$

Then the error covariance becomes

$$\begin{aligned}\mathbf{P}(k|k) = \text{cov}\Big[\mathbf{W}(k) - \Big(\hat{\mathbf{W}}(k|k-1) &+ \mathbf{G}(k)\mathbf{X}(k)\left(\mathbf{X}^T(k)\mathbf{W}(k|k-1)\right.\\ +v(k) - \mathbf{X}^T(k)\hat{\mathbf{W}}(k|k-1)\Big)\Big)\Big] &= \text{cov}\Big[\left[\mathbf{W}(k) - \hat{\mathbf{W}}(k|k-1)\right]\\ -\mathbf{G}(k)\mathbf{X}(k)\mathbf{X}^T(k)\left[\mathbf{W}(k) - \hat{\mathbf{W}}(k|k-1)\right] &- \mathbf{G}(k)\mathbf{X}(k)v(k)\Big]\end{aligned} \tag{15.33}$$

Now define the *apriori* error covariance matrix of the weights as $\mathbf{P}(k|k-1)$ where

$$\mathbf{P}(k|k-1) = \text{cov}\left[\mathbf{W}(k) - \hat{\mathbf{W}}(k|k-1)\right] \tag{15.34}$$

Then (15.33) becomes

$$\mathbf{P}(k|k) = \text{cov}\left[\left[\mathbf{I} - \mathbf{G}(k)\mathbf{X}(k)\mathbf{X}^T(k)\right]\mathbf{P}(k|k-1) - \mathbf{G}(k)\mathbf{X}(k)v(k)\right] \tag{15.35}$$

Expanding (15.35) and noting that the covariance matrix of the regressors is

$$\mathbf{R}(k) = E[\mathbf{X}(k)\mathbf{X}^T(k)] \tag{15.36}$$

$$\mathbf{P}(k|k) = [\mathbf{I} - \mathbf{G}(k)\mathbf{R}(k)]\mathbf{P}(k|k - 1)[\mathbf{I} - \mathbf{G}(k)\mathbf{R}(k)]^T$$
$$+ \sigma_v^2 \mathbf{G}(k)\mathbf{R}(k)\mathbf{G}^T(k) \tag{15.37}$$

Now we minimise by using the gradient with respect to the $\mathbf{G}$ matrix

$$\nabla_\mathbf{G} \text{trace}(\mathbf{P}(k|k)) = 0 \tag{15.38}$$

After a little algebra (differentiating the trace of a matrix with respect to a matrix) we get the optimal time-varying gain matrix as

$$\mathbf{G}(k) = \mathbf{P}(k|k - 1)[\mathbf{R}(k)\mathbf{P}(k|k - 1) + \sigma_v^2 \mathbf{I}]^{-1} \tag{15.39}$$

With

$$\mathbf{P}(k|k) = (\mathbf{I} - \mathbf{G}(k)\mathbf{R}(k))\mathbf{P}(k|k - 1) \tag{15.40}$$

Our modified LMS algorithm is hence

$$\hat{\mathbf{W}}(k|k) = \hat{\mathbf{W}}(k|k - 1) + \mathbf{P}(k|k - 1)[\mathbf{R}(k)\mathbf{P}(k|k - 1) + \sigma_v^2 \mathbf{I}]^{-1} \mathbf{X}_k e_k \tag{15.41}$$

Clearly the error covariance matrix is updated from (15.40). The error is found from $e(k) = y(k) - \mathbf{X}^T(k)\hat{\mathbf{W}}(k|k - 1)$ where $y(k)$ is the noisy measurement.

15.5.1   Newtons Method

Now we must analyse (15.40) to see what we can glean from this result. The first result is interesting. Consider the case when the measurement noise is zero. Set $\sigma_v^2 = 0$ in (15.41). We get

$$\hat{\mathbf{W}}(k|k) = \hat{\mathbf{W}}(k|k - 1) + \mathbf{R}^{-1}(k)\mathbf{X}(k)e(k) \tag{15.42}$$

This equation is recognizable as the Newton method of LMS. It has fast convergence but require the inverse of the regressor covariance matrix. It is well documented in the literature. For example [9] considers the Newton method for nonstationary environments.

At first sight Eq. (15.41) looks original and different from previous results in this area. We can proceed further by defining a gain vector in (15.41). Let

$$\hat{\mathbf{W}}(k|k) = \hat{\mathbf{W}}(k|k-1) + \mathbf{K}^g(k)e_k \tag{15.43}$$

where

$$\mathbf{K}^g(k) = \mathbf{P}(k|k-1)\left[\mathbf{R}(k)\mathbf{P}(k|k-1) + \sigma_v^2\mathbf{I}\right]^{-1}\mathbf{X}_k \tag{15.44}$$

Is a column vector of length n, instead of the earlier matrix.

Despite the fact it looks a little different, it can be shown with a little more algebra [10] that (15.44) is just another form of the Kalman gain vector. Hence what we have done is convert the LMS algorithm into a Kalman filter. In other words, the LMS algorithm with optimal time-varying step size is a Kalman filter.

15.5.2 White Driving Noise Case [11]

If and only if the driving noise to the LMS algorithm is white, a great simplification occurs. The states of the Kalman filter become decoupled and only one step size is needed just as with ordinary LMS. We no longer need the gain vector in (15.44). There is a major difference however in that the step size is time-varying and no longer a constant. The algorithm becomes

$$e(k) = y(k) - \mathbf{X}^T(k)\hat{\mathbf{W}}(k) \tag{15.45}$$

$$\hat{\mathbf{W}}(k) - \hat{\mathbf{W}}(k-1) + \mu(k)\mathbf{X}(k)e(k) \tag{15.46}$$

$$\mu(k) = \frac{p(k|k-1)}{\sigma_u^2 p(k|k-1) + \sigma_v^2} \tag{15.47}$$

$$p(k|k) = \left[1 - \mu(k)\sigma_u^2\right]p(k|k-1) \tag{15.48}$$

In the above σ_u^2 and σ_v^2 are the variances of the reference regressor white noise and additive white measurement noise respectively. The step size will start at a high value and reduce to a low value near the optimum.

As an example, consider white noise of unit variance passed through a second order transfer function $y(k) = \frac{1}{1-1.5z^{-1}+0.8z^{-2}}\xi(k)$ and uncorrelated white noise of unit variance added at the output. $s(k) = y(k) + v(k)$. The recursive algorithms will estimate this system as an FIR system which is effectively the inverse. Figure 15.7 shows the LMS algorithm working with a fixed step size estimating the first 10 weights whilst Fig. 15.8 shows the same example with the LMS-Kalman approach. The first true values of the weights become (by long division) {1, 1.5, 1.45, 0.975, 0.3025, −0.3263, −0.7314 ...}.

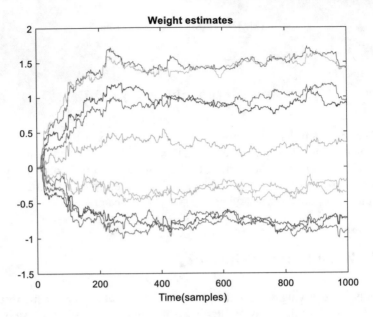

Fig. 15.7 LMS estimate of weights for a fixed step size

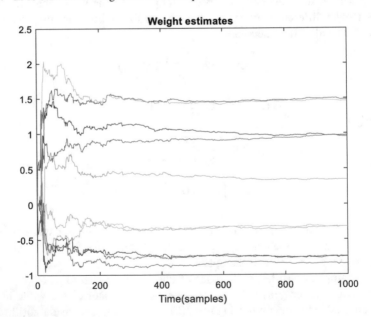

Fig. 15.8 LMS-Kalman estimated weights using variable step size

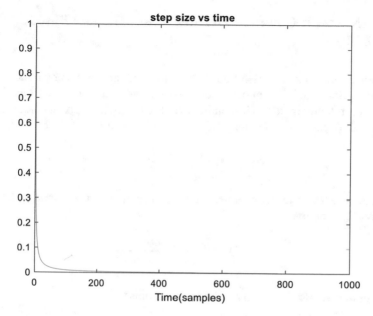

Fig. 15.9   Step size versuss time for LMS-Kalman approach

With the fixed step size, it is hard to find a value of step size that gives smooth estimates without the convergence taking too long. Figure 15.9 shows the step size vs time for the LMS-Kalman approach.

The step size is 1 initially reducing asymptotically towards zero. It is also interesting to examine the equation for the scalar error covariance $p(k|k) = [1 - \mu(k)\sigma_u^2]p(k|k-1)$ to see that for stability i.e. so that the error covariance dies out with time rather than grows larger, we must have that

$$\left|1 - \mu(k)\sigma_u^2\right| < 1 \tag{15.49}$$

Giving

$$0 < \mu(k) < \frac{2}{\sigma_u^2} \tag{15.50}$$

Although we have a better LMS algorithm than ordinary LMS, this is only the case for white driving noise through the unknown system. If coloured (correlated) noise is passed through the system, then the Kalman filter is the only estimator which has optimal time-varying gains. For the coloured driving noise case LMS is sub-optimal. From experience we know it still converges with a fixed step size, but the results will not be as good as the Kalman estimator.

15.6   LMS with Coloured Driving Noise. Toeplitz Based LMS

The LMS algorithm is successful due to its simplicity and low computational cost. However, it does suffer from slow convergence under certain conditions. This is when the correlation matrix is ill-conditioned. For example, suppose coloured noise is generated by passing white noise through an AR colouring filter of the form

$$u(k) = W_c(z^{-1})\xi(k) \tag{15.51}$$

where $\xi(k)$ is zero mean white noise with variance σ_ξ^2, and the colouring filter has the Z transfer function

$$W(z^{-1}) = \frac{1}{1 + a_1 z^{-1} + a_2 z^{-2}} \tag{15.52}$$

If a regression vector is created for the coloured noise

$$\mathbf{X}(k) = \begin{bmatrix} u(k) & u(k-1) & \cdots & u(k-p) \end{bmatrix}^T \tag{15.53}$$

Then the correlation matrix

$$\mathbf{R} = E\begin{bmatrix} \mathbf{X}(k)\mathbf{X}^T(k) \end{bmatrix} \tag{15.54}$$

plays a major role in the least-squares estimate of the system the coloured noise is trying to estimate. The reason for this is because the least-squares estimate involves the inverse of this matrix. The LMS algorithm does not of course require a matrix inverse (though the Newton method does), but the rate of convergence of the LMS algorithm is nevertheless still governed by the matrix itself. Haykin [12] discusses this problem at some length. One of the problems which is quoted is when

$$W(z^{-1}) = \frac{1}{1 - 1.911z^{-1} + 0.95z^{-2}} \tag{15.55}$$

The poles are at $0.955 \pm j0.1924$ which are close to the unit circle. This is narrowband noise with most of the power concentrated within a narrow band of frequencies and little lese outside of it. It results in a correlation matrix which is "stiff". That is, the ratio of the largest to smallest eigenvalue is big. This ratio is known as the *condition number* of the matrix. Consider an example where this coloured noise is driving an FIR system and we try and identify the system recursively using LMS. Let the system be $H(z^{-1}) = 1 - z^{-1} + 0.5z^{-2}$. The problem when we apply LMS is that the step size cannot be made too big or the algorithm will go unstable. Figure 15.10 shows the convergence with the larges practical step size.

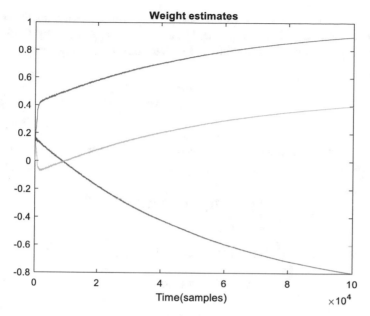

Fig. 15.10 Convergence of LMS algorithm with coloured noise. Shows slow convergence

After 100,000 samples the algorithm is getting close to convergence, but the step size cannot be increased, or instability will result.

One possible solution is to use a lower triangular Toeplitz (LTT) method [13]. We have studied LTT matrices in previous chapters. This is a related topic whereby instead of the matrix being populated by process parameters, it is populated by noise regressors instead. For an nth order system, we represent its output by the $(n + 1)$ order vector $\mathbf{y}(k)$ where

$$\mathbf{y}(k) = \mathbf{T}\mathbf{W}(k) \tag{15.56}$$

where the $(n + 1)$ LTT matrix $\mathbf{T}$ is time-varying

$$\mathbf{T} = \begin{bmatrix} u(k) & 0 & 0 & 0 & 0 & 0 \\ u(k+1) & u(k) & 0 & 0 & 0 & 0 \\ u(k+2) & u(k+1) & u(k) & 0 & 0 & 0 \\ . & u(k+2) & u(k+1) & u(k) & 0 & 0 \\ . & . & . & . & u(k) & 0 \\ u(k+n) & u(k+(n-1)) & u(k+(n-2)) & . & . & u(k) \end{bmatrix} \tag{15.57}$$

and $\mathbf{W}(k)$ is the $(n + 1)$ vector of true weights. This represents an FIR system using the convolution matrix.

The main difference here is that we are now dealing with a vector output as opposed to scalar for ordinary LMS. We proceed in the same way as ordinary LMS to find the optimal estimator. We must minimise a vector norm instead however.

For an error vector

$$\mathbf{e}(k) = \mathbf{y}(k) - \mathbf{T}\hat{\mathbf{W}}(k) \tag{15.58}$$

where $\hat{\mathbf{W}}(k)$ is the estimated weight vector. We minimize the vector norm by differentiating the scalar cost function

$$J = \frac{1}{2}E \left\| \mathbf{y}(k) - \mathbf{T}\hat{\mathbf{W}}(k) \right\|_2^2 \tag{15.59}$$

with respect to the vector $\hat{\mathbf{W}}(k)$. This gives us the gradient

$$\nabla_w J = -E\left[\mathbf{T}^T \mathbf{e}(k)\right] = 0 \tag{15.60}$$

The optimal solution for the weight vector is readily found,

$$\mathbf{W}^* = \mathbf{R}_c^{-1} \mathbf{P}_c \tag{15.61}$$

where a new *convolution* correlation matrix $\mathbf{R}_c$ is defined as

$$\mathbf{R}_c = E\left[\mathbf{T}^T \mathbf{T}\right] \tag{15.62}$$

and the *convolution* cross-correlation vector is defined as

$$\mathbf{P}_c = E\left[\mathbf{T}^T \mathbf{y}(k)\right] \tag{15.63}$$

The steepest descent algorithm is used to find the new update of the weight vector. It becomes

$$\hat{\mathbf{W}}(k+1) = \hat{\mathbf{W}}(k) + \mathbf{T}^T \mathbf{e}(k) \tag{15.64}$$

where the error vector is given by (15.58). The algorithm uses blocks of data since the measurement is now a vector instead of a scalar.

The difference now is that the covariance matrix $\mathbf{R}_c$ differs from the normal LMS one and its condition number is not so high. This allows for faster convergence. The algorithm converges in around 500 blocks of data as shown in Fig. 15.11. This is 1500 samples and considerably faster than ordinary LMS.

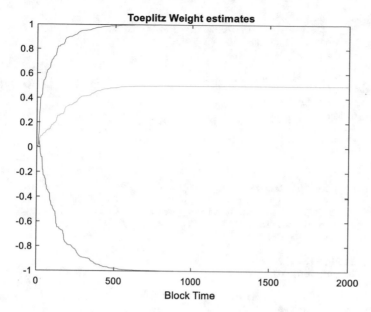

Fig. 15.11 Convergence of the same problem using the LTT LMS approach

The MATLAB code that generated Fig. 15.11 is shown below.

MATLAB Code 15.3

```
%LT Toeplitz Least-mean squares (LMS) algorithm
%Driven by coloured noise
clear
close all
%Number of samples in simulation
Npoints=100000;
%FIR system coefficients
c=[1 -1 0.5];
%Number of weights
Nweights=length(c);
%Step size
K=0.0001;

%Unit variance driving noise
sq=1;
%Random white Noise length Npoints Variance rd
rd=sqrt(sq)*randn(Npoints,1);
%Generate coloured driving noise
num=1;
den=[1 -1.911 0.95];
yd=filter(num,den,rd);
%yd=rd;
%This array v is a plotting index only
v=zeros(Nweights,1);
for j=1:Nweights
 v(j)=j;
end
%Now the Lower Triangular Toeplitz method
%Define Wt
Wt=zeros(Nweights,1);
%error vector;
ev=zeros(Nweights,1);
yp=zeros(Nweights,1);
np=zeros(Nweights,1);
x=zeros(Nweights,1);
%Define Matrix T
T=zeros(Nweights,Nweights);
icount=0;
Nmax=Npoints/50;
ws2=zeros(Nmax,Nweights);

for i=1:Nweights:Nmax*Nweights
        icount=icount+1;
   %Put Nweights worth of coloured input data in a separate vector np
for k=1:Nweights
        if(i+k-1)>0
        np(k)=yd(i+k-1);
        else
        end
 end
%Filter np to get the output data yp
        %Convolution
     yp=filter(c,1,np);
```

```
            %Create Lower Triangular Toeplitx matrix
        %From coloured noise vector np.
        m=Nweights;
    tr=zeros(1,m);
    tr(1)=np(1);
    nul=zeros(1,(m-length(np)));
    tc=horzcat(np,nul);
    T=toeplitz(tc,tr);
    %
     %error vector
    ev=yp-T*Wt;
    Wt=Wt+K*T'*ev;
    %Store weights
    for i1=1:Nweights
    ws2(icount,i1)=Wt(i1);
    end
    end
    Wt
    figure
    plot(ws2(:,v))
    title('Toeplitz Weight estimates ')
    set(gcf,'color','w');
    xlabel('Block Time index')
```

References

1. K.J. Astrom, B. Wittenmark, On self tuning regulators. Automatica **9**(3), 185–199 (1973)
2. L. Ljung, On positive real transfer functions and the convergence of some recursive schemes. IEEE Trans. Autom. Control **22**(4), 539–551 (1977). https://doi.org/10.1109/TAC.1977.110 1552
3. L. Ljung, T. Soderstrom, *Theory and practice of recursive identification:* (Signal processing,optimization and control). Camb. MA, USA: MIT Press, 1983
4. R.E. Plackett, Some therems on least-squares. Biometrika **37**, 149 (1950)
5. R.E. Kalman, A new approach to linear filtering and prediction problems. Trans. ASME. J. Basic Eng. **82**(Series D), 35–45 (1960)
6. T.J. Moir, *Feedback* (Springer Nature, Berlin, 2020)
7. J.F. Barrett, T.J. Moir, A unified approach to multivariable discrete-time filtering based on the Wiener theory. Kybernetika **23**(3), 177–197 (1987)
8. L. Guo, Estimating time-varying parameters by the Kalman filter based algorithm: stability and convergence. IEEE Trans. Autom. Control **35**(2), 141–147 (1990). https://doi.org/10.1109/9. 45169
9. M.L.R. de Campos, P.S.R. Diniz, A. Antoniou, Performance of LMS-Newton adaptation algorithms with variable convergence factor in nonstationary environments, in *1993 IEEE International Conference on Acoustics, Speech, and Signal Processing*, 27–30 Apr 1993, vol. 3, pp. 392–395 (1993). https://doi.org/10.1109/ICASSP.1993.319517
10. T.J. Moir, A new look at the LMS algorithm with correlated driving noise, in *2016 3rd International Conference on Signal Processing and Integrated Networks (SPIN)*, Amity University, Noida India, 11–12 Feb 2016 (IEEE, pp. 24–27). https://doi.org/10.1109/SPIN.2016.7566656

11. P.A.C. Lopes, G. Tavares, J.B. Gerald, A new type of normalized LMS algorithm based on the Kalman Filter, in *2007 IEEE International Conference on Acoustics, Speech and Signal Processing—ICASSP '07*, 15–20 Apr 2007, vol. 3, pp. III-1345–III-1348 (2007). https://doi.org/10.1109/ICASSP.2007.367094
12. S. Haykin, *Adaptive Filter Theory* (Prentice Hall, New Jersey USA, 1996)
13. T.J. Moir, FIR system identification for correlated noise using the convolution matrix, an investigation. Signal Image Video Proc. **10**(6), 1049–1054 (2016). https://doi.org/10.1007/s11760-015-0857-2

Printed in the United States
by Baker & Taylor Publisher Services